Y0-BZB-309

HYDRONICS TECHNOLOGY

Justin Duncan

BNP Business News Publishing Company
Troy, Michigan

Library of Congress Cataloging in Publication Data

Duncan, Justin.
Hydronics technology / Justin Duncan.
p. cm.
ISBN 0-912524-87-1
1. Hot-water heating. 2. Steam-heating. 3. Air conditioning.
I. Title.
TH7467.D86 1995 94-23005
697'.4--dc20 CIP

Editor: Joanna Turpin
Art Director: Mark Leibold
Copy Editor: Carolyn Thompson

Printed in the United States of America
7 6 5 4 3 2 1

Dedication

I dedicate this book to my adopted country, the United States of America. I truly am proud of the opportunities which this great land offers. Combined with hard and dedicated work, accomplishments follow. This book is one of mine.

Acknowledgments

Special thanks to editors Joanna Turpin and Carolyn Thompson, whose active cooperation and effort made this book possible.

Table of Contents

Chapter 1

Introduction to Hydronics

Hydronics is part of the large sphere of hydraulics, which in turn, is a part of fluid mechanics. Hydronics uses water as a medium for producing heating, cooling, or other environmentally useful work. Hydronics may also use steam to convey heat to consumers.

Webster's dictionary defines hydronic as: "relating to or being a system of heating or cooling that involves transfer of heat by circulating fluid (as water or steam) in a closed system of pipes." While this definition is correct, it is somewhat incomplete. A more complete definition of a hydronic system is: a system that removes heat (cooling) or adds heat (heating) to an area by circulating water in a closed system using specialized equipment and pipes.

Hydronics is the science or knowledge of cooling and heating with a fluid. Hydronic heating systems have been used for well over 100 years. About 50 years ago, automatic controls such as zone control were added. Other improvements have also been made over the years, resulting in better systems.

The hydronic cooling system usually transfers the heat from an area to the outdoor air, Figure 1-1. Heat is first transferred from the indoor ambient air to the chilled water, then from the chilled water to the refrigerant, and finally either to the atmosphere or to another water system (cooled and recirculated or wasted). The recirculated system uses a cooling tower and does not waste water (only a minimal of 2% to 4% is lost through evaporation and mist). A once-through water-cooled system uses large quantities of water which is wasted and therefore not recommended.

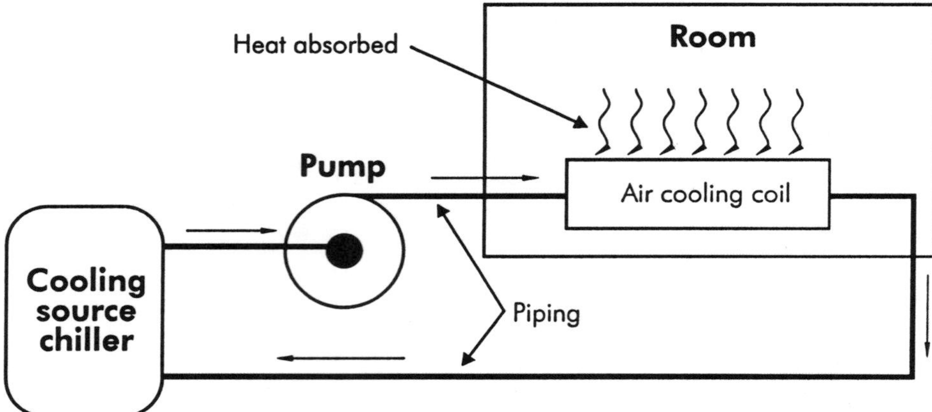

Figure 1-1. Basic hydronic cooling loop

The hydronic heating system involves the controlled combustion of a solid fuel (coal, wood), liquid fuel (oil), or gas (natural) in a boiler to produce steam or hot water, Figure 1-2. The heat generated is distributed through a network of pipes to radiators, convectors, etc., to heat various spaces. Energy in the form of heat may be converted from a combustible fuel to water or steam in a boiler or from water to water (or steam to water) in a heat exchanger. The heated water then circulates in a closed system and releases the heat in a controlled manner (water to air) to the space to be heated.

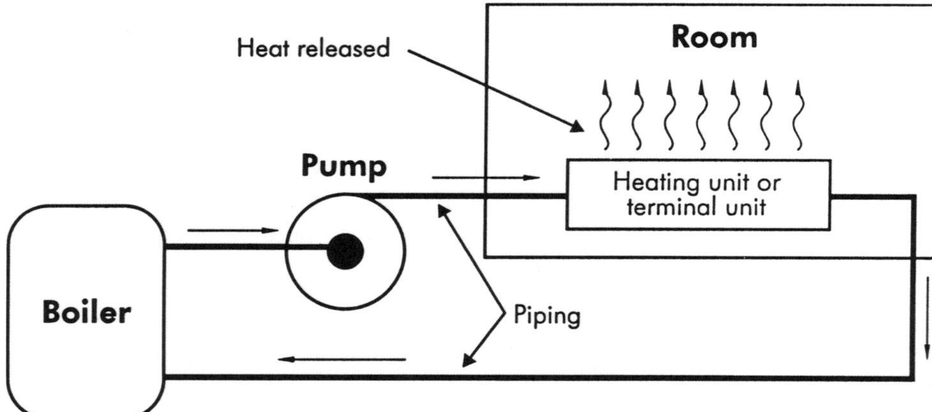

Figure 1-2. Basic hydronic heating loop

The elements that determine heat transfer from one fluid to another are the:

- temperature difference between the two fluids;
- area of the exchanger;
- fluid velocity within the shell and within the coil(s).

The designer and owner aim to have a heating and/or cooling installation that performs well continuously and without trouble. Water is an excellent medium for conveying coolness or heat from the equipment that produces it (chiller or boiler, respectively) to the point where it is absorbed or released, respectively. In heating, once the heat is released from the terminal unit, the water with lower temperature returns to the heat exchanger to be reheated. The operation is reversed for cooling.

When hot water is introduced into a tank above a level of cold water, the two mix very little without outside influence. This is called *stratification*. However, when the cold water is introduced above hot water the two will naturally mix. A pump (also called a circulator) usually helps water to circulate. This type of system is called a *forced circulation type* and has the following advantages:

- Smaller pipe sizes
- Better control of the amount of water circulated and the amount of heat released
- Easier and better selection of the distribution type to be used

To design and size a hydronic system that will perform as expected, heat loss (heat gain for cooling) calculations must be prepared. This book will explain the elements of such calculations. The examples included illustrate the types of systems available and emphasize their advantages. The understanding of a hydronic installation must combine a theoretical base with training and skills that are obtained through specialized schooling programs. All these programs and courses must be coupled with practical work.

Hydronics is a specialty, and it is necessary to acquire the knowledge of design and installation by learning and working. Numerous problems arise in every working day of the hydronic specialist. In order to solve them properly, it is necessary to understand how a hydronic system works.

The aim of this book is to familiarize the reader with hydronic technology. It is not the intent of this book to give instructions on how to install a boiler, chiller, or cooling tower, or how to fix a hydronic functioning problem. The intent is to set a foundation or augment existing basic hydronic knowledge and/or form a solid starting base. Understanding and experience are required in order to undertake the design and installation of a hydronic system, which is where this book will help. The knowledge gained by studying this book will hopefully contribute to better installations and more economical, trouble-free hydronic systems.

ASSOCIATIONS

One association designed to help hydronic specialists is the Hydronics Institute, Inc. (H.I.), which has radiation testing laboratories, hydronic system classes, and numerous other services to ensure progress and growth of the hydronic industry. H.I. was founded in 1915, and its headquarters are located at 35 Russo Place, Berkeley Heights, New Jersey 07922.

In addition, the Hydronics Institute prepares statistics and continuously disseminates information to members and other individuals upon demand. Standards for various categories of heating equipment are developed and published continuously. The Hydronics Institute includes a carefully scheduled training program known as I=B=R (Institute of Boiler and Radiator Manufacturers) schools.

Another hydronic related organization that publishes information for this trade is the Better Heating-Cooling Council (BHC). The Council prepares and publishes pamphlets on topics related to heating and cooling systems. Their address is the same as that for the Hydronics Institute.

Chapter 2

The Properties of Water and Hydraulics

Hydronics is a practical application of hydraulics. Hydronics deals with the flow of heating and cooling fluids in pipes, as well as with the equipment that produces the heating and cooling. In order to fully understand hydronics, it is necessary to become familiar with some hydraulic principles and the various properties of water.

Properties of Water

In its natural cycle, water evaporates from the surface of natural bodies of water, then clouds form, and rain or snow develops. Upon reaching the ground, water percolates very slowly through the upper layers of the earth's crust, which is called permeable strata. During this slow, natural, downward movement, water comes in contact with various naturally occurring substances (minerals). This process filters the water while adding minerals, mostly in the form of salts. At some point, equilibrium is reached, and water accumulates in natural underground reservoirs.

Dissolved minerals in ground water may affect its potential usage. If the concentration of a certain mineral is excessive for the specific water usage, it may need to be removed during water treatment. Due to natural filtration, most ground water contains virtually no bacteria and no suspended matter, which means most underground water needs little treatment to become domestic (potable).

In order to be used in a hydronic system, water must be treated. Treatment usually involves air elimination, prevention of algae growth, antifreeze for chilled water, etc. The addition of chemicals is required to help destroy harmful and non-harmful organic bacteria that generally exist and grow in a closed water system. If not destroyed, the bacteria or algae may cause an increased resistance to flow and even stoppages. Non-harmful bacteria encourage algae growth and slime, which discolor the water and produce odors.

Neither the water occurring in nature nor the water used for domestic purposes is pure. Water contains a number of naturally dissolved materials and chemicals (generally in very small amounts), as well as certain chemicals that are purposely added through treatment for hydronic systems.

The natural properties of water are: solubility, hardness, specific electrical conductance, hydrogen-ion concentration (pH), dissolved carbon dioxide, and dissolved solids. Figure 2-1 shows the pH value of solutions and recommended pump construction materials to be used.

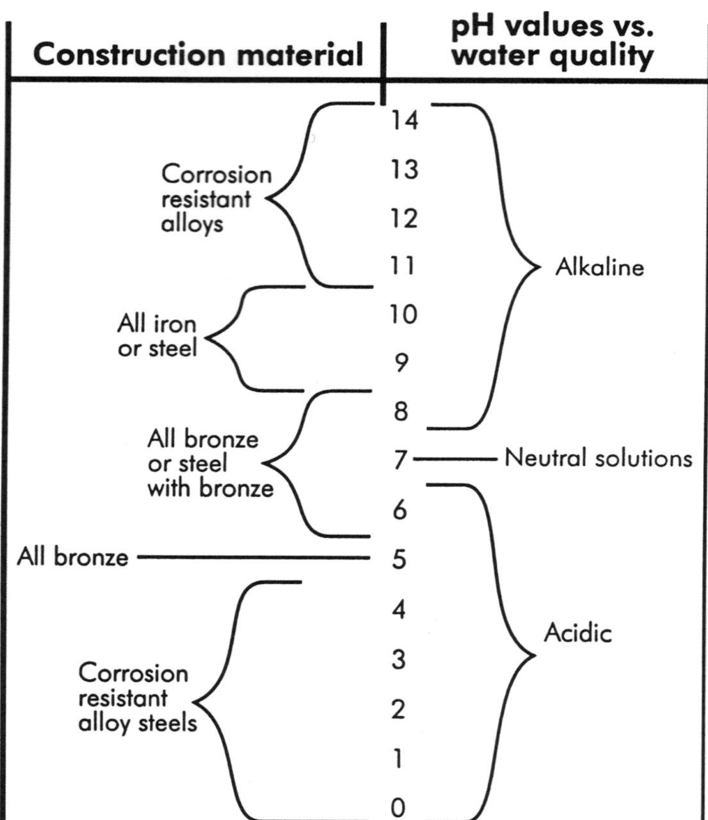

Figure 2-1. Recommended pump construction materials in relation to pH value of solutions

Some of the physical properties of water are: density, viscosity, noncompressibility, boiling point, and freezing point. It is important to recognize these properties because in one way or another, they influence the flow of water in a hydronic system.

Density

By definition, density is the ratio of mass (weight) of a substance to the volume it occupies. Density is given in pounds per cubic foot, which may be written as lb/ft^3 or lb/cu ft. In hydronic calculations, water density is usually considered to be 62.3 lb/ft^3. This value represents the density of water at a temperature of 70°F (room temperature). Water density varies slightly with the temperature; the warmer the water, the less dense it becomes, as shown below:

Water temperature (°F)	Water density (lb/ft^3)
32	62.416
50	62.408
70	62.300
100	61.998
150	61.203
210	59.843

Viscosity

Viscosity, as applied to hydronics, concerns the friction of water molecules among themselves, as well as along the walls of the pipes and fittings. It is the physical property that directly influences the flow of water in pipes. The forces at work between the water molecules themselves are called *cohesion* and *adhesion*. These forces can be measured in the laboratory.

Cold water is more cohesive than warm water, thus its viscosity is greater. The greater viscosity increases the friction of the flow of cold water through pipes. Warm water flows somewhat more easily through pipes because it is not as cohesive. However, the actual difference in viscosity between cold and hot water is so small that it is considered insignificant for practical purposes and is negligible in calculations.

Viscosity is measured in centistokes or centipoise. At 60°F, water has an absolute viscosity equal to 1.12 centistokes, which corresponds to a measurement of kinematic viscosity of 0.00001216 ft^2/sec.

Specific Gravity

Specific gravity (SG) is a number that has no dimensions or units of measurement. The specific gravity of water is interrelated to its specific volume and is used as a reference for other liquids. The specific gravity of a substance is expressed as the ratio of its weight to the weight of an equal volume of water, at 39°F. The density of water at 70°F is 62.3 lb/ft^3, so the specific gravity is:

$$SG = \frac{d}{d_w} = \frac{d}{62.3}$$

where: d = density of substance (lb/ft^3)
d_w = density of water at 70°F (62.3 lb/ft^3)

The value of specific gravity of water changes slightly with temperature. The specific volume reference for water is at 60°F. To calculate the specific gravity (SG) at a temperature other than 60°F, use the following example:

$$SG_{100} = \frac{v_{60}}{v_{100}} = \frac{0.016033}{0.016130} = 0.9939$$

Table 2-1 shows the properties of water at temperatures between 32° and 640°F.

Temp	Specific volume		Density		Kinematic viscosity centistokes
F	ft³/lb	gal/lb	US ft³/lb	SI g/cm³	
32	0.016022	0.1199	62.414	0.9998	1.79
33	0.016021	0.1198	62.418	0.9999	1.75
34	0.016021	0.1198	62.418	0.9999	1.72
35	0.016020	0.1198	62.420	0.9999	1.68
36	0.016020	0.1198	62.420	0.9999	1.66
37	0.016020	0.1198	62.420	0.9999	1.63
38	0.016019	0.1198	62.425	1.0000	1.60
39	0.016019	0.1198	62.425	1.0000	1.56
40	0.016019	0.1198	62.425	1.0000	1.54
41	0.016019	0.1198	62.426	1.0000	1.52
42	0.016019	0.1198	62.426	1.0000	1.49
43	0.016019	0.1198	62.426	1.0000	1.47
44	0.016019	0.1198	62.426	1.0000	1.44
45	0.016020	0.1198	62.42	0.9999	1.42
46	0.016020	0.1198	62.42	0.9999	1.39
47	0.016021	0.1198	62.42	0.9999	1.37
48	0.016021	0.1198	62.42	0.9999	1.35
49	0.016022	0.1198	62.41	0.9998	1.33
50	0.016023	0.1199	62.41	0.9998	1.31
51	0.016023	0.1199	62.41	0.9998	1.28
52	0.016024	0.1199	62.41	0.9997	1.26
53	0.016025	0.1199	62.40	0.9996	1.24
54	0.016026	0.1199	62.40	0.9996	1.22
55	0.016027	0.1199	62.39	0.9995	1.20
56	0.016028	0.1199	62.39	0.9994	1.19
57	0.016029	0.1199	62.39	0.9994	1.17
58	0.016031	0.1199	62.38	0.9993	1.16
59	0.016032	0.1199	62.38	0.9992	1.14
60	0.016033	0.1199	62.37	0.9991	1.12
62	0.016036	0.1200	62.36	0.9989	1.09
64	0.016039	0.1200	62.35	0.9988	1.06
66	0.016043	0.1200	62.33	0.9985	1.03
68	0.016046	0.1200	62.32	0.9983	1.00
70	0.016050	0.1201	62.31	0.9981	0.98
75	0.016060	0.1201	62.27	0.9974	0.90
80	0.016072	0.1202	62.22	0.9967	0.85
85	0.016085	0.1203	62.17	0.9959	0.81
90	0.016099	0.1204	62.12	0.9950	0.76
95	0.016114	0.1205	62.06	0.9941	0.72
100	0.016130	0.1207	62.00	0.9931	0.69
110	0.016165	0.1209	61.98	0.9910	0.61
120	0.016204	0.1212	61.71	0.9886	0.57
130	0.016247	0.1215	61.56	0.9860	0.51
140	0.016293	0.1219	61.38	0.9832	0.47

Temp	Specific volume		Density		Kinematic viscosity centistokes
F	ft³/lb	gal/lb	US ft³/lb	SI g/cm³	
150	0.016343	0.1223	61.19	0.9802	0.44
160	0.016395	0.1226	60.99	0.9771	0.41
170	0.016451	0.1231	60.79	0.9737	0.38
180	0.016510	0.1235	60.57	0.9703	0.36
190	0.016572	0.1240	60.34	0.9666	0.33
200	0.016637	0.1245	60.11	0.9628	0.31
210	0.016705	0.1250	59.86	0.9589	0.29
212	0.016719	0.1251	59.81	0.9580	
220	0.016775	0.1255	59.61	0.9549	
230	0.016849	0.1260	59.35	0.9507	
240	0.016926	0.1266	59.08	0.9484	
250	0.017006	0.1272	58.80	0.9420	0.24
260	0.017089	0.1278	58.52	0.9374	
270	0.017175	0.1285	58.22	0.9327	
280	0.017264	0.1291	57.92	0.9279	
290	0.01736	0.1299	57.60	0.9228	
300	0.01745	0.1305	57.31	0.9180	0.20
310	0.01755	0.1313	56.98	0.9128	
320	0.01766	0.1321	56.63	0.9071	
330	0.01776	0.1329	56.31	0.9020	
340	0.01787	0.1337	55.96	0.8964	
350	0.01799	0.1346	55.59	0.8904	0.17
360	0.01811	0.1355	55.22	0.8845	
370	0.01823	0.1364	54.84	0.8787	
380	0.01836	0.1374	54.47	0.8725	
390	0.01850	0.1384	54.05	0.8659	
400	0.01864	0.1394	53.65	0.8594	0.15
410	0.01878	0.1404	53.25	0.8530	
420	0.01894	0.1417	52.80	0.8458	
430	0.01909	0.1428	52.38	0.8391	
440	0.01926	0.1441	51.92	0.8317	
450	0.01943	0.1453	51.47	0.8244	0.14
460	0.01961	0.1467	50.99	0.8169	
470	0.01980	0.1481	50.51	0.8090	
480	0.02000	0.1496	50.00	0.8010	
490	0.02021	0.1512	59.48	0.7926	
500	0.02043	0.1528	48.95	0.7841	0.13
510	0.02067	0.1546	48.38	0.7750	
520	0.02091	0.1564	47.82	0.7661	
530	0.02118	0.1584	47.21	0.7563	
540	0.02146	0.1605	46.60	0.7465	
550	0.02176	0.1628	45.96	0.7362	0.12
560	0.02207	0.1651	45.31	0.7258	
570	0.02242	0.1677	44.60	0.7145	
580	0.02279	0.1705	43.88	0.7029	
590	0.02319	0.1735	43.12	0.6908	
600	0.02364	0.1768	42.30	0.6776	0.12
610	0.02412	0.1804	41.46	0.6641	
620	0.02466	0.1845	40.55	0.6496	
630	0.02526	0.1890	39.59	0.6342	
640	0.02595	0.1941	38.54	0.6173	

Table 2-1. Water properties at various temperatures (Reprinted from *Cameron Hydraulic Data* book)

Compressibility

Water at ambient temperature is considered noncompressible for all practical purposes.

Boiling/Freezing Points

Water boils at 212°F (100°C) at sea level (atmospheric pressure). If the pressure varies, the boiling temperature point will also vary. The lower the pressure exerted upon the surface, the lower the boiling point. For example, atmospheric pressure is lower on top of a mountain, so water boils at a lower temperature. The changes in the boiling point as a function of pressure are as follows:

Absolute pressure (psi)	Water boiling point (°F)
1	101.8
6	170.1
14.7 (atmospheric)	212.0

Figure 2-2 shows boiling temperature in relation to absolute pressure (psia). To determine the boiling point of water as a function of gauge pressure (psig), the practical formula for 1 psig and higher is as follows:

$$\text{Boiling temperature} = 14\sqrt{P} + 198$$

where: P = gauge pressure exerted

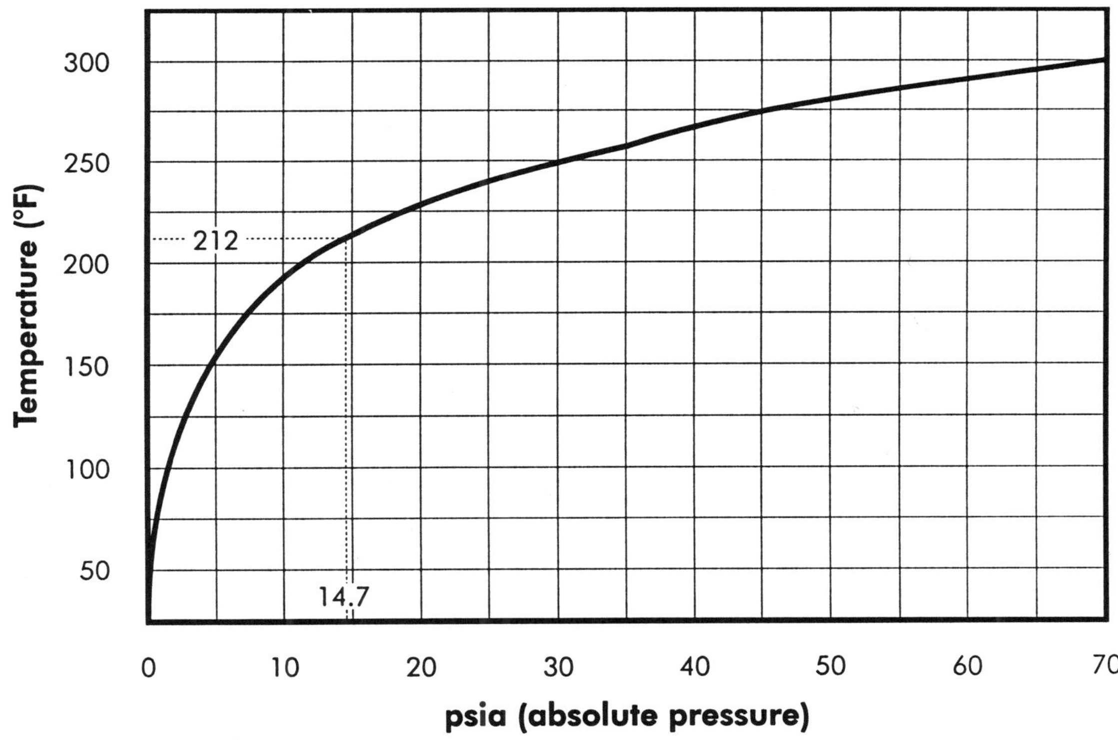

Figure 2-2. Boiling point of water at various absolute pressures

The freezing point for water is 32°F (0°C). At this temperature, water turns into a solid and will not flow. When heated or frozen, the volume of water increases.

Water Flow

The flow of water can be characterized as *laminar* or *turbulent*. In laminar flow, streams of water molecules flow naturally parallel to each other up to a certain velocity. Above that velocity, the flow becomes turbulent. This characteristic was demonstrated by Osborne Reynolds, who developed a simple formula to determine the ***Reynolds number*** (R), which classifies the flow as laminar or turbulent. If the Reynolds number (R) is less than 2000, the flow is laminar. The simplified formula is as follows:

$$R = \frac{VD\ \rho}{\upsilon} = \frac{\text{velocity x diameter x density}}{\text{absolute viscosity}}$$

where:
- V = water velocity (ft/sec)
- R = Reynolds number (no unit of measurement)
- D = pipe diameter (ft)
- υ = absolute or dynamic viscosity (lb/ft-sec)
- ρ = density of fluid (lb/cu ft)

Replacing values for water in the above formula it becomes:

$$R = (7740)\left(\frac{VD}{v}\right)$$

where:
- V = water velocity in pipe (ft/sec)
- D = diameter of pipe (inches)
- v = kinematic viscosity (centistokes)
- 7740 = coefficient

For water, the kinematic viscosity at 68°F is approximately 1 centistoke.

To satisfy the laminar flow, water velocity must be less than 1 ft/sec, which never happens in practical application. The flow of water in pipes at "normal" velocities of 4 to 8 ft/sec is always turbulent. It is also true that the velocity of flowing water in a cross section of the pipe is not uniform. There is no adverse influence of a turbulent flow in a hydronic design.

The velocity of water is greatest at the center of the pipe. More friction exists along the walls, where water molecules rub against pipe walls, Figure 2-3. The average velocity represents 80% of the maximum velocity at the center of the pipe.

Tables 2-2 and 2-3 show the recommended water velocity for various services and the recommended water velocity to limit erosion.

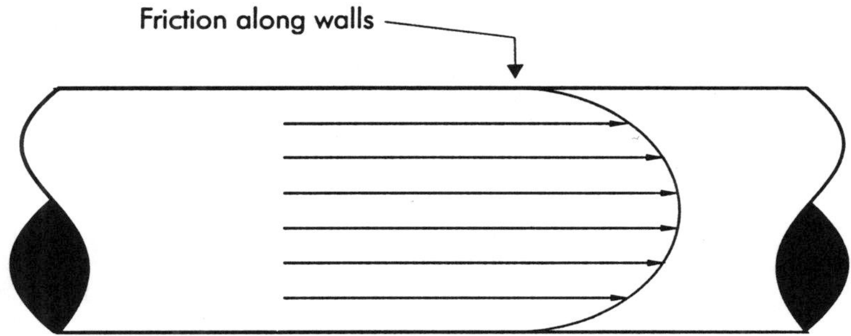

Figure 2-3. Flow of water in pipes

Type of operation	Water velocity (fps)
General service	5 to 10
Header	4 to 15
Riser	3 to 10
City water	3 to 7
Pump discharge	8 to 12
Pump suction	4 to 7
Drain line	4 to 7

Table 2-2. Recommended water velocity for various services (Courtesy, *Carrier, Handbook of Air Conditioning System Design*, 1965, reprinted with permission by McGraw-Hill, Inc.)

Hours of operation (hr/yr)	Water velocity (fps)
1500	12
2000	11.5
3000	11
4000	10
6000	9
8000	8

Table 2-3. Recommended water velocity to limit erosion (Courtesy, *Carrier, Handbook of Air Conditioning System Design*, 1965, reprinted with permission by McGraw-Hill, Inc.)

The basic formula for fluid flow is:

$$Q = AV$$

where:
Q = flow (ft^3/sec)
A = cross section of pipe (ft^2)
V = water velocity in pipe (ft/sec)

Another item often used in hydronic calculations is the *velocity head*. This is defined as the decrease in head (or the loss of pressure), which corresponds to the velocity of flow. The formula for velocity head is:

$$h_v = \frac{V^2}{2g}$$

where: h_v = velocity head (ft)
V^2 = velocity in pipe $(ft/sec)^2$
g = acceleration of gravity (ft/sec^2)

As a dimensional verification:

$$h_v = \left(\frac{ft^2}{sec^2}\right)\left(\frac{sec^2}{ft}\right) = ft$$

Water flows from a high pressure or high elevation to a lower one. The amount of potential energy due to the elevation of water above a certain reference point is called *static head*. Static head is measured in feet of water and can be converted into psi (pounds per square inch).

When used by itself, the term *head* is rather misleading. It is commonly taken to mean the difference in elevation between the suction level and the discharge level of the liquid being pumped. Although this is partially correct, it does not take into account all of the conditions that should be included to give an accurate description. The different definitions of head are listed below:

- *Friction head* - the pressure expressed in lb/in^2 or ft of liquid needed to overcome the resistance to the flow in the pipe and fittings.
- *Suction lift* - the source of liquid supply that is below the center line of the pump.
- *Suction head* - the source of liquid supply that is above the center line of the pump.
- *Static suction lift* - the vertical distance from the center line of the pump down to the free level of the liquid source.
- *Static suction head* - the vertical distance from the center line of the pump up to the free level of the liquid source above.
- *Static discharge head* - the vertical elevation from the center line of the pump to the point of free discharge.
- *Dynamic suction lift* - the sum of static suction lift, friction head loss, and velocity head.
- *Dynamic suction head* - the static suction head minus the sum of friction head loss and velocity head in the suction side of the pump.
- *Dynamic discharge head* - the static discharge head plus friction head plus velocity head in the discharge side of the pump.
- *Total dynamic head* - the dynamic discharge head plus dynamic suction lift.

HYDRAULICS

Hydraulic principles are based on and intimately connected with the chemical, physical, and mechanical properties of water. These properties were discussed earlier and include water density, viscosity, and the type of flow (laminar or turbulent) as a function of velocity, temperature, and pressure.

Hydronics deals with two types of fluids: gas and liquid. The main difference between these two fluids is that gas is compressible. Liquids, mainly water in hydronic systems, are basically noncompressible. Water flows in pipes by gravity from a higher to a lower elevation. If the water is required to flow from a lower to a higher elevation or even to flow through pipes in the same horizontal plan, the flow must be assisted (pushed) by a mechanical device such as a pump.

A piping system or network includes pipes, fittings, and valves. Examples of fittings are: elbows (standard, 30, 45, and 90 degrees), tees, close return bends, sudden pipe enlargements or contractions, etc. Examples of valves are: gate, globe, angle, check, butterfly, etc. Valves used in hydronic systems are segregated into two major groups: stop valves (on-off) and regulating (throttling) valves. Throttling valves may also serve as balancing valves. Water flowing through pipes, fittings, valves, and equipment (e.g., boilers, chillers, water heaters, etc.) produces friction or a loss of pressure, Figure 2-4. This pressure loss or resistance to flow occurs because the molecules of water "rub" against the walls of the pipes and fittings. The types of energy involved in the flow of water include kinetic and potential energy.

Hydraulic calculations are important to ensure pipes are sized correctly. If the pipes are not sized correctly and the flow of water stops suddenly, the dynamic force may produce water hammer, shock, or noise. If the velocity is constantly too high when water is flowing, erosion might also occur in the pipes. The general hydraulic problems used in this chapter will include a pumped open system, a gravity open system and a closed system, and a closed tank with a heat exchanger.

Example 2-1. What is the outlet pressure for water flowing in a straight pipe having the following characteristics:

Pipe material: Schedule 40 black steel
Pipe diameter: 2"
Flow: 40 gpm
Length of pipe: 50 ft
Inlet water pressure 25 ft = 10.82 psi

Solution 2-1. Remember that when water flows along the pipe, friction, loss of head, or loss of power result. Also realize that while data is provided for calculations in this book, in an actual hydronic application, this data must either be calculated (e.g., required flow) or selected (e.g., piping material, limiting water velocity in pipes, etc.).

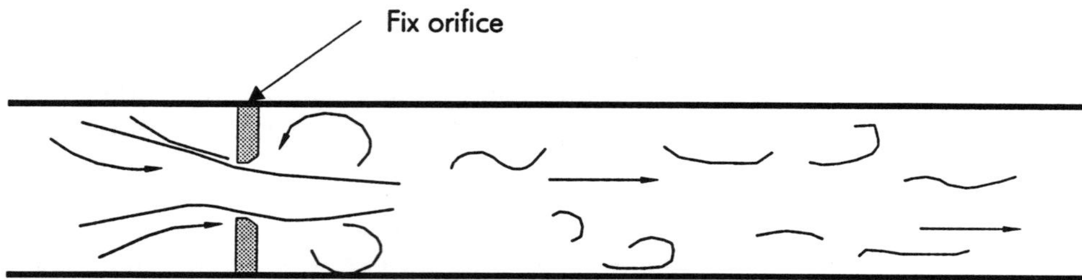

A: How an orifice influences the flow in pipe

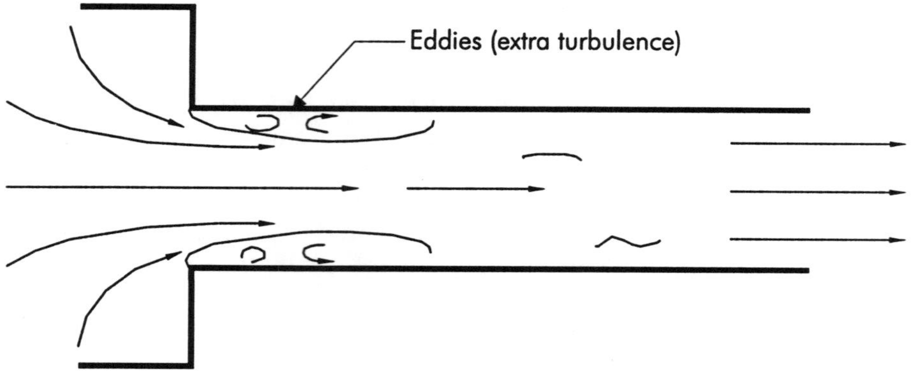

B: Turbulence in the water flow at a sharp contraction

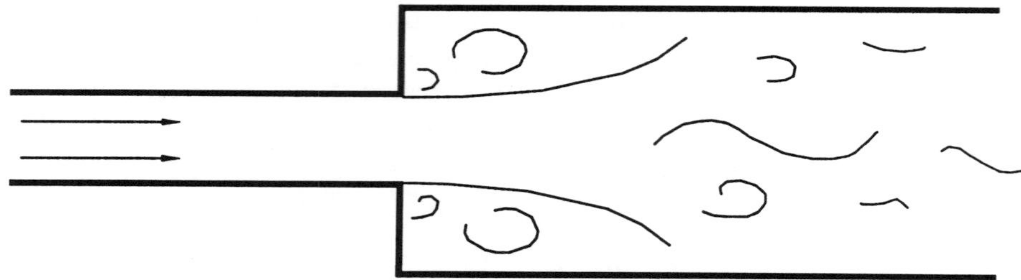

C: Turbulence in the water flow at a sudden expansion

Figure 2-4. Friction in fittings

Based on the information in Appendix B, at a flow of 40 gpm, the velocity of water is 3.82 ft/sec and the head loss is 3.06 ft per 100 ft of Schedule 40 steel pipe. The pipe length in this problem is only 50 ft, so the head loss is:

$$(50 \text{ ft})\left(\frac{3.06 \text{ ft}}{100 \text{ ft}}\right) = 1.53 \text{ ft}$$

Since the pressure at the pipe entrance is given as 25 ft, the pressure available at the pipe exit is 25 feet - 1.53 feet = 23.47 ft, or 10.16 psi, Figure 2-5.

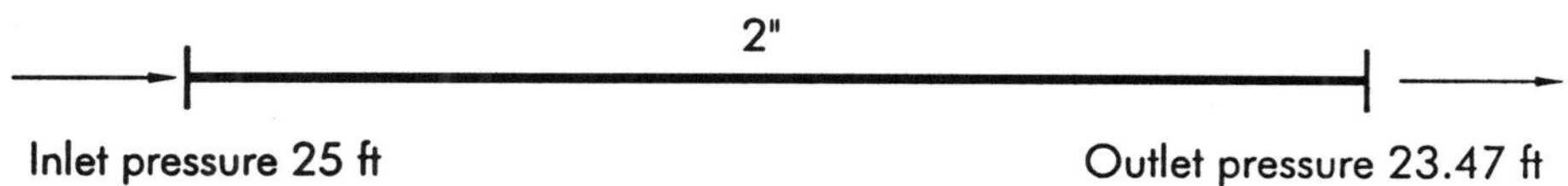

Figure 2-5. Straight pipe

Example 2-2. Based on the data given in Example 2-1, what is the outlet pressure for water flow in pipe with fittings, Figure 2-6? (There is no difference in elevation between pipe entrance and exit.)

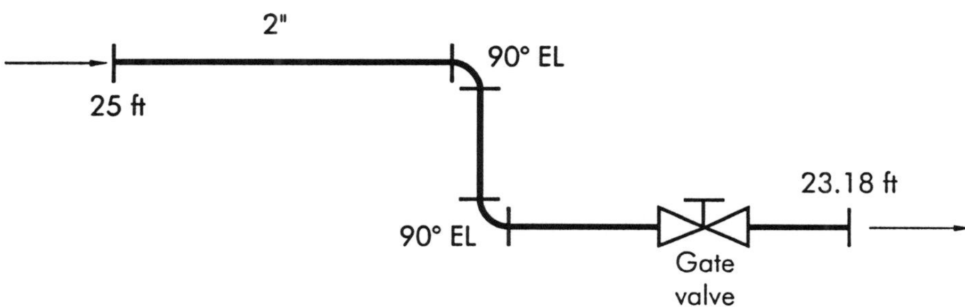

Figure 2-6. Pipe with fittings

Solution 2-2. The velocity head, which may be found in Appendix B, is calculated as follows:

$$hv = \frac{V^2}{2g} = \frac{(3.82)^2}{(2)\ (32.17)} = 0.227 \text{ ft}$$

The friction in fittings and valve (hf) can be found in Appendix B and is based on the velocity head calculation. The K factor is a resistance coefficient, which helps calculate the friction in fittings. It represents the coefficients for each individual fitting, the sum of which equals total friction in fittings (e.g., K1 + K2 + ..., where K represents each fitting) when multiplied with the velocity head. For the problem at hand, the K value for a 2-inch, 90 degree standard elbow is 0.57. The K value for a 2-inch gate valve is 0.15. Thus, the friction loss in fittings is as follows:

$$hf = [K1\ (\text{elbow}) + K2\ (\text{elbow}) + K3\ (\text{valve})]\left(\frac{V^2}{2g}\right)$$

$$= (0.57 + 0.57 + 0.15)\ (0.227) = 0.293 \text{ ft}$$

The friction in this length of pipe was already calculated in Example 2-1, so just add it to the friction in fittings and valve:

$$1.53 + 0.293 = 1.823 \text{ ft}$$

The pressure available at the pipe exit is now:

25 ft - 1.823 ft = 23.177 ft (round to 23.18), or 10.03 psi

In this example, the available pressure is lower because of the added friction in the fittings and valve.

Example 2-3. Based on the data given in Examples 2-1 and 2-2, what is the outlet pressure for water flow in a pipe with a vertical portion, Figure 2-7?

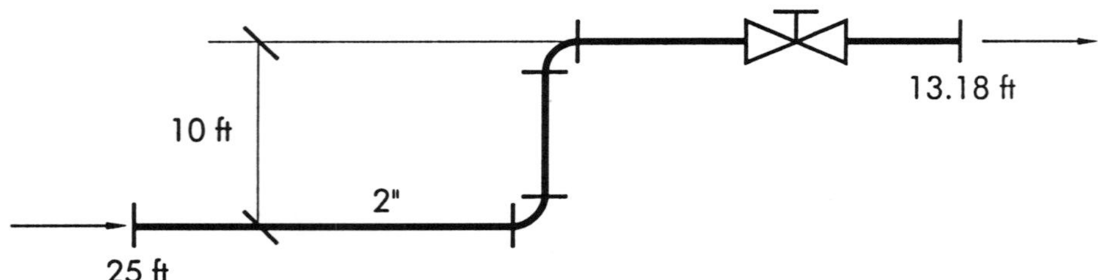

Figure 2-7. Pipe with vertical portion (difference in elevation between the two pipe ends)

Solution 2-3. The calculations made in Example 2-2 remain unchanged since the data is the same. However, there is now some pressure loss due to a 10-ft vertical section.

The friction in pipe and fittings remains the same at 1.823 ft. From the inlet pressure, deduct the difference in elevation:

25 ft - 1.823 ft - 10 ft = 13.177 ft (round to 13.18), or 5.70 psi

Example 2-4. Water must be transferred from a holding tank located at a low elevation to a reservoir located at a higher elevation, Figure 2-8. Calculate the total head loss in this open system in order to select the appropriate pump size to do the job. Use the following data for this calculation:

Pipe material: Schedule 40, standard weight, steel
Pipe diameter: 3"
Flow: 130 gpm
Water velocity: 5.64 ft/sec (Table 2-4)
Friction: 3.9 ft per 100 ft of pipe (Table 2-4)
Limiting velocity: 6 fps (see footnote 1)

Solution 2-4. The system shown in Figure 2-8 can be divided into two sections: the suction side of the pump and the discharge side of the pump. The head loss in the suction side of the pump includes: friction loss in the pipe; friction loss in fittings and valves; and static head (since the flow is

Friction in 3 inch pipe
Asphalt-dipped Cast Iron and New Steel Pipe
(Based on Darcy's Formula)

Flow U.S. gal. per min.	Asphalt-dipped cast iron 3.0" inside dia.			Standard wt. steel - sch 40 3.068" inside dia.			Extra strong steel - sch 80 2.900" inside dia.			Schedule 160 steel 2.624" inside dia.		
	Velocity ft. per sec.	Velocity head-ft.	**Head loss ft. per 100 ft.**	Velocity ft. per sec.	Velocity head-ft.	**Head loss ft. per 100 ft.**	Velocity ft. per sec.	Velocity head-ft.	**Head loss ft. per 100 ft.**	Velocity ft. per sec.	Velocity head-ft.	**Head loss ft. per 100 ft.**
10	0.454	0.000	0.042	0.434	0.003	0.038	0.49	0.00	0.050	0.593	0.005	0.080
15	0.681	0.010	0.088	0.651	0.007	0.077	0.73	0.01	0.101	0.89	0.012	0.164
20	0.908	0.010	0.149	0.868	0.012	0.129	0.97	0.02	0.169	1.19	0.022	0.275
25	1.13	0.02	0.225	1.09	0.018	0.192	1.21	0.02	0.253	1.48	0.034	0.411
30	1.36	0.03	0.316	1.3	0.026	0.267	1.45	0.03	0.351	1.78	0.049	0.572
35	1.59	0.04	0.421	1.52	0.036	0.353	1.70	0.04	0.464	2.08	0.067	0.757
40	1.82	0.05	0.541	1.74	0.047	0.449	1.94	0.06	0.592	2.37	0.087	0.933
45	2.04	0.06	0.676	1.95	0.059	0.557	2.18	0.07	0.734	2.67	0.111	1.16
50	2.27	0.08	0.825	2.17	0.073	0.676	2.43	0.09	0.86	2.97	0.137	1.41
55	2.50	0.10	0.990	2.39	0.089	0.776	2.67	0.11	1.03	3.26	0.165	1.69
60	2.72	0.12	1.17	2.6	0.105	0.912	2.91	0.130	1.21	3.56	0.197	1.99
65	2.95	0.14	1.36	2.82	0.124	1.06	3.16	0.15	1.4	3.86	0.231	2.31
70	3.18	0.16	1.57	3.04	0.143	1.22	3.40	0.18	1.61	4.15	0.268	2.65
75	3.40	0.18	1.79	3.25	0.165	1.38	3.64	0.21	1.83	4.45	0.307	3.02
80	3.63	0.21	2.03	3.47	0.187	1.56	3.88	0.23	2.07	4.75	0.35	3.41
85	3.86	0.23	2.28	3.69	0.211	1.75	4.12	0.26	2.31	5.04	0.395	3.83
90	4.08	0.26	2.55	3.91	0.237	1.95	4.37	0.29	2.58	5.34	0.443	4.27
95	4.31	0.29	2.83	4.12	0.264	2.16	4.61	0.33	2.86	5.63	0.493	4.73
100	4.54	0.32	3.12	4.34	0.293	2.37	4.85	0.36	3.15	5.93	0.546	5.21
110	4.99	0.39	3.75	4.77	0.354	2.84	5.33	0.44	3.77	6.53	0.661	6.25
120	5.45	0.46	4.45	5.21	0.421	3.35	5.81	0.52	4.45	7.12	0.787	7.38
130	5.90	0.54	5.19	5.64	0.495	3.90	6.30	0.62	5.19	7.71	0.923	8.61
140	6.35	0.63	6.00	6.08	0.574	4.50	6.79	0.71	5.98	8.31	1.07	9.92
150	6.81	0.72	6.87	6.51	0.659	5.13	7.28	0.82	6.82	8.90	1.23	11.3
160	7.26	0.82	7.79	6.94	0.749	5.80	7.76	0.93	7.72	9.49	1.40	12.8

Table 2-4. Friction of water (Reprinted from *Cameron Hydraulic Data* book)

against gravity, it is considered a loss). Static head must be added to the friction loss, since the pump must overcome both of these factors to push the column of water up through the pipes.

The friction loss in the pipe (given in ft/100 ft) can be calculated with Darcy's formula, which will be discussed later in this chapter. However, for the purposes of this calculation, the corresponding values will be taken directly from Table 2-4.

On the suction side of the pump, the head loss due to friction (hp) in an 8-ft length of pipe is calculated as follows:

$$hp = (8\text{ ft})\left(\frac{3.9\text{ ft}}{100\text{ ft}}\right) = 0.312\text{ ft}$$

To calculate the friction loss in fittings, use the same formula used in Solution 2-1; that is:

$$hf = (K)\left(\frac{V^2}{2g}\right)$$

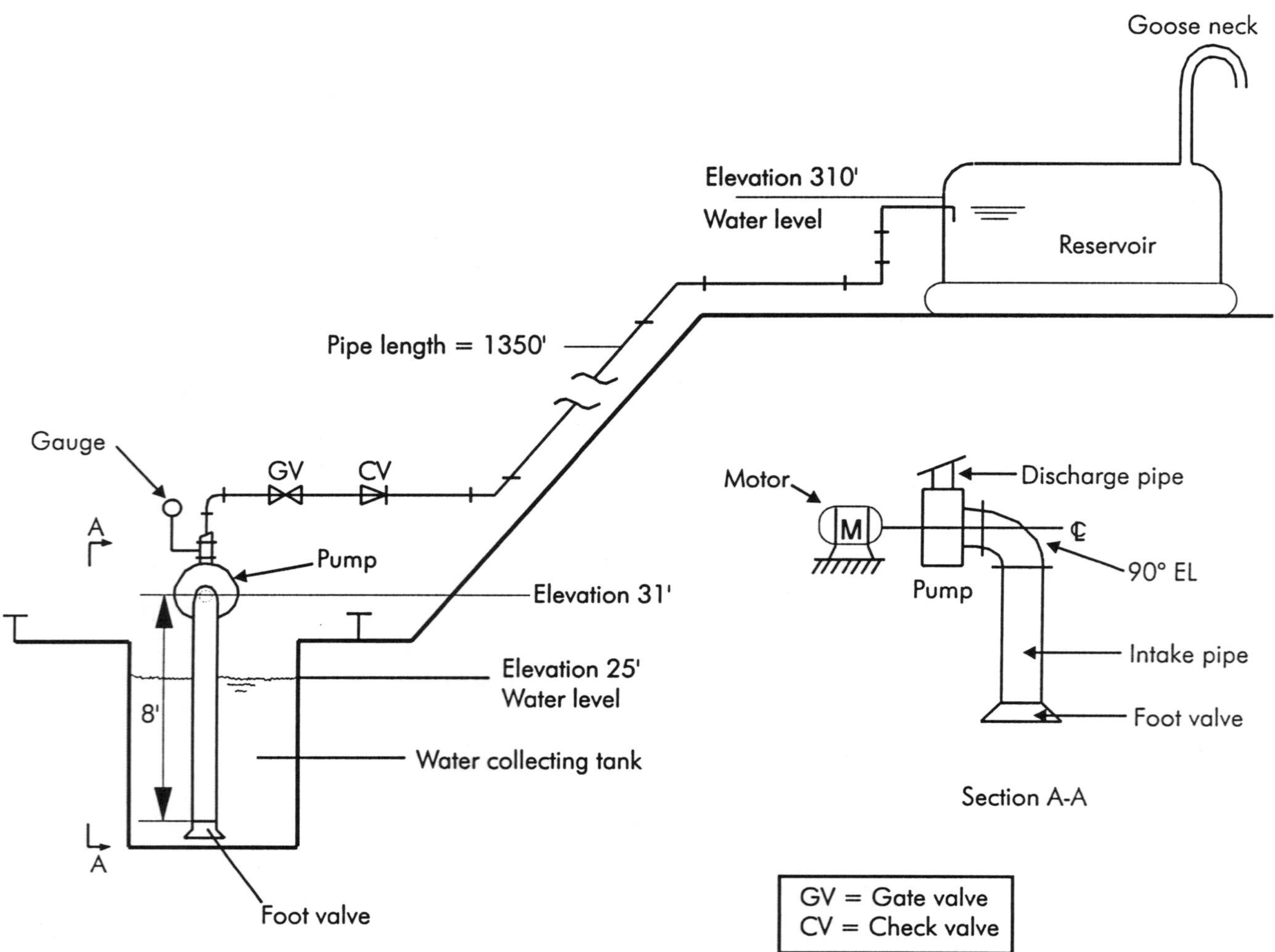

Figure 2-8. Transfer of water from a lower elevation to a higher elevation

where:
- hf = friction in fitting (ft)
- V = water velocity in pipe (5.64 ft/sec per Table 2-4)
- g = the acceleration of gravity (32.174 ft/sec^2)
- K = resistance coefficient for each fitting

therefore: $$hf = (K)\left(\frac{(5.64)^2}{(2)\ (32.174)}\right) = (K)\ (0.494)$$

On the suction side of the pump, the fittings are a foot valve (a type of check valve) and a 90 degree elbow at the entrance of the pump (see Section A-A in Figure 2-8). Based on the tables listing the values of resistance coefficients included in Appendix B, the applicable value of these K coefficients is 1.4 for the foot valve and 0.54 for the 90 degree elbow. Insert these values to complete the equation above:

$$hf = (1.4 + 0.54)\ (0.494) = 0.958 \text{ ft}$$

To calculate the total head loss on the suction side, add the static head to the friction losses (pipes and fittings). Remember that the static head is the difference in elevation between the center line elevation of the pump and the water level[2] in the holding tank. (Since we are calculating the losses by dividing the system into suction and discharge sides of the pump, the pump center line becomes the reference point between the two ends of the system in this case.)

From Figure 2-8, it is possible to calculate the different in elevation:

$$31 \text{ ft} - 25 \text{ ft} = 6 \text{ ft}$$

The total suction head or pressure loss, which is measured in ft, becomes:

6 ft (static head) + 0.312 ft (friction in pipe) + 0.958 ft (friction in fittings) = 7.27 ft

Now it is necessary to calculate the friction or head loss in the discharge side of the pump. For the 1350-ft length of 3" diameter discharge pipe, the friction in the pipe is:

$$(1350 \text{ ft})\left(\frac{3.9 \text{ ft}}{100 \text{ ft}}\right) = 52.65 \text{ ft}$$

Before calculating the friction loss in fittings, first tabulate the applicable K-coefficient values from the following list:

Fitting	Value of K	Pipe diameter
45 degree elbow	0.29	3"
90 degree standard elbow	0.54	3"
Gate valve	0.14	3"
Swing check valve	1.80	3"
Pipe exit	1.00	sharp edge

It is now possible to calculate the friction loss in fittings:

$$hf = (K)\left(\frac{V^2}{2g}\right)$$

$$hf = (0.54 + 0.14 + 1.8 + 0.29 + 0.29 + 0.54 + 0.54 + 1.0)\left(\frac{V^2}{2g}\right)$$

$$hf = (5.14)\ (0.494) = 2.54 \text{ ft}$$

The static head, or difference in elevation, is 279 ft (310 ft - 31 ft). The static head, in this case, is considered a loss. (As mentioned earlier, the lower reference point in this case is the pump center line.)

Since the flow is against gravity and the static head must be overcome, the total discharge head loss (h_d) is:

h_d = 52.65 ft (loss in pipe) + 2.54 ft (loss in fittings) + 279 ft (static head) = 334.19 ft

Adding together the losses on both sides of the pump, the total pressure loss in the system (H_T) is as follows:

H_T = 7.27 ft (suction head) + 334.19 ft (discharge head) = 341.46 ft

Table Verification

In Example 2-4, we obtained the pipe friction value from a neat and orderly table. However, these tables are the result of a great deal of calculation based on the ***Darcy-Weisbach formula.*** Another similar formula is called the *Hazen* ***and Williams empirical formula.*** The Hazen and Williams formula is referred to as empirical because it is based on laboratory and field observations. The Hazen and Williams pressure loss formula is:

$$hf = (0.002083)(L)\left(\frac{100}{C}\right)^{1.85}\left(\frac{Q^{1.85}}{d^{4.8655}}\right)$$

where:
- hf = friction loss in pipe in ft/100 ft
- 0.002083 = an empirically determined coefficient
- L = length of pipe in ft (in this case, it is 100 ft)
- C = roughness coefficient based on the pipe material (Table 2-5)
- Q = flow in gallons per minute (gpm)
- d = pipe diameter in inches

Example 2-5. To ensure the values given in the tables for Example 2-4 are correct, calculate the friction loss using the Hazen and Williams formula and the following data:

L = 100 ft
C = 150
d = 3"
Q = 130 gpm

Solution 2-5. Plug the data into the Hazen and Williams formula as follows:

$$hf = (0.002083)(100)\left(\frac{100}{150}\right)^{1.85}\left(\frac{130^{1.85}}{3^{4.8655}}\right) = 3.8 \text{ ft}$$

Pipe material	Values of C		
	Range	Average value	Normally used value
Bitumastic-enamel-lined steel centrifugally applied	160-130	148	140
Asbestos-cement	160-140	150	140
Cement-lined iron or steel centrifugally applied	-	150	140
Copper, brass, or glass as well as tubing	150-120	140	130
Welded and seamless steel	150-80	140	100
Wrought iron, Cast iron	150-80	130	100
Tar-coated cast iron	145-50	130	100
Concrete	152-85	120	100
Full riveted steel (projecting rivets in girth and horizontal seams)	-	115	100
Corrugated steel	-	60	60

Value of C	150	140	130	120	110	100	90	80	70	60
For (100/C) at 1.85 power is	0.47	0.54	0.62	0.71	0.84	1	1.22	1.5	1.93	2.57

Table 2-5. Values of the constant C used in Hazen and Williams formula

The previous value used from the table was 3.90 ft per 100 ft, which is very close to the one just calculated (3.80 ft/100 ft). This exercise demonstrates that for all practical purposes, the table may be used with confidence.

Alternative Solution

There is an easier and faster way to solve the hydraulic problem given in Example 2-4. It involves using the *equivalent length for fittings*, Table 2-6. This easier alternative is defined as an equivalent length of straight pipe that has the same friction loss as the respective fitting or valve.

Various piping books and publications may indicate slightly different equivalent length values for the same fitting. These differences are usually small and therefore negligible. If a certain type of fitting cannot be found in an available table, an approximate value should be estimated based on a similar fitting. The equivalent value (length) can be obtained from the manufacturer of the fitting, Table 2-7.

Example 2-6. Solve the same problem given in Example 2-4, but use the equivalent length for fittings and valves and make the calculation for the total head loss for the entire system (suction and discharge). The system data remains the same.

Solution 2-6. The developed length of pipe, or the actual measured length is:

1350 ft (discharge) + 8 ft (suction) = 1358 ft

Nominal pipe size	Gate valve -- full open	Globe valve -- full open	Butterfly valve	Angle valve -- full open	Swing check valve -- full open	90° elbow	Long radius 90° & 45° std elbow	Close return bend	Standard tee -- through flow	Standard tee -- branch flow	Mitre bend 45°	Mitre bend 90°
1/2	0.41	17.6		7.78	5.18	1.55	0.83	2.59	1.0	3.1		
3/4	0.55	23.3		10.3	6.86	2.06	1.10	3.43	1.4	4.1		
1	0.70	29.7		13.1	8.74	2.62	1.40	4.37	1.8	5.3		
1 1/4	0.92	39.1		17.3	11.5	3.45	1.84	5.75	2.3	6.9		
1 1/2	1.07	45.6		20.1	13.4	4.03	2.15	6.71	2.7	8.1		
2	1.38	58.6	7.75	25.8	17.2	5.17	2.76	8.61	3.5	10.3	2.6	10.3
2 1/2	1.65	70.0	9.26	30.9	20.6	6.17	3.29	10.3	4.1	12.3	3.1	12.3
3	2.04	86.9	11.5	38.4	25.5	7.67	4.09	12.8	5.1	15.3	3.8	15.3
4	2.68	114	15.1	50.3	33.6	10.1	5.37	16.8	6.7	20.1	5.0	20.1
5	3.36	143	18.9	63.1	42.1	12.6	6.73	21.0	8.4	25.2	6.3	25.2
6	4.04	172	22.7	75.8	50.5	15.2	8.09	25.3	10.1	30.3	7.6	30.3
8	5.32	226	29.9	99.8	33.3	20.0	10.6	33.3	13.3	39.9	10.0	39.9
10	6.68	284	29.2	125	41.8	25.1	13.4	41.8	16.7	50.1	12.5	50.1
12	7.96	338	34.8	149	49.7	29.8	15.9	49.7	19.9	59.7	14.9	59.7
14	8.75	372	38.3	164	54.7	32.8	17.5	54.7	21.8	65.6	16.4	65.6
16	10.0	425	31.3	188	62.5	37.5	20.0	62.5	25.0	75.0	18.8	75.0
18	16.9	478	35.2	210	70.3	42.2	22.5	70.3	28.1	84.4	21.1	84.4
20	12.5	533	39.2	235	78.4	47.0	25.1	78.4	31.4	94.1	23.5	94.1
24	15.1	641	47.1	283	94.3	56.6	30.2	94.3	37.7	113	28.3	113
30	18.7					70	37.3	117	46.7	140	35	140
36	22.7					85	45.3	142	56.7	170	43	170
42	26.7					100	53.3	167	66.7	200	50	200
48	30.7					115	61.3	192	76.7	230	58	230

Table 2-6. Equivalent length for pipe fittings

Size of fittings	1/2		3/4		1		1 1/4		1 1/2		2		2 1/2		3		4		5	6	8	10	12
Material & fitting	C.I.	COP.	C.I.	COP.	C.I.	COP.	C.I.	COP.	C.I.	COP.	C.I.	COP.	C.I.	COP.	C.I.	COP.	C.I.	COP.	C.I.	C.I.	C.I.	C.I.	C.I.
Radiator Valve angle	← 5.0 →																	33.6	31.2	37.2	50.4	62.4	74.4
Boiler	4.0	5.0	5.0	7.0	6.5	9.0	9.0	12.0	9.0	12.0	12.6	16.8	15.6	20.8	18.6	24.8	25.2	33.6	31.2	37.2	50.4	62.4	74.4
Sq. Head Cock Open	← 1.5 →								2.5	2.5	3.2	3.2	3.9	3.9	4.7	4.7	6.1	6.1					
Radiator C.I.	← 7.5 →																						
Std. Venturi			22.5	18	13.5	14	13.5	9.5	13.5	10.5	14.5		14		14								
Super Venturi			37.0	31.5	27.5	28.5	33	21	32.5	26	30		29.5		28.5				31.2	37.2	50.4	62.4	74.4
Air Scoop			2.0	2.0	2.7	2.7	4.0	4.0	4.8	4.8	6.8	6.8	8.0	8.0	13.3	13.3	15.0	15.0	31.2	37.2	50.4	62.4	74.4

C.I. = Cast Iron
COP = Copper

Table 2-7. Excerpts from a manufacturer's guide for equivalent lengths of straight pipe for specialty fittings (Courtesy, Taco, Inc.)

For the application, and based on Table 2-6, the equivalent length for fittings and valves are listed below:

Fittings	Quantity	Equivalent length each fitting (ft)	Total equivalent length (ft)
3" foot valve	1	25.50	25.50
90 degree elbow	4	7.67	30.68
45 degree elbow (long radius)	2	4.09	8.18
3" gate valve	1	2.04	2.04
3" swing check valve	1	25.50	25.50
Sharp pipe exit (estimate)	1	17.50	17.50
			109.40

Now add the total equivalent length for fittings to the actual (developed) pipe length as follows:

$$1358 \text{ ft} + 109.4 \text{ ft} = 1467.4 \text{ ft} = \text{total equivalent length of pipe}$$

Therefore, the total friction loss in pipe, fittings, and valves is:

$$(1467.4 \text{ ft})\left(\frac{3.9 \text{ ft}}{100 \text{ ft}}\right) = 57.22 \text{ ft}$$

The total difference in elevation of the water levels (static head) is equal to 285 ft (310 ft - 25 ft); therefore, the total head (pressure) loss is 342.22 ft (57.22 ft + 285 ft).

Compare the previous value of 341.46 ft, which was obtained in Example 2-4 using the K resistance coefficients, to the value of 342.22 ft, which was obtained using the equivalent length method. Both results are close; therefore, it is easier to use the equivalent length method when solving hydronic problems. Keep in mind that these results are based on an engineering judgment, in which case minor approximations are acceptable.

These examples have shown that by knowing the flow, pipe diameter, system configuration, and pipe material, it is possible to calculate the head or friction loss in the system. When calculating a problem like this, consider that the piping system will age, so add in a safety factor of 10% to 15% to establish the acceptable value when selecting the pump. Thus, a value or approximately 380 ft is required for the selection of the pump head (342.22 ft + 37.78 ft). The pump head value and the pump flow (previously given as 130 gpm) are used in the selection of the pump.

The objective of the examples given earlier was to determine the correct pump size for a system. A pump is a mechanism that is used to push a liquid with a specific force to overcome system friction loss and any existing difference in elevation. The pump produces this force with the help of a motor or driver and consumes energy in the process.

Example 2-7. Select a pump for a closed hydraulic system, which is very similar to a hydronic system. The system is shown in Figure 2-9, and the data is as follows:

Pipe material: Schedule 40 standard steel
Flow: 130 gpm
Water velocity: 5.6 ft/sec
Friction coefficient in pipe: 3.90 ft/100 ft
Developed pipe length or actual length: 420 linear ft
Pipe diameter: 3"

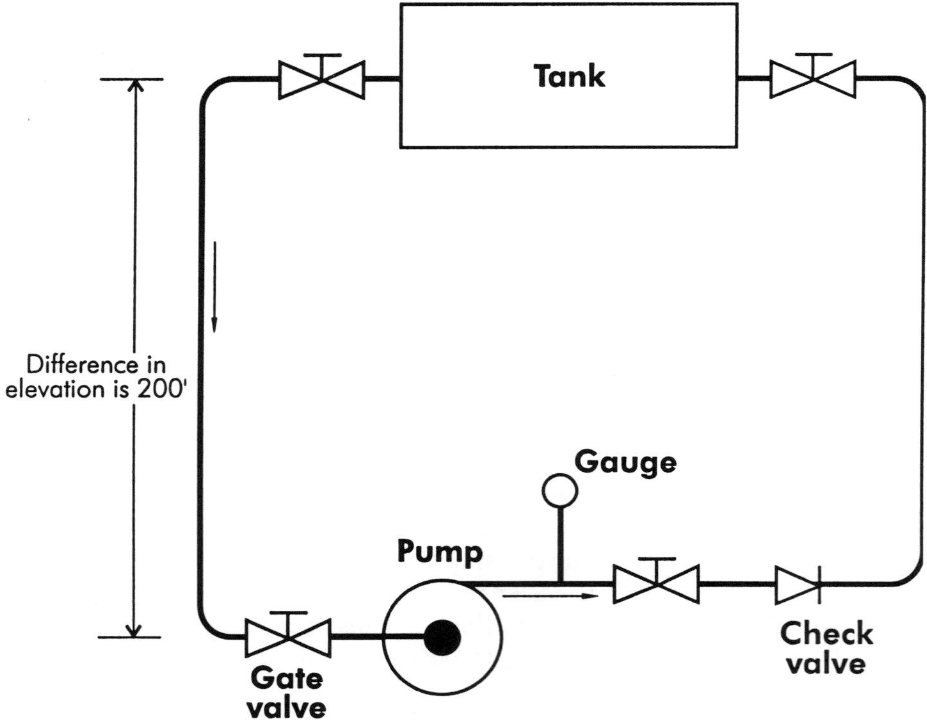

Figure 2-9. Closed hydraulic system

Solution 2-7. Using Table 2-6, it is possible to determine the equivalent length for the fittings in this system:

Fittings	Quantity	Equivalent length each fitting (ft)	Total equivalent length (ft)
90 degree elbow	4	7.67	30.68
3" gate valve	4	2.04	8.16
Swing check valve	1	25.50	25.50
Sharp pipe exit	1	17.50	17.50
Sharp pipe entrance	1	17.50	17.50
Gauge	1	0	0
			99.34

Add the equivalent length for fittings and valves to the actual (developed) pipe length:

$$420 \text{ ft} + 99.34 \text{ ft} = 519.34 \text{ ft}$$

The total friction loss in pipe, fittings, and valves is:

$$(519.34 \text{ ft})\left(\frac{3.9 \text{ ft}}{100 \text{ ft}}\right) = 20.25 \text{ ft}$$

In this case, the static head at the pump suction is equal to the static head at the pump discharge, so they cancel each other out. In other words, the energy required to force the water in a closed hydronic loop is not affected by the static pressure. The pump is selected solely on the friction loss of the total equivalent length of pipe and the pressure loss in apparatus. This is valid if the system is primed or full of water when it starts.

The pump for this system must be selected for a flow of 130 gpm and a head of:

$$20.25 \text{ ft} + 2.75 \text{ ft (safety factor)} = 23 \text{ ft}$$

Example 2-8. Calculate the outlet pressure in the system shown in Figure 2-10, in which water flows by gravity. The technical data is as follows:

Water flow (Q): 50 gpm
Pipe diameter: 2"
Pipe material: Type K copper tubing
Pipe length: 380 linear feet (developed length)
Fittings: Two gate valves; one sudden contraction (from the tank into the pipe); one sudden enlargement (discharge open to the atmosphere); one 30 degree elbow; one 45 degree elbow
Difference in elevation: 150 ft (static head)

Solution 2-8. From Appendix B, the flow of 50 gpm water in a 2" diameter Type K copper tubing (pipe) has a velocity of 5.32 ft/sec and a friction loss of 5.34 ft per 100 ft.

Using Table 2-6, it is possible to determine the equivalent length for the fittings in this system:[3]

Fittings	Quantity	Equivalent length each fitting (ft)	Total equivalent length (ft)
45 degree elbow	1	2.76	2.76
2" gate valve	2	1.38	2.76
30 degree elbow[4]	1	2.76	2.76
Sudden contraction[5]	1	10.30	10.30
Sudden enlargement[6]	1	10.30	10.30
			28.88, or round to 29

The total equivalent length of the pipe and fittings is:

$$380 \text{ ft} + 29 \text{ ft} = 409 \text{ ft}$$

Friction loss in the pipe and fittings is:

$$(409 \text{ ft})\left(\frac{5.34 \text{ ft}}{100 \text{ ft}}\right) = 21.84 \text{ ft}$$

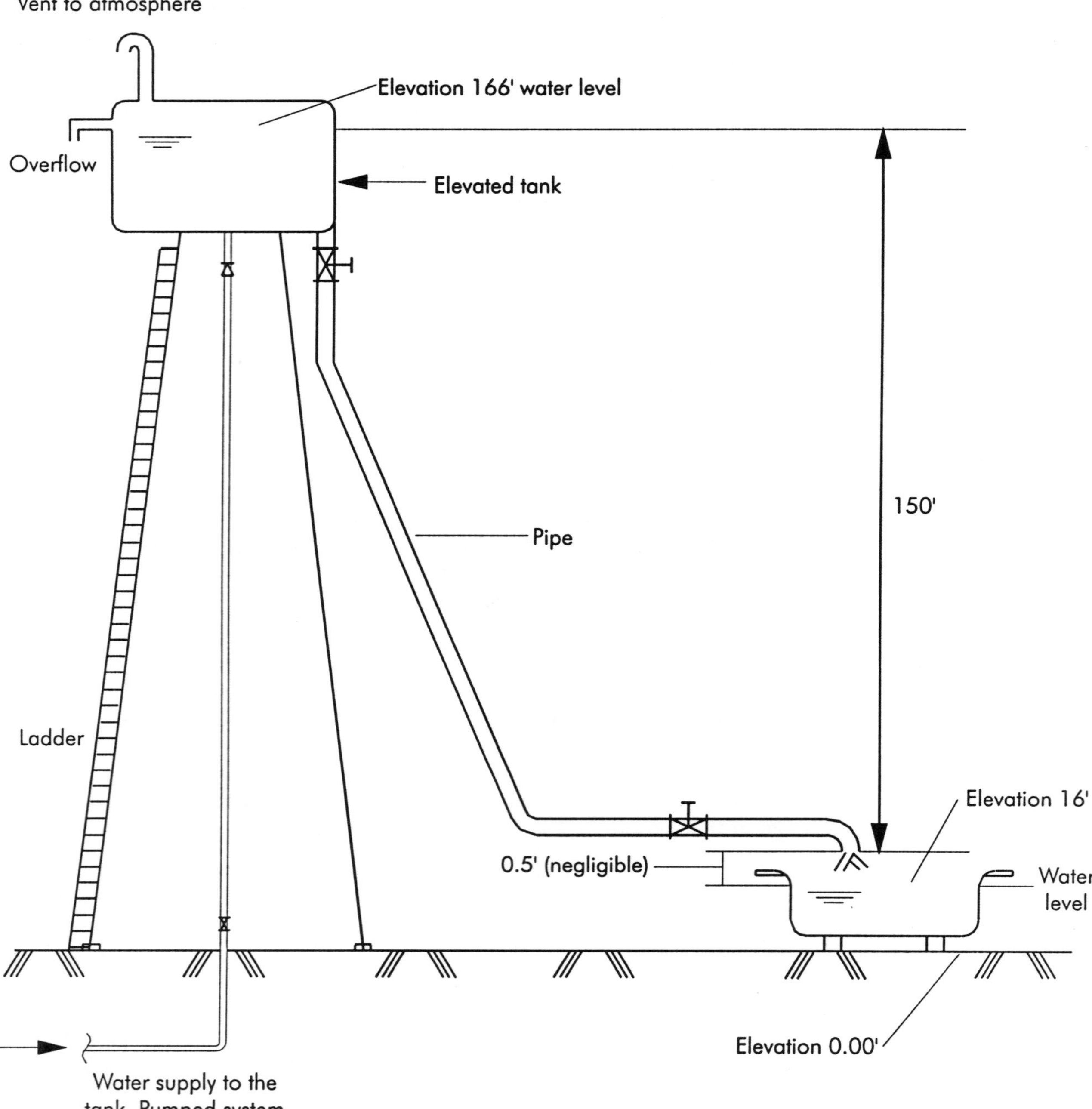

Figure 2-10. Gravity system

Given the difference in the elevation of 150 feet, the outlet pressure is:

$$150 \text{ ft} - 21.84 \text{ ft} = 128.16 \text{ ft}$$

$$\frac{128.16 \text{ ft}}{2.31 \text{ ft/psi}} = 55.48 \text{ psi}$$

The difference in elevation (static head) in this gravity flow type example (or in any other downhill flow) assists the flow, because the weight of the water column pushes the water down toward the discharge. This is the reason why the friction is *deducted* from the static head.

Example 2-9. Calculate the same problem given in Example 2-8, only this time with a flow of 200 gpm. All other data remains the same. Based on Appendix B, the velocity is 21.3 ft/sec, and the friction is 65.46 ft per 100 ft. *Note: A velocity of 21.3 ft/sec is unacceptable, but it is used here for illustrative purposes.*

Solution 2-9. From Example 2-8, we determined that the equivalent length of pipe and fittings is 409 ft. Therefore, the friction loss in pipe and fittings is:

$$(409 \text{ ft})\left(\frac{65.46 \text{ ft}}{100 \text{ ft}}\right) = 267.73 \text{ ft}$$

Given the difference in the elevation of 150 ft, the outlet pressure has a negative value:

$$150 \text{ ft} - 267.73 \text{ ft} = -117.73 \text{ ft}$$

The result of the calculation means that 200 gpm cannot flow through the system, because the pipe diameter is too small and the friction is too high for such flow. Water will flow but only at a maximum rate of 145 gpm. This value can be mathematically calculated as follows:

$$\left(\frac{X \text{ ft}}{100 \text{ ft}}\right)(409 \text{ ft}) = 150 \text{ ft static head}$$

$$X = \frac{(150 \text{ ft})(100 \text{ ft})}{409 \text{ ft}} = 36.67 \text{ ft}$$

From Appendix B, for a 2" Type K copper tube at the friction calculated above the corresponding flow is approximately 145 gpm.

Example 2-10. Calculate the required pump head in a pumping system that has equipment located at different elevations, Figure 2-11. In this system, the pump transfers water at 100°F from an open tank through a heat exchanger where water is heated to 160°F into a closed tank, which is under a constant pressure of 20 psi. The problem data is as follows:

Flow rate (water): = 74,600 lb/hr
Pipe material: Schedule 40 standard steel
Pipe diameter: 3"

Solution 2-10. First calculate the specific gravity of water entering the heat exchanger at 100°F and leaving the heat exchanger at 160°F. To make the calculation, a reference temperature point is required. This reference point is usually the density of water at 60°F.[7] The specific gravity of water at 100°F is:

$$\frac{\rho\ 100°F}{\rho\ 60°F} = \frac{v\ 60°F}{v\ 100°F} = \frac{0.016033}{0.016130} = 0.994$$

where: ρ = water density
v = specific volume

The specific density at a higher temperature is inverse to the specific volume. The specific gravity of water at 160°F is (see Table 2-1):

$$\frac{\rho\ 160°F}{\rho\ 60°F} = \frac{v\ 60°F}{v\ 160°F} = \frac{0.016033}{0.016395} = 0.978$$

The next step is to determine the gpm. Since water volume increases with temperature, it is necessary to determine the gpm for 100° and 160°F. Only the lb/hr is given, but it is easy to convert to gpm using the following formula:

$$gpm = \frac{lb/hr}{(60\ min/hr)(8.33\ lb/gal)(specific\ gravity)}$$

For 100°F:

$$gpm = \frac{74{,}600\ lb/hr}{(60\ min/hr)(8.33\ lb/gal)(0.994)} = 150.2$$

For 160°F:

$$gpm = \frac{74{,}600\ lb/hr}{(60\ min/hr)(8.33\ lb/gal)(0.978)} = 152.6$$

From an engineering point of view, the difference in gpm between the two calculations is small and may be ignored, because a safety factor of 10% to 15% will eventually be added to pump capacity. The basic flow of 150 gpm may be used.

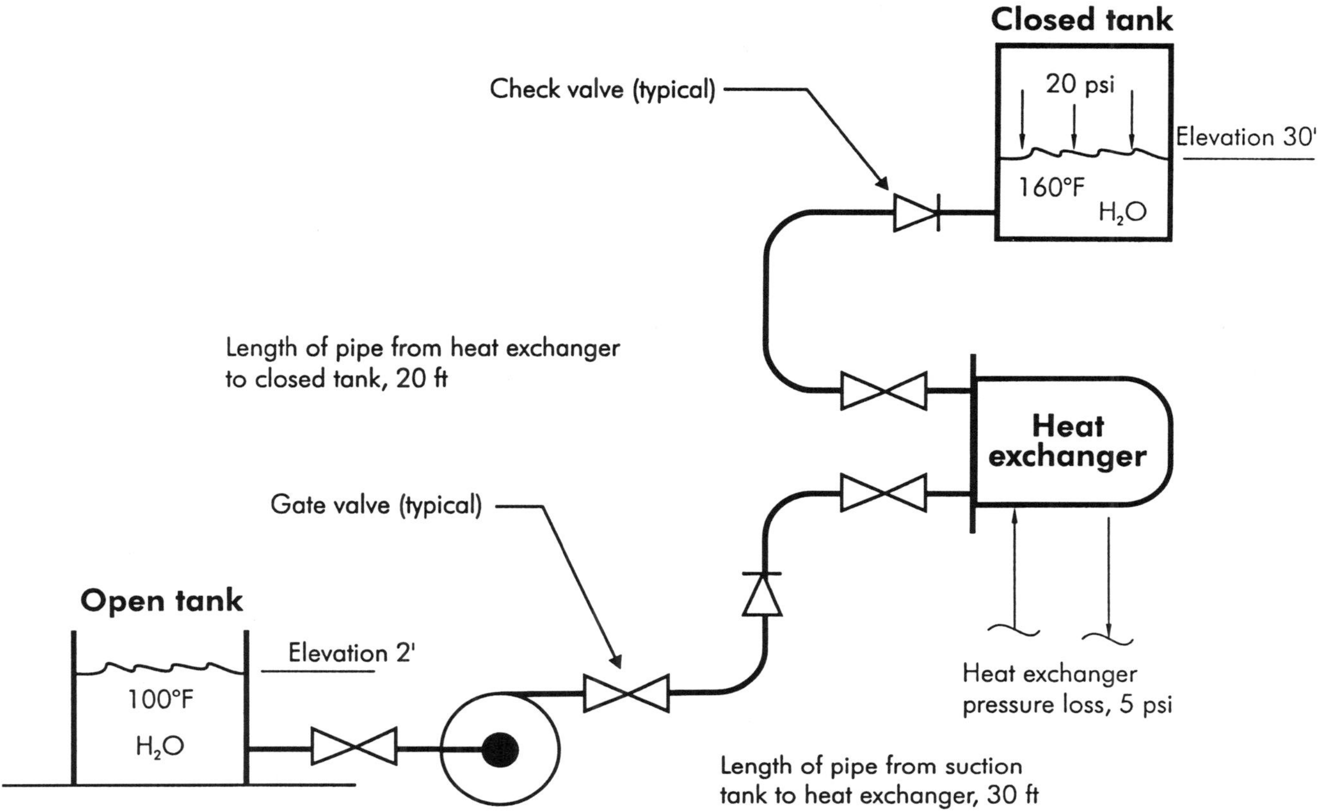

Figure 2-11. Pumping system with different elevations

Find the average specific gravity in the heat exchanger by taking the measurements found earlier and averaging them:

$$\frac{0.994 + 0.978}{2} = .986$$

Based on manufacturer's information, the heat exchanger head loss is 5 psi.

The next step is to add the measured length of pipe (developed length) to the equivalent length for fittings. The total developed length of pipe is 50 ft (30 ft + 20 ft). Using Table 2-6, it is possible to determine the equivalent length for the fittings in this system:

Fittings	Quantity	Equivalent length each fitting (ft)	Total equivalent length (ft)
90 degree elbow	5*	7.67	38.25
Gate valve	4	2.04	8.16
Entering/exit loss	2	15.30	30.60
Swing check valve	2	25.50	51.00
			128.01

*The fifth elbow is at the pump inlet.

The total equivalent length of pipe plus fittings and valves is 178.01 ft (50 ft + 128.01 ft). We can round the number up to 180 ft.

The total friction loss (see Appendix B for friction in ft per 100 ft in a 3" Schedule 40 steel pipe when flowing 150 gpm) is:

$$(180 \text{ ft})\left(\frac{5.13 \text{ ft}}{100 \text{ ft}}\right) = 9.23 \text{ ft}$$

The head loss is:

$$(30 - 2) + 9.23 + 11.71 + \left(\frac{(20)(2.31)}{0.978}\right) = 96.17 \text{ (97) ft}$$

The head loss in the heat exchanger is determined as follows:

$$\frac{(5 \text{ psi})(2.31 \text{ ft/psi})}{\text{specific gravity}} = \frac{11.55}{0.986} = 11.71 \text{ ft}$$

To determine the pump head, add the 10% safety factor to the total head loss:

$$97 \text{ ft} + 9.7 \text{ ft} = 106.7 \text{ ft (round to 107.0)}$$

The final chosen pump shall have a capacity of 150 gpm and a head of 107 ft. Consequently, the pump characteristics shall be 150 gpm capacity, @ 107 ft head.

Using the specific gravity modified for the supply water temperature throughout this head loss calculation introduces a very small error. Table 2-8 illustrates the error introduced by the temperature difference.

where:
- CHW = Chilled water
- LTW = Low temperature hot water
- MTW = Medium temperature hot water
- HTW = High temperature hot water
- HHTW = High-high temperature hot water
- M = Median temperature
- S = Supply temperature

The conclusion is that the positive nature of the error justifies the simplification obtained from basing all calculations on the nominal supply of water temperature.

System	Temperatures				Calculation method*			Error, %		
	Supply		Return							
	°F	°C	°F	°C		°F	°C	Pipe friction	Fitting loss	Total head loss
CHW	40	4	60	16	M	50	10	-0.1	0	-0.1
					S	40	4	1.3	-0.1	1.2
LTW	180	82	140	60	M	160	71	-0.1	0	-0.1
					S	180	82	0.1	0.7	0.1
MTW	300	149	140	60	M	220	104	-1.0	-0.6	-1.0
					S	300	149	2.1	3.4	2.2
MTW	300	149	220	104	M	260	127	-.02	-1.0	-0.2
					S	300	149	1.6	2.0	1.6
HTW	400	204	240	116	M	320	160	-0.8	-0.7	-0.8
					S	400	204	4.2	4.8	4.3
HHTW	440	227	240	116	M	340	171	-1.4	-1.2	-1.3
					S	440	227	5.8	6.5	5.9

*M = median temperature, S = supply temperature

Table 2-8. Temperature calculated method error (Courtesy, E. Hansen, *Hydronic System Design and Operation*, 1985, reprinted with permission by McGraw-Hill, Inc.)

MEASUREMENTS

The following is a short list of some useful units of measurement:

Acceleration of gravity = 32.2 ft/sec^2
1 ft^3 H_20 = 62.3 lb
1 gal = 0.1337 ft^3 H_20
1 gal H_20 = 8.33 lb (at 70°F)
1 ft^3 = 7.48 gal H_20

Therefore:
1 ft^3 of H_20 = (7.48) (8.33) = 62.3 lb
1 atm = 14.696 ~ 14.7 lb/in^2 or psi = 29.92 in. Hg (mercury)
1 atm = 33.96 ft of water
1 psi = 2.31 ft of water
1 ft of H_20 = 0.433 psi (see footnote 8)

NOTES

[1] In most piping applications, a velocity of 6 fps or less is advisable. Pipe sizes based on this velocity usually give years of trouble-free operation.

[2] The difference in elevation (static head) is normally from water level to water level or to a water discharge outlet elevation.

[3] More (and possibly slightly different) equivalent length values are available from other specialty tables. There is no problem with minor differences because at the end of the calculation a safety factor is added.

[4] Estimated to be the same as the 45 degree elbow.

[5] Estimated length as a standard tee through the branch.

[6] Ibid

[7] The specific gravity for water at various temperatures may be found in Appendix C.

[8] The conversion value of 0.433 is derived from the following:

$$0.433 = \frac{14.7 \text{ psi}}{33.96 \text{ ft}}$$

Reciprocal: 1psi = 2.31 ft of H_20.

Chapter 3

Heat Load Calculations

In order to size a hydronic system for either a cooling or heating operation, it is necessary to determine or roughly calculate the load required for cooling and/or heating. These calculations of heat gains and losses are related to hydronic systems; however, they are part of the larger sphere of hvac (heating, ventilating, and air conditioning) systems. Such calculations involve the use of numerous tables and formulas, which will be discussed briefly in this chapter but are not an integral part of this book. For reference, heat loss and gain blank calculation sheets are included in Appendix C.

A cooling system is designed to remove sensible and latent heat from a space. A heating system is designed to add sensible heat to a space. Both cooling and heating systems provide a better ambient environment for people working, living, or resting in temperature-controlled areas as well as required when specialized equipment is in operation. In other words, they improve working and living conditions.

Water System Classification

To facilitate the understanding of heating and cooling loads, it is first necessary to learn some of the terms used in hydronic systems.[1]

Low Temperature Water System (LTW)

This type of hot water heating system operates within the pressure and temperature limits of the ASME boiler construction code for low pressure heating boilers. The maximum allowable working pressure for low pressure heating boilers is 160 psi (1100 kPa) with a maximum temperature limitation of 250°F (121°C). The usual maximum working pressure for boilers for LTW systems is 30 psi (200 kPa), although boilers specifically designed, tested, and stamped for higher pressures may frequently be used with working pressures up to 160 psi. Steam-to-water or water-to-water heat exchangers are also often used to produce LTW.

LTW systems are used in buildings ranging in size from small single dwellings to very large and complex structures. Terminal heat transfer units include convectors, cast iron radiators, baseboards, commercial finned-tubes, fan-coil units, unit heaters, unit ventilators, and multizone air-handling units. Modern LTW systems are of the forced type and use a pump to circulate water.

A heat transfer coil inside or outside of the boiler is often used to supply hot water directly to the domestic water system or to a storage tank. A large storage tank may be included in the system to store energy in the form of hot water to be used when heat input devices such as the boiler or a solar energy collector are not supplying energy.

Medium Temperature Water System (MTW)

This type of hot water heating system operates at temperatures of 350°F (175°C) or less, with pressures not exceeding 150 psi (1030 kPa). The usual design supply temperature is approximately 250° to 325°F (120° to 160°C), with a usual pressure rating for boilers and equipment of 150 psi (1030 kPa).

High Temperature Water System (HTW)

This type of hot water heating system operates at temperatures over 350°F (175°C), with usual pressures of about 300 psi (2100 kPa). The maximum design supply water temperature is 400° to 450°F (200° to 230°C) with a pressure rating for boilers and equipment of about 300 psi (2100 kPa). The pressure-temperature rating of each component must be checked against the design characteristics of the system.

High-High (or Very High) Temperature Water System (HHTW)

A hot water heating system operating at temperatures of 420° to 455°F (215° to 235°C). Characteristics: high pressure, widest TD (temperature difference).

Chilled Water System (CW)

This type of chilled water-cooling system operates with usual design supply water temperatures of 40° to 55°F (4.4° to 13°C) and normally operates within a pressure range of 125 psi (860 kPa). Antifreeze or brine solutions may be used for systems (usually process applications) that require temperatures below 40°F (4.4°C). In some cases, even well water, which supplies a temperature of 60°F (15°C), can be used.

Dual-Temperature Water System (DTW)

This type of combination water heating and cooling system circulates separate hot and/or chilled water with common return piping and separate terminal heat transfer apparatus. They operate within the pressure and temperature limits of LTW systems, with usual winter design supply water temperatures of about 100° to 150°F (38° to 66°C) and summer supply water temperatures of 40° to 55°F (4.4° to 13°C).

Condenser Water System

This type of condenser water system removes the heat from water-cooled refrigerant condensers, usually in connection with cooling towers or city or well water services.

Steam System Classification

Temperature is the thermal state of both liquid and vapor at any given pressure. The vapor temperature can be raised by adding more heat, resulting in a superheated system. A superheated system is used in the following situations:

- Where higher temperatures are required
- In large distribution systems to compensate for heat losses and to ensure that steam is delivered at the desired saturated pressure and temperature
- To ensure that the steam is dry and does not contain entrained liquid, which could damage the turbine in some turbine-driven equipment

Some of the terms used in steam systems include the following:

- *Enthalpy of the liquid (sensible heat)* - the amount of heat in Btu (kJ) required to raise the temperature of a pound (kg) of water from 32°F (0°C) to the boiling point at the pressure indicated.
- *Enthalpy of evaporation (latent heat of vaporization)* - the amount of heat required to change a lb (kg) of boiling water at a given pressure to a lb (kg) of steam at the same pressure. This same amount of heat is released when the vapor condenses back to a liquid.
- *Enthalpy of steam (total heat)* - the combined enthalpy of liquid and vapor, which represents the total heat above 32°F (0°C) in the steam.
- *Specific volume* - the reciprocal of density and is the volume of unit mass, which indicates the volumetric space that 1 lb (1 kg) of steam or water will occupy.

Some of the unique properties and advantages of steam are as follows:

- Most of the heat content of steam is stored as latent heat, which permits large quantities of heat to be transmitted efficiently with little change in temperature. Since the temperatures of saturated steam are pressure-dependent, a negligible temperature reduction occurs from a reduction in pressure caused by pipe friction losses as steam flows through the system. This occurs regardless of insulation efficiency as long as the boiler maintains the initial pressure. The steam traps remove the condensate.
- Steam, as all fluids, flows from areas of high pressure to areas of low pressure and is able to move throughout a system without an external energy source (no pump required). Heat dissipation causes the vapor to

condense, creating a reduction in pressure caused by the dramatic change in specific volume (1600:1 at atmospheric pressure).

- As steam gives up its latent heat at the terminal equipment, the condensate that forms is initially at the same pressure and temperature as the steam. When this condensate is discharged to a lower pressure (as when a steam trap passes condensate to the return system), the condensate contains more heat than necessary to maintain the liquid phase at the lower pressure. This excess heat causes some of the liquid to vaporize or "flash" to steam at the lower pressure. The amount of liquid that flashes to steam can be calculated.

BASICS

It is important to remember that heat is not destroyed but merely transferred from one place to another. One of the basic rules of physics states that energy can neither be created nor destroyed. Heat is a form of energy. The intensity of sensible heat can be measured with a thermometer in degrees Fahrenheit (°F) or Celsius (°C). Appendix D contains information concerning changing the degrees from Fahrenheit to Celsius or vice versa.

Latent heat is the energy added to or taken from a substance, which does not change the temperature of the substance. Instead, the heat energy causes the substance to change its state. An example of this is when water changes from a liquid to a solid (ice) or from a liquid to steam. *Sensible heat* is the thermal energy that changes the temperature of a substance or object. Sensible heat can be measured with a thermometer.

The British thermal unit (Btu) is the unit of measurement that indicates the quantity of heat content or heat released. The definition of a Btu is the amount of heat needed to raise the temperature of one pound (lb) of water one °F. For cooling purposes, it is the amount of heat that must be removed to lower the temperature of one pound of water one °F. The most common unit of time used is Btu per hour (Btuh), which is the rate at which heat is transferred in one hour. A larger unit of heat removal or transfer from a warm area to the outdoors is a *refrigeration ton*, which is equal to 12,000 Btuh.

Heat is always conveyed from a higher temperature to a lower one. This means that to remove heat from a room, equipment must be used that is capable of circulating a fluid with a lower temperature than that of the room. The refrigeration process uses refrigerant to accomplish this task. Well water or ice may do the same job, but they are not practical.

Heat is transferred by convection, conduction, and radiation.

COOLING SYSTEMS

A refrigeration system employs equipment that removes heat. Such a system removes heat from an enclosed place and releases it to the outdoor atmosphere or a body of water. Refrigerant is used to transfer the heat (both latent and sensible) in the refrigeration cycle.

Two of the most common refrigerants used in refrigeration systems are R-12 and R-22. These refrigerants are in the process of being phased out, because they have the ability to deplete the protective ozone layer in the stratosphere. New refrigerants are being developed that will take their place.

A chiller produces chilled water, which is then pumped to finned coils. The lower temperature of water circulated in the finned coils reduces or lowers the temperature in the space where they are located by absorbing the heat for ambient comfort or industrial/commercial purposes. Heat accumulated by the refrigerant in the chiller is transferred to water and then normally rejected to the atmosphere by an open or closed cooling tower. A closed cooling tower has finned coils that are cooled by air and is called an indirect tower system.

The amount of chilled water that must be circulated through the cooling coils to satisfy the installation must be calculated. When making these calculations, the psychrometric chart is often useful (see Appendix C). Conditions that are plotted on a psychrometric chart include certain properties of air: dry bulb temperature, wet bulb temperature, dew point temperature, and relative humidity. These help define the selection of proper refrigeration equipment. Other information required for the calculation include indoor design conditions, outdoor design conditions (based on the geographic location), and heat released from various sources. This information may be found in specialty books or through associations such as the American Society of Heating, Refrigerating and Air-Conditioning Engineers, Inc.

Major sensible heat gains in an area can either be external or internal. The external heat gains include heat from the sun and infiltration. Internal heat gains include people, heat-producing equipment, electric motors, appliances, lighting, food, and hot pipes, ducts, etc. The formulas used to calculate these heat gains may be found in ASHRAE books. The formulas shown here for cooling load calculation procedures are reprinted with permission from the ASHRAE *Handbook--Fundamentals*, 1989.

Cooling Load Calculation Procedures

The following formulas are for external heat gains.

The formula for the **roof** is:

$$q = UA\,(CLTD)$$

where:

- q = the amount of heat gain in Btuh applicable in each case
- U = roof assembly heat transfer coefficient
- A = area calculated from building plans or field measured for a remodeling
- CLTD = cooling load temperature difference (roofs)

Note: Adjust CLTD for (a) latitude-month correction, (b) exterior surface color, (c) indoor design temperature, (d) outdoor design temperature, (e) attic conditions, (f) U-values, and (g) insulation.

The formula for the **walls** is:

$$q = UA\,(CLTD)$$

where:
U = wall design heat transfer coefficient
A = area calculated from building plans
CLTD = cooling load temperature difference (walls)

Note: Adjust CLTD for same factors as roofs.

The formula for **glass** is:

$$q\ \text{conduction} = UA\,(CLTD)$$

where:
U = glass design heat transfer coefficient
CLTD = cooling load temperature difference (glass)

Note: Adjust CLTD for (a) inside design temperature, (b) outside design temperature, and (c) daily range.

$$q\ \text{solar} = A(SC)(SHGF)(CLF)$$

where:
SC = shading coefficient
SHGF = maximum solar heat gain by orientation, latitude, and month
CLF = cooling load factor with no interior shade or with shade

The formula for **partitions**, **ceilings**, and **floors** is:

$$q = UA\Delta t$$

where:
U = design heat transfer coefficient
A = areas calculated from building plans or fields measured for a remodeling
Δt = design temperature difference, from an adjacent unconditioned area to room

The following formulas are for internal heat gain.

The formula for **people** is:

$$q\ \text{sensible} = N(\text{Sensible heat gain})(CLF)$$
$$q\ \text{latent} = N(\text{Latent heat gain})$$

where: N = number of people in space from best available source
Sensible and latent heat gain = calculated from occupancy
CLF = cooling load factor, people, by hours of occupancy

Note: CLF = 1.0 with high occupancy density.

The formula for **lights** is:

$$q = (\text{Input})(\text{CLF})$$

where: Input = rating from electrical plans or lighting fixture data, Btuh
CLF = cooling load factor, lights, by use of schedule and hours
CLF = 1.0 with 24 hour operation

The formula for **power** is:

$$q = (\text{Heat gain})(\text{CLF})$$

where: CLF = 1.0 with 24 hour operation

The formula for **appliances** is:

$$q\ \text{sensible} = (\text{Heat gain})(\text{CLF})$$

where: Sensible and latent heat gain = appliance manufacturer's data
CLF = cooling load factor by scheduled hours (1.0 with 24 hour operation)

Note: Set latent heat to 0 if appliance is under exhaust hood.

For **ventilation** and **infiltration air,** the formulas are:

$$q\ \text{sensible} = 1.10Q\Delta t$$
$$q\ \text{latent} = 4840Q\Delta W$$
$$q\ \text{total} = 4.5Q\Delta H$$

where: Q = ventilation cfm, ASHRAE Standard 62; add infiltration cfm
Δt = outside/inside air temperature difference, °F
ΔW = outside/inside air humidity ratio difference, lb water/lb dry air
ΔH = outside/inside air enthalpy difference, Btu/lb dry air

There are quite a few tables from which particular data may be obtained to be included in the various formulas listed above. These tables include the following:

- Solar gain through glass
- Solar and transmission gains through roof and walls
- Internal heat gain:
    - — number of people
    - — power equipment
    - — number of lights
    - — number of appliances

By adding all of the individual heat gains, the sensible heat can be determined. The next step is the latent heat calculation, which includes tables indicating the latent heat gain from the following sources:

- Air infiltration
- People
- Steam produced
- Appliances
- Other heat gains
- Vapor transmission

A certain amount (10% to 20%) of fresh air is continuously introduced into the space, which must also be cooled. The rest of the air is recirculated. During summer, this fresh air contributes to the increase of the load so the following items must be added:

- Sensible air heat
- Latent air heat
- Recirculated air heat gain

All of the elements discussed so far must be considered in the calculation of all air conditioning systems. While these calculations are not considered to be part of hydronics proper, they are a subdiscipline, which requires specialized understanding and professional skill.

The heat gains discussed earlier may be clarified by the following statements:

- Solar heat gain is reduced by shading devices, overhangs, reveals, and adjacent buildings.
- Solar gain must be carefully selected, because it does not always peak as normally expected.
- Outside air temperature used to calculate the heat inflow is listed in climate tables, which are based on records over a long period of time. Various state codes indicate values to be used for various cities.
- For indoor heat gain from people, include convection, respiration, and perspiration. The amount depends on the physical activity and the surrounding conditions.
- Power for electric motors is a significant load found mostly in industrial applications. The motor and equipment driven may or may not be located within the same enclosure.

- Certain tanks containing hot liquids are open to the atmosphere, so evaporation takes place. The heat released, sensible and latent, must be determined and accounted for.

After the heat gain or loss is estimated, the appropriate equipment can be selected to offset the respective load. An example based on a heat gain summary and the detailed hydronic calculation and equipment selection is included in Appendix G.

The equipment must be sized for the peak load occurrence. The load estimate starts with data obtained based on the area survey. Once the data is gathered, a load estimating sheet (see Appendix C) must be completed. The survey is based on architectural drawings and includes the following:

- Orientation of building - point of compass
- Building usage (residence, commercial, hospital, laboratory, industrial, etc.)
- Space dimensions (based on architectural drawings available)
- Construction of building components, walls, roofs, windows, etc.
- Number of occupants
- Lighting
- Electric motors operating
- Appliances and business equipment
- Fresh air introduced per person

HEAT LOAD CALCULATION

In the winter, a heat loss calculation will determine the heating load required. Based on this load, the heating equipment and the system as a whole may be selected. The heat loss is based on the structure (e.g., walls, windows, roof, partitions, slab), as well as infiltration of outside air and fresh outside air purposely introduced for ventilation.

Structural transmission coefficients are required in order to calculate heat losses. In addition, outdoor design temperatures (based on available tables, U.S. Weather Bureau information, or indicated in the state code) must be known, and the indoor space design temperature must be established. This information may be obtained from special technical books.

To calculate a heating load, prepare the following information connected to the building design and weather data at design conditions:[2]

- Select outdoor design weather conditions: temperature, wind direction, and wind speed. Winter climatic data can be found in ASHRAE guides or obtained from the U.S. Weather Bureau or state code.
- Select the indoor air temperature to be maintained in each space during the cold weather (check state code energy conservation rules).
- Estimate temperatures in adjacent unheated spaces.
- Select or compute heat transfer coefficients for outside walls and glass; inside walls, non-basement floors, and ceilings, if these are next to unheated spaces; and the roof.

- Determine net area of outside wall, glass, and roof. These determinations can be made from building plans or from the actual building, using inside dimensions.
- Compute heat transmission losses for each kind of wall, glass, floor, ceiling, and roof in the building by multiplying the heat transfer coefficient in each case by the area of the surface and the temperature difference between indoor and outdoor air or adjacent unheated space.
- Compute heat losses from basement or grade-level slab floors.
- Select unit values and compute the energy associated with infiltration of cold air around outside doors, windows, and other openings. These unit values depend on the kind or width of crack, wind speed, and the temperature difference between indoor and outdoor air. An alternative method is to use air changes.
- When positive ventilation using outdoor air is provided by an air-heating or air-conditioning unit, the energy required to warm the outdoor air to the space temperature must be provided by the unit. The type of calculation for this load component is identical to that for air infiltration. Besides the heat required for the outdoor air drawn in by the unit, the unit must also provide heat to compensate for natural infiltration losses. If no mechanical exhaust is used and the outdoor air supply equals or exceeds the amount of natural infiltration, some reduction in infiltration heat loss may be considered.
- The total heat load is represented by the sum of the transmission losses or heat lost through the confining walls, floors, ceiling, glass, and other surfaces, plus the heat loss associated with cold air entering by infiltration and that required to replace mechanical exhaust.
- In buildings with a sizable, reasonable, steady internal heat release from sources other than the heating system, compute and deduct this heat release under design conditions from the total heat losses computed above.
- Consider using pick-up loads that may be required in intermittently heated buildings or in buildings using night thermostat setbacks. Pick-up loads frequently require an increase in heating equipment capacity to bring the temperature of structure, air, and material contents to the specified temperature. At the end of the calculation, 5% to 15% of the heat loss calculated must be added as a safety factor.

Because of the difference between outside and inside temperatures combined with heat flowing from high to low temperatures, structure heat loss occurs from within a heated building. The hvac engineer must determine the rate at which the heat is lost. The basic formula is:

$$H = AU\Delta t$$

where:
H = heat loss, Btuh
A = area of the surface, ft^2
U = heat transfer coefficient, Btuh per ft^2 per °F
Δt = temperature difference between outdoor air and indoor air design temperature

The rate at which heat will travel through a construction material depends on the nature of the construction material and its thickness. This rate is called the *U factor* (heat transmission coefficient) of a material. It indicates how many Btu will flow through a material with one-square-foot surface in one hour for each degree of temperature difference between inside and outside.

The construction and insulating materials are rated based on the R factor (construction material resistance to heat flow). To calculate the U factor (heat transmission coefficient), use the following formula:

$$\frac{1}{R} = U$$

To determine the heat load necessary to compensate for infiltration and/or outdoor air supply, use the following formula:

$$Q = (cfm)\ (1.08)\ (\Delta t)$$

where:
- Q = heat load, Btuh
- cfm = cubic feet of air per minute
- 1.08 = coefficient
- Δt = temperature difference between outdoor air and indoor air design temperature

To calculate the number of air changes per hour, use the following formula:

$$AC = \frac{(cfm)(60)}{V}$$

where:
- AC = number of air changes per hour
- cfm = cubic feet of air per minute
- V = space volume, ft^3

Appendix C contains a table that will help to determine heat losses quickly and easily.

A heating calculation allows for proper selection of heating equipment (e.g., boiler in the hydronic system), as well as the pipe network and appurtenances. In order to provide a safety factor, indoor heat released from people, certain small appliances, light, etc., should not be deducted from the heat loss calculation unless they are steady gains. Tables are available for recommended indoor temperatures for various types of occupancies, summer and winter.

SAFETY FACTORS

In order to determine the required capacity of a heating or cooling system, it is common practice to apply a factor or series of factors to the heat gain or heat loss calculations. This factor is a safety factor, and there is little guidance available concerning how to apply a safety factor to heat gain calculations.

Safety factors are generally applied either as individual factors used to account for the unknown within specific parts of a calculation or as an overall safety factor applied to the total nominal heat gain or heat loss. Sometimes both approaches are used, which may result in larger-than-necessary components being selected. Some basic rules of thumb concerning safety factors are as follows:

- Safety factors for preliminary and final heat gains should be within the appropriate ranges (5% to 10%).
- Selection of a particular safety factor value should be the judgment of the technician or engineer based upon the level of uncertainty in a calculation.
- The safety factor range for piping must apply to both insulated and uninsulated pipes and include all allowances for supports, hangers, valves, etc.

NOTES

[1] These definitions are reprinted by permission of the American Society of Heating, Refrigerating and Air-Conditioning Engineers, Atlanta, Georgia, from the 1987 ASHRAE *Handbook--HVAC Systems and Applications.*

[2] Reprinted by permission of the American Society of Heating, Refrigerating and Air-Conditioning Engineers, Atlanta, Georgia, from the 1985 ASHRAE *Handbook--Fundamentals.*

Chapter 4

Piping, Pumps, and Accessories

Mankind has used piping to convey liquids (mostly water) for thousands of years. Archaeologists have discovered piping made from hollowed out logs, clay pipes, and bronze pipes throughout different stages of human evolution. This chapter will include information concerning modern pipe materials, arrangements, supports, sizing, as well as radiators, coils, and pumps.

Piping

The hot and chilled water produced in a boiler or chiller must reach the users by way of a network of pipes, valves, and other fittings. Pipe sizes must be correct in order to achieve the necessary requirements. In addition to flow and pressure, there must be reduced noise levels and appropriate velocity to prevent water hammer. (Water hammer is produced when the water velocity in pipes is high and a valve is suddenly closed.)

At the same time, economical pipes sizes must be selected to prevent overdesign and unwarranted costs. The most economical pipe sizes are the ones accurately determined or selected from tables when the flow and limiting velocity are known.

Professional skill is continuously accumulated, but practical experience and knowledge are required in the selection of pipe sizes and the installation of a good working hydronic system. In summary, the pipes in a hydronic system must be:

- the correct size;
- well supported;
- able to accommodate the expansion and contraction due to hot water circulation;
- able to accommodate minor vibration;
- tight (no leakage).

When installing the water piping system indoors, there are some practical rules that should be respected:

- Water pipes must be installed parallel to the walls in an organized and geometrically regular system.
- Pipes must be pitched to allow for drainage and prevent air accumulation.
- Pipes may not be installed in stairwells, hoistery, elevator shafts, or in front of windows, doors, or any other wall opening.
- Vertical pipes are usually located in chases. Enough depth of the chase is necessary to accommodate pipes and fittings.
- The location of horizontal pipes installed above ceilings must be checked (first on the drawings) to ensure they do not interfere with other pipe ducts, lights, beams, etc., located above the ceiling.
- In specialty rooms (e.g., telephone rooms, electrical rooms, control rooms, etc.), water pipe installation is usually not permitted. However, special construction of pipe within pipe or concrete encasement may be necessary.
- Pipes may not be installed in areas where they may be subject to freezing or corrosion without special protection like heat tracing plus insulation, corrosion resistant paint, etc.
- Pipe location must not prevent access to equipment, panels, motor control centers, etc.
- Connections between piping and equipment (e.g., boilers, chillers, pumps, water heaters, etc.) must include shut-off valves and a disconnectible union or flange for easy dismantling.
- Manual or automatic air vents must be installed at high points.
- A drain valve or dirt trap must be installed at the lowest point in a piping system.
- All water heating pipe mains must be installed level or with a maximum pitch of 1 in. per 100 ft in the flow direction.
- Branches of mains must be as short and uniform as possible.
- Steam pipes must be pitched in the direction of the flow to allow for condensate to flow toward the traps.
- Hot water, steam, and chilled water pipes must be insulated.

When installing the water piping system outdoors, the following practical rules apply:

- Underground pipes must be installed a minimum of 3 ft from the building foundation (except for those that penetrate the foundation).
- Pipes penetrating through the building foundation must be installed with insulation within steel sleeves. The space between the pipe plus

insulation (if any) and the sleeve must be filled with a waterproof elastic caulking material to prevent building settlement from shearing the pipe.

- Underground pipes must be installed below the frost level and laid in firm prepared beds.
- Pipes must be reasonably protected from injury or aggressive environments.
- Hot water, steam, and chilled water pipes must be insulated.

Pipe Materials

Pipes may be fabricated of ferrous (iron base metals such as various forms of steel) or non-ferrous (copper, brass, aluminum, plastic) materials. Ferrous pipes used in hydronic systems include seam-welded steel and seamless steel, which may be black or galvanized metal. Pure metals are seldom used, because alloys are more resistant to the environment. Naturally, each type of material has unique characteristics that make it more or less desirable for a particular application. When the proper material is selected for the piping system, few problems are anticipated. Incorrect piping material could present many problems that may render the system vulnerable to hazardous situations.

As knowledge and technology advance, certain materials may be replaced with better and safer materials. An example of this is the use of lead pipe and lead containing solder. Both materials are now prohibited in domestic water systems. Studies showed that lead dissolves in the water and is poisonous to humans. Certain chemicals used in hot water treatment to prevent deposits on pipe walls are also no longer used, because it was determined that they, too, are harmful to people and the environment.

There are a number of factors contributing to piping material selection for a hydronic installation. Some of them are applicable to overhead pipes, some to underground pipes, and some to both. These factors include the following:

- Initial material cost - pipes and fittings
- Installation cost
- Expected service life
- Weight
- Ease of making joints
- Chemical resistance
- Susceptibility to corrosion
- Thermal expansion
- Friction losses coefficient
- Pressure rating
- Rigidity
- Resistance to crushing
- Ease of replacement (availability in the area)

Piping materials commonly used for various piping system applications include the following:

Application	Pipe material
Hot water heating	Black steel, hard copper
Steam	Black steel (extra heavy for high pressure)
Condensate	Black steel, hard copper
Chilled water	Black or galvanized steel, hard copper
Domestic hot water	Galvanized steel, copper

Copper Piping

Copper piping is widely used in hydronic systems. The type of copper piping (tubing) used in hydronic systems is usually Type K, which has the heaviest (thickest) pipe wall. Copper pipe is lighter and more resistant to corrosion than steel pipe; however it costs nearly twice as much as the black steel pipe per linear foot.

There are two kinds of copper tubing: *flexible* (soft temper) and *rigid.* Flexible copper tubing accepts only compression fittings or flare fittings. To make a compression fitting, slide the nut and then tighten the nut to the fitting. The soft metal of the compression ring will be pressed against the fitting and pipe to form a water-tight and long-lasting seal. Flexible-type copper pipes can usually be shaped by hand; however, a special bending tool may also be used.

If soldering a connection to a fitting that has other tubing already soldered to it, wrap the finished joints with wet rags to protect them from heat. Before using a torch, drain all water from pipes.

Steel Piping

One of the ways steel pipe is classified is by its wall thickness. This is called the *pipe schedule*, and it has been used in the U.S. for the last 60 years. Steel pipes that are designated as Schedule 10 have a thin wall and a low resistance to pressure. Schedule 160 pipes have very heavy, thick walls. Another type of steel pipe has even heavier wall construction and is designated as XX Strong. The most common type of steel pipe used is standard steel Schedule 40.

Black steel pipe is strong, durable, and relatively inexpensive. It is available in standard 21-ft lengths, but it is also available in a large variety of precut, prethreaded sizes or may be cut to order. When pipe is cut to order, it can often be threaded at the same time.

When connecting pipes, fittings, or valves, apply joint-sealing compound or plastic joint-sealing tape to the outside threads. Tighten joints with a wrench to avoid straining the rest of the pipes. Do not over-tighten joints, and never screw a pipe more than one turn after the last thread is no longer visible.

Table 4-1 indicates the characteristics for Schedule 40 and Schedule 80 steel pipes. Schedule 30 is also included for sizes over 10 inches in diameter.

Nominal pipe size (in.)	Nominal metric size (mm)	Schedule no.*	Outside diameter (in.)	Inside diameter (in.)	Wall thickness (in.)	Trans. area (sq in.)
1/2	13	40(S)	0.840	0.622	0.109	0.3040
		80(X)	0.840	0.546	0.147	0.2340
3/4	19	40(S)	1.050	0.824	0.113	0.5330
		80(X)	1.050	0.742	0.154	0.4330
1	25	40(S)	1.315	1.049	0.133	0.8640
		80(X)	1.315	0.957	0.179	0.7190
1-1/4	32	40(S)	1.660	1.380	0.140	1.495
		80(X)	1.660	1.278	0.191	1.283
1-1/2	38	40(S)	1.900	1.610	0.145	2.036
		80(X)	1.900	1.500	0.200	1.767
2	50	40(S)	2.375	2.067	0.154	3.355
		80(X)	2.375	1.939	0.218	2.953
2-1/2	64	40(S)	2.875	2.469	0.203	4.788
		80(X)	2.875	2.323	0.276	4.238
3	76	40(S)	3.500	3.068	0.216	7.393
		80(X)	3.500	2.900	0.300	6.605
4	100	40(S)	4.500	4.026	0.237	12.73
		80(X)	4.500	3.826	0.337	11.50
5	125	40(S)	5.563	5.047	0.258	20.01
		80(X)	5.563	4.813	0.375	18.19
6	150	40(S)	6.625	6.065	0.280	28.99
		80(X)	6.625	5.7610	0.432	26.07
8	200	40(S)	8.625	7.981	0.322	50.0
		80(X)	8.625	7.625	0.500	45.6
10	250	40(S)	10.750	10.020	0.365	78.9
		(X)	10.750	9.750	0.500	74.7
		80	10.750	9.564	0.593	71.8
12	300	30	12.750	12.090	0.330	115.0
		(S)	12.750	12.000	0.375	113.1
		40	12.750	11.938	0.406	111.9
		(X)	12.750	11.750	0.500	108.0
		80	12.750	11.376	0.687	101.6
14	350	30(S)	14.000	13.250	0.375	138.0
		40	14.000	13.125	0.438	135.3
		(X)	14.000	13.000	0.500	133.0
		80	14.000	12.5000	0.750	122.7
16	400	30(S)	16.000	15.250	0.375	183.0
		40(X)	16.000	15.000	0.500	176.7
		80	16.000	14.314	0.843	160.9
18	—	(S)	18.000	17.250	0.375	234.0
		(X)	18.000	17.000	0.500	227.0
		40	18.000	16.874	0.562	224.0
		80	18.000	16.126	0.937	204.2
20	—	30(S)	20.000	19.000	0.500	284.0
		40	20.000	18.814	0.593	278.0
		80	20.000	17.938	1.031	252.7

* S = standard X = strong pipe

Table 4-1. Steel pipe characteristics (Reprinted by permission from *Heating/Piping/Air Conditioning*, October 1990)

More details regarding piping material alternatives and their specific application in hydronic systems may be found in the sample specification included in the appendix.

Pipe Installations

Pipes may be connected to pipes, fittings, or valves in any of the following ways depending on the material listed:

- Flanged (cast steel or cast iron)
- Welded (forged steel)
- Screwed (black or galvanized steel)
- Brazed (copper or brass)

Pipes are fabricated in certain standardized lengths, which must be assembled to form an installation. Steel pipe may have threaded ends, so the pieces and fittings are screwed together. Copper tubes are soldered together.

When hot water circulates through pipes, the pipes first expand and then contract as the water cools. Expansion and contraction must be calculated and the piping arrangement made so that movement is limited to a maximum of 1-1/2". Hydronic pipes are not installed in a straight, continuous line in ordinary buildings, so expansion loops are usually not required, because every bend helps to absorb small pipe movements. However, if the length is large or if the water (or steam) has high temperature, expansion calculations are required to determine the need and size of expansion loops. These are dependent on pipe length, fluid temperature, and the piping material. Calculations for expansion are based on formulas and specialized tables, which detail shape, size, and other installation elements. These tables are not included in this book, but they can be found in other specialized piping manuals.

Hydronic pipes that convey cold and hot water must be insulated. Cold pipes are insulated to avoid condensation, and hot water pipes are insulated to reduce heat loss. Both are insulated to be protected against freezing, if such a possibility exists. Piping insulation is called thermal insulation. There is a large selection of materials to be used for insulation, and the hydronic specialist must make an appropriate choice. Insulation material and its thickness must be compared to cost. Table 4-2 lists the common types of insulation that may be used in hydronic, as well as other systems. The sample specification in the appendix gives practical details on the insulation types and their installation.

Pipe Sizing

As illustrated in the examples in Chapter Two, the sizing of pipes in a hydronic system is based on the flow and the selection of (a limiting) velocity. It should be clear that the friction loss, head loss, or loss of pressure along the pipe is dependent on the flow velocity. To select the correct and most economical pipe size, a calculation should be performed.

Type of insulation	Applicable ASTM standard	Forms of insulation	Service temperature range (°F)	Thermal conductivity,* Btu in./(hft² °F) mean temperature	Common application
Calcium silicate	C-533, type I	Pipe, block	Above ambient to 1200	0.41	High-pressure steam, hot water, condensate; load bearing
Cellular glass	C-552	Pipe, block	-450 to 800	0.38	Dual temperature, cold water, brine; load bearing
Cellular elastomer, flexible	C-534	Pipe, sheet	-40 to 200	0.27	Dual temperature to 200°F, cold water, runouts, non-load bearing
Cellular polystyrene	C-578, type I	Board (pipe)	-65 to 165	0.25	Dual temperature to 165°F, cold water, hot water to 165°F; limited load bearing
Cellular polyurethane	C-591, type I	Board (pipe)	-40 to 225	0.16	Dual temperature to 200°F, cold water, hot water to 225°F; limited load bearing
Mineral fiber; fiberglass	C-547, class 1	Pipe	-20 to 450	0.23	Dual temperature to 450°F, steam, condensate, hot and cold water; non-load bearing
Mineral fiber; rock or slag	C-547, class 3	Pipe	Above ambient to 1200	0.23	Steam, condensate, hot and cold water; non-load bearing

*The ASTM standards listed give maximum thermal conductivity values, which are approximately 10% higher than the nominal values given by most insulation manufacturers.

Table 4-2. Various types of thermal insulation

Every engineer and designer wants to provide an economical installation that will cost less to install and operate. A balance must be established between the cost of the pipe and the cost of power during operation. If a larger diameter pipe is used, the installation will cost more. However, the larger pipe will have less friction and conversely less power consumption.

Example 4-1. Determine the most economical pipe size for a hydronic system in which a pump must supply 200 gpm water through a 500-ft equivalent length of steel pipe Schedule 40 (the total equivalent length includes the actual pipe length plus the equivalent length of fittings and valves). The following data is known:

Pump efficiency: 70%
Price per kilowatt-hour (kWh): $0.10
Hours of operation: 2000 hours/year

Other pertinent information is as follows:

Pipe size (inches)	Friction (ft/100 ft)	Cost of pipe installed ($)*	Interest & repairs @ 14% year
3	8.90	9775	1368
3-1/2	4.26	11,500	1610
4	2.25	13,000	1820
5	.728	20,000	2800
6	.304	26,000	3640

*Data obtained from R.S. Means Estimating Manual.

Solution 4-1. To determine the power consumption, use the following formula:

$$\text{kWh} = \left(\frac{(\text{Q gpm})(\text{Friction})}{3960}\right)\left(\frac{1}{\text{EFF}}\right)(0.746 \text{ kW})(\text{hr of operation/year})$$

where: EFF = pump efficiency
0.746 kW = 1 hp
Q = flow in gpm
3960 = coefficient

Replace the known values in the above formula:

$$\text{kWh} = \left(\frac{(200 \text{ gpm})(\text{Friction})}{3960}\right)\left(\frac{1}{.70}\right)(0.746\text{kW})(2000 \text{ hr/year}) = (107.75)(\text{friction})$$

Calculate the cost of power for each pipe size:

Pipe size (inches)	Friction (ft/500 ft)	Cost of power ($)*	Total cost per year power, repairs, interest
3	44.50	479.5	1847.5
3-1/2	21.30	229.5	1839.5
4	11.25	121.2	1941.2
5	3.64	39.2	2839.2
6	1.52	16.4	3656.4

* Cost of power is: (107.75) (friction loss in the equivalent pipe length) ($0.10/kWh).

This comparison calculation shows that the 3-1/2" diameter pipe is the most economical to be installed. This is a trial and error calculation based on the pipe size, friction loss, and pump horsepower. It renders the most economical pipe size based on initial cost and operation cost. In order for a system to function well, it is desirable to maintain a near constant velocity in the pipes of various dimensions. For additional and/or reduced flow in branches, sizing pipes based on a velocity that is too high might generate unwanted noise and erosion. Smaller diameter pipes cost less, but for a small job, the difference is insignificant. For a large job, a balance between initial cost, power expenditure, good management, and maintenance requirements must be determined. The cost must include the power consumption, because higher velocity triggers higher pressure loss and higher power consumption.

As shown in the examples included in Chapter Two, equivalent length of fittings, valves, expansion joints, etc., must be added to the developed (measured) pipe length to obtain the total equivalent length. When the total equivalent length is multiplied by the friction coefficient, the result is the total friction or head loss in a system. The pressure loss or friction loss calculations (the base for pipe sizing) may be done using either the Darcy-Weisbach or Hazen and Williams formula. Charts and tables listing pipe diameters, pipe materials, flow, friction coefficients, and velocities are also available. Most of these tables are calculated based on a water temperature of 60°F. When higher temperatures are dealt with, some errors are introduced. These are tolerable considering the safety factors that are normally included at the end of each calculation.

Velocity

Pressure drop depends on the flow velocity. The pressure drop range for small pipes in a hydronic system should be between 1 and 4 ft/100 ft with an acceptable average of 2.5 ft/100 ft. ASHRAE recommends velocities of 4 fps for pipes up to 2" diameter. Larger pipe size velocities may be slightly greater (up to a maximum of 6 to 6.5 fps).

Table 4-3 shows the velocities and gpm ranges for various steel and copper pipe sizes that are recommended by one manufacturer.

Another important element included in the pipe sizing calculation is the water flow rate. In a hydronic system, this can be determined after the selection of the water temperature entering and returning from the system. The water flow rate is based on the amount of heat load. The formula to calculate the heat load is:

$$q = M_w Cp(t_1 - t_2)$$

where:
- q = the total system heat load, or the heat load for a particular heat distributing unit (Btuh)
- M_w = mass of water circulated (lb/hr)
- Cp = specific heat of water usually taken as 1
- $(t_1 - t_2)$ = design water temperature difference (TD) between supply and return measured in °F, for the entire system or for a single heat-distributing unit within the system (as applicable)

The above equation may be easily rearranged to yield the mass flow of water (M_w), which is normally done for high temperature water (HTW) systems (as applicable). Most calculations for low temperature water (LTW) systems use flow rate in gallons per minute. Therefore M_w in lb/hr may be converted to gpm by the following relationship:

$$M_w = (gpm)\ (60)\ (8.17) = 490\ gpm$$

where:
- 60 = minutes per hour
- 8.17 = weight measured in pounds of 1 gal of water at 180°F

	STEEL PIPE				COPPER TUBING			
Nominal Pipe or Tubing Size Inches	Minimum GPM	Velocity Ft/Sec	Maximum GPM	Velocity Ft/Sec	Minimum GPM	Velocity Ft/Sec	Maximum GPM	Velocity Ft/Sec
1/2"	---	---	1.8	1.9	---	---	1.5	2.1
3/4"	1.8	1.1	4.0	2.4	1.5	1.0	3.5	2.3
1"	4.0	1.5	7.2	2.7	3.5	1.4	7.5	3.0
1 1/4"	7.2	1.6	16.	3.5	7.	1.9	13.	3.5
1 1/2"	14.	2.2	23.	3.7	12.	2.1	20.	3.6
2"	23.	2.3	45.	4.6	20.	2.1	40.	4.1
2 1/2"	40.	2.7	70.	4.7	40.	2.7	75.	5.1
3"	70.	3.0	120.	5.2	65.	3.0	110.	5.2
3 1/2"	100.	3.2	170.	5.4	90.	3.1	150.	5.2
4"	140.	3.5	230.	5.8	130.	3.5	210.	5.6
5"	230.	3.7	400.	6.4	---	---	---	---
6"	350.	3.8	610.	6.6	---	---	---	---
8"	600.	3.8	1200.	7.6	---	---	---	---
10"	1000.	4.1	1800.	7.4	---	---	---	---
12"	1500.	4.3	2800.	8.1	---	---	---	---

Table 4-3. Velocities and gpm ranges for various pipe sizes (Courtesy, *Guide for Hydronic Engineers,* Taco, Inc.)

If the average water temperature in the system is much different than 180°F, a new value should replace the 8.17 in the equation to adjust for water density. By replacing the value in the previous equation, it becomes:

$$q = (490)\,(gpm)\,(t_1 - t_2)$$

or

$$gpm = \frac{q}{(490)(t_1 - t_2)}$$

where: 490 = the value of M_w

$t_1 - t_2$ = difference in water temperature (supply vs return)

There are different rules that apply when sizing steam pipes, and these rules are governed by other equations that depend on the following:

a) Initial pressure of steam and the total pressure drop between the steam source and the user (radiators, convectors, etc.)
b) Maximum velocity of steam allowable for quiet and dependable operation
c) Equivalent length of pipe from the boiler to the furthest user or consumer

For quiet operation, steam velocity should be between 8000 and 12,000 ft/min.

Piping Supports

Pipes are commonly supported by hanger rods, which connect to the structure above or are clamped to building support beams. Standard hardware (clamps) is available for making connections between support rods, beams, and piping, as well as the building structure.

In addition to supporting the pipe, pipe supports must also allow for expansion movements without introducing restraining forces, which result in unplanned stresses either in the pipe or in the support structure. Pipes must be supported around equipment connections in such a manner as to avoid transmitting forces from piping to the equipment. Such forces could cause deformation and misalignment within the machinery, reduced reliability, and provoke premature failures. The equipment should not be expected to support the pipe weight.

Supports for insulated pipe cannot bear directly on the insulation. Typically, heavy gauge sheet-steel half-sleeves (called shields) are placed between the hanger and the insulation to spread the load. Rigid insulation with the same thickness as the primary insulation is then placed between the shield and the pipe.

Pipe support hardware should be carefully protected against corrosion. Pipe support failures may cause more damage than failure of the pipe itself. Protective coating or plating of support hardware is recommended wherever surrounding atmospheric conditions are corrosive.

Particular care and expertise in designing pipe support systems are required in the following situations:

- Where higher pressures are used
- Where water hammer can occur
- In high rise buildings
- When pipe sizes exceed 12 inches in diameter
- When large anchor loads or external forces bear on the pipes

Pipe anchors must be provided at regular intervals to control and contain piping movements. A pipe anchor should be constructed so that the entire pipe circumference is clamped. A half-pipe clamp can cause line misalignment, which introduces unbalanced stresses that may cause the anchor to distort or fail.

Valves

One of the most important fittings in a hydronic system is the valve. In general, valves control (start, stop, regulate, or prevent reversal) the flow of water in pipes, equipment, or at terminal units. Valve components include:

- the body;
- a disk, which influences the flow;
- a stem, which controls the disk;
- a seat for the disk;
- a bonnet to hold the stem;
- end connections.

There are numerous types of valves, and some are better suited for a specific task than others. There is not one type of valve which can be used for all functions. The selection depends on the pressure drop requirements, water temperature, replacement ease and the control accuracy required.

Valves are connected to pipes with threaded, flanged, welded, grooved, or soldered ends. Solder is usually used for connections to copper tubing.

Most valves are rated by their capability to withstand pressure (measured in psi). Valve ratings also correspond to a maximum operating temperature. Table 4-4 lists the valve types and their primary functions.

PUMPS

A pump is a mechanism used to push a liquid with a specific force to overcome system friction loss and any existing difference in elevation (static pressure). Liquid traveling through piping, equipment, and apparatus creates friction, which results in pressure loss or *head loss*. Pumps are installed in systems to overcome this loss of pressure.

There are many types of pumps. Those usually used in hydronic systems are centrifugal pumps (called circulators in small systems). The centrifugal pump is a mechanism that adds energy in the form of velocity to a flowing liquid. The pump produces this velocity with the help of a motor or driver and consumes energy in the process.

The pump's housing is referred to as the casing. The casing encloses the impeller, which is equipped with blades or vanes. The liquid enters at the center or the "eye" of the impeller. The impeller rotates, adding velocity and causing the centrifugal force to push the liquid out. This force (centrifugal) acts on the mass of liquid, moving it in a circular route away from the center of rotation.

This centrifugal force pushes the liquid out of the pump housing through the outlet located at the edge of the volute. Figure 4-1 shows the basic components of a centrifugal pump. The impeller's blades are bent slightly backwards to minimize the shock action. The velocity is the greatest at the impeller's periphery, where the liquid is discharged through a spiral shaped passage called the volute. This shape is designed so that there is an equal velocity of the liquid at all circumference points. The capacity (Q) of a pump is the rate of fluid flow delivered, which is generally expressed in gallons per minute (gpm).

The head or pressure furnished is the energy per unit weight of a liquid. In general, the pump is selected to overcome the total head (e.g., friction in pipes, fittings, valves, and apparatus) in a closed hydronic system. The head developed by a centrifugal pump is a function of the impeller diameter and the speed of rotation (rpm). The information obtained for the head loss and flow in a system are the two elements used in pump selection.

Valve	Subdivisions	Function
Gate valve	Solid wedge disks Split wedge disks Double disk parallel sit Rising stem Nonrising stem Quick-opening type	Stops or starts fluid flow. The "gate" is raised to open and lowered to close the valve. Used either fully open or closed. Used for high temperatures and/or corrosive materials.
Globe valve	Standard angle Angle Plug	Throttles (partially open) or controls the flow quantity. (Frequent operation of flow adjusting.)
Ball valve	Full port size Reduced port size Standard port size (one size smaller than the pipe)	Same as globe valve.
Butterfly valve		Same as globe valve.
Check valve	Swing type Lift type Foot valve	Prevents backward flow in pipes. (Constructed so that reversing the flow closes the valve.)
Safety or relief valve (adjustable)	Pressure Temperature Pressure and temperature	Relieves the pressure and/or temperature by allowing steam or water to escape.

Table 4-4. Valve types

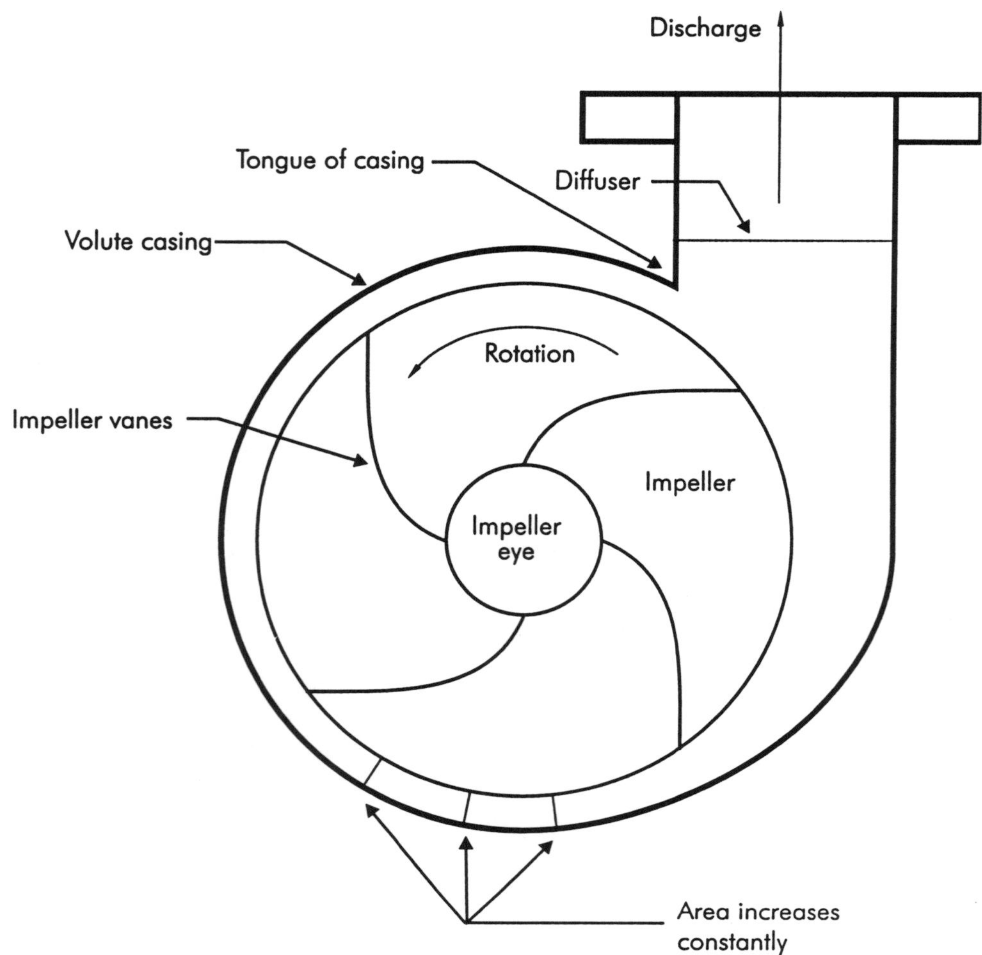

Figure 4-1. Centrifugal pump

Manufacturers fabricate pumps and motors in discrete increments. Figure 4-2 shows pump curve charts from one manufacturer, which are used to make a pump selection. Consider what centrifugal pump to choose when working with a system that requires a pump to deliver 130 gpm against a head of 380 feet (380 ft x 0.433 psi/ft = 164 psi).

On this pump curve, the flow delivery capacity in gallons per minute (gpm) is located on the horizontal axis (abscissa). The pump's efficiency percentage is shown on parallel curves on the upper part of the diagram. In this case, the efficiency of the selected pump is close to 60%. The necessary electric motor horsepower (hp) to do the work is marked on the lines slanted down to the right. Since the value falls between two lines, choose the higher hp value of 30 hp. The impeller diameter along the vertical line helps select a 7-9/16" diameter impeller. The possibility that more capacity may be required at a later date must be considered in the pump selection. This implies the possibility of changing the impeller in the future with a larger one. The pump curve sheet also indicates the number of rpm (revolutions per minute), which is shown on top of the pump data page. In Figure 4-2, the value is 3500 rpm.

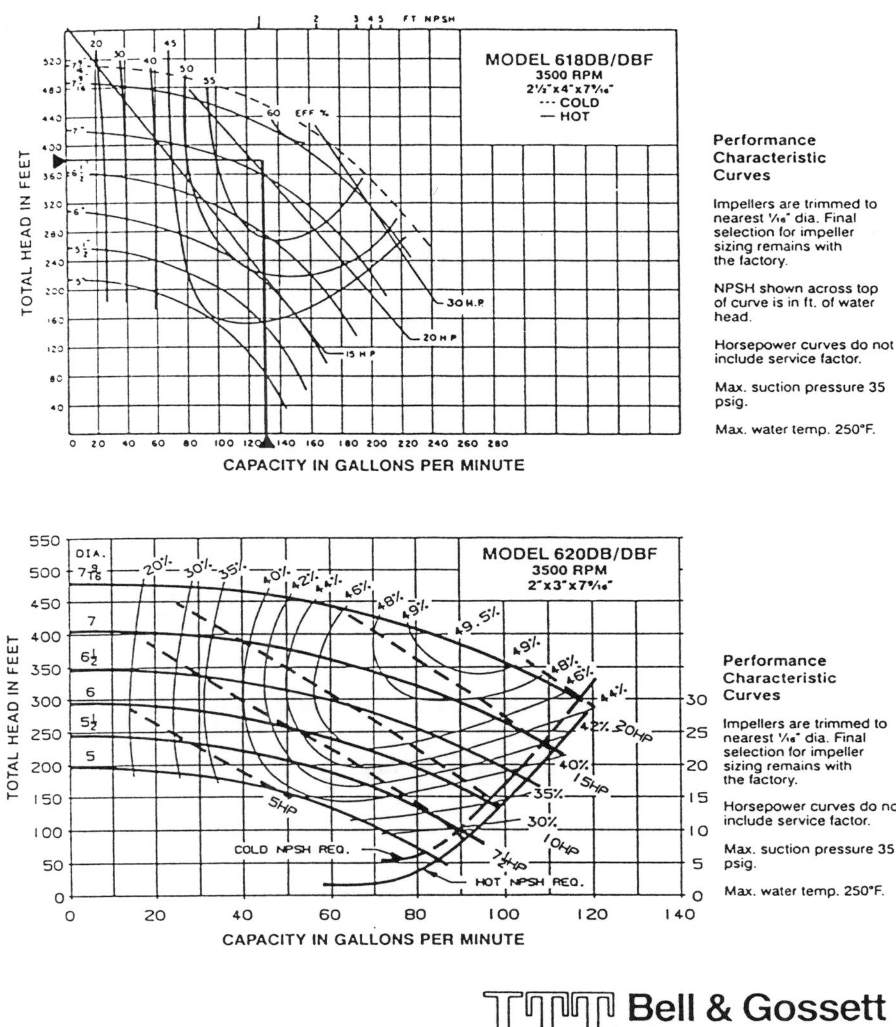

Figure 4-2. Pump curves (Courtesy, ITT Fluid Handling)

Pump Efficiency

One important element for the pump is its efficiency. The efficiency decreases as the flow increases or decreases. The pump efficiency percentage is calculated as follows:

$$EFF = \left(\frac{WHP}{BHP}\right)(100)$$

where: WHP = water horsepower
BHP = brake (or shaft) horsepower

Example 4-2. Calculate the efficiency percentage of a pump with flow of 150 gpm, head of 65 ft, and brake horsepower of 3.6.

Solution 4-2. First calculate the output hp (water hp) as follows:

$$\frac{(\text{gpm})(\text{head})}{3960} = \frac{(150 \text{ gpm})(65 \text{ ft})}{3960} = 2.5 \text{ hp}$$

Then apply the efficiency equation:

$$\text{EFF} = \left(\frac{\text{WHP}}{\text{BHP}}\right)(100) = \left(\frac{2.5 \text{ hp}}{3.6 \text{ hp}}\right)(100) = 69\%$$

The efficiency of the pump is 69%.

The horsepower information may be found on the motor nameplate. The BHP represents the motor horsepower multiplied by the efficiency. Horsepower is the ability to do useful work at a rate of 33,000 ft-lb/minute. Water horsepower refers to the horsepower developed by pounds of water (W) falling from a height (h) measured in feet and divided by 33,000.

The horsepower output of a pump moving water may be calculated using the following formula:

$$\text{hp} = \frac{(\text{gpm})(\text{total dynamic head, in ft})(8.33 \text{ lb/gal})}{33{,}000 \text{ ft-lb/minute}}$$

Note: Water weighs 8.33 lb/gal at 70°F.

Another similar formula to calculate output power is:

$$\text{hp} = \frac{(Q)(H)(S)}{3960}$$

where:
Q = flow in gpm
H = total head in ft
S = specific gravity of the liquid pumped
3960 = coefficient

Pump Performance

Pump performance is established by the pump manufacturer and based on tests executed under strictly controlled conditions. Pump operation is governed by three basic rules:

1. Pump capacity varies directly proportional to the speed or the impeller's diameter change.
2. Pump head (pressure) delivered varies with the square power of the speed or the impeller's diameter.

3. Brake horsepower varies with the third (cube) power of the speed of the impeller's diameter.

These rules may be interpreted as follows:

- At a constant impeller diameter:
  - — capacity varies directly with the speed;
  - — head varies as the square of the speed;
  - — horsepower varies as the cube of the speed.
- At a constant speed:
  - — capacity varies directly with the impeller diameter;
  - — head varies as the square of the impeller diameter;
  - — horsepower varies as the cube of the impeller diameter.

Example 4-3. As an example of capacity vs speed, what is the new capacity of a pump when the rotation is increased from 1150 rpm to 1750 rpm? The initial capacity is 200 gpm.

Solution 4-3.

$$\frac{1150 \text{ rpm}}{1750 \text{ rpm}} = \frac{200 \text{ gpm}}{X \text{ gpm}} \qquad X \text{ gpm} = \frac{(1750 \text{ rpm})(200 \text{ gpm})}{1150 \text{ rpm}} = 304.4 \text{ gpm}$$

Example 4-4. As an example of head vs speed, what is the new head of a pump when the rotation is increased from 1150 rpm to 1750 rpm. Initial pump head is 50 ft.

Solution 4-4.

$$\frac{(1150 \text{ rpm})^2}{(1750 \text{ rpm})^2} = \frac{50 \text{ ft}}{X \text{ ft}} \qquad X \text{ ft} = \frac{(1750 \text{ rpm})^2(50 \text{ ft})}{(1150 \text{ rpm})^2} = 115.8 \text{ ft}$$

Example 4-5. As an example of horsepower vs speed, what is the new horsepower required for a pump when the rotation is increased from 1150 rpm to 1750 rpm. Initial pump horsepower is 10 hp.

Solution 4-5.

$$\frac{(1150 \text{ rpm})^3}{(1750 \text{ rpm})^3} = \frac{10 \text{ hp}}{X \text{ hp}} \qquad X \text{ hp} = \frac{(1750 \text{ rpm})^3(10 \text{ hp})}{(1150 \text{ rpm})^3} = 35.24 \text{ hp}$$

Other Pump Information

An important element to consider when establishing the characteristics of a pump is the safety factor, which must be added to the total loss calculation. This safety factor usually is 10% of the calculated pressure loss.

Another element in pump selection is the net positive suction head (NPSH). This element is related to the pump priming (see Appendix B). If necessary, the pump manufacturer can offer help with pump selection, usually free of charge. For such assistance, contact the local representative and relay all pertinent data and installation details.

To select the correct pump to do the job, an assortment of pump catalogs is required.

When installing a pump, consider the following requirements:

- Pump must be rigidly connected to its support and concrete foundation (pad) to eliminate vibration.
- Pump should not be used to support the connected pipes.
- Pump and motor should be correctly aligned to prevent vibration.
- Supply pipe must be correctly sized and connected straight into the pump eye so that the flow is uniform.

In-line pumps are used primarily in residential and small commercial heating systems and have a relative low head (maximum 40 ft or 17.3 psi). They can handle an approximate flow up to 170 gpm. They cannot accept different impeller diameter replacements, so a valve must be installed in the discharge line for any adjustment purpose.

Pumps may be installed in parallel or in series. Pumps working in parallel deliver the same maximum head even if one may be capable of delivering a higher head. Pumps working in series form a system in which pressure is raised in steps. Pump discharge heads are additive, but the capacity is of a single pump.

On the discharge side of the pump check valves must be installed to prevent reverse flow.

Table 4-5 illustrates some of the problems that might develop when pumps are used and their solutions.

TERMINAL UNITS

There are four types of heat distributing units: baseboard and floor vectors; convectors; radiators; and fan coil units. Any of these units may transfer heat from the media to the surroundings. Their function is to maintain a controlled temperature in the area in which they are installed.

Based on tests and practical experience, heat distributing units work best when located under a window or along outside walls. The unit size must be based on heat loss calculations — not on the length of the wall available.

The heat output rating is measured in Btuh. Larger systems are measured in MBtuh, which is 1000 Btuh. When necessary, this measurement may be converted to EDR (equivalent direct radiation). The rate for steam is 1 ft^2 EDR = 240 Btuh with 1 psig steam condensing in the heat transfer unit. For hot water with an average of 170°F, 1 ft^2 EDR = 150 Btuh.

The recommended working pressures for the four types of terminal units are based on both the type of system and the heat distributing unit, which has built-in features that determine water temperature drop. The following guidelines are suggested:

Not Enough Water Is Delivered

Problem	Solution
1. Pump not primed.	Fill up pump and suction pipe with water.
2. Discharge head too high	Are valves wide open? Check pipe friction losses. Larger piping may correct condition.
3. Speed too low	Check whether motor is directly across-the-line and receiving full voltage. Current frequency may be too low, motor may have an open phase.
4. Wrong direction of rotation	Symptoms are an overloaded drive and reduced pump capacity. Could occur with wrong electrical connection. Compare turning of motor with directional arrow on pump casing.
5. Impeller clogged	Remove top of pump casing and clean impeller.
6. Air leaks into suction piping	Test flanges for leakage. Seal leaks.
7. Defective wearing rings	Inspect. Replace if worn.
8. Defective impeller	Inspect. Replace if damaged or blade eroded.
9. Defective packing	Replace packing and sleeves, if worn.
When Not Enough Pressure Is Delivered	
10. Speed too low	See Item 3.
11. Mechanical defects	See Items 7, 8, and 9.
12. Too small impeller diameter	Check with pump manufacturer to see if a larger impeller can be used. Otherwise, cut pipe losses, increase speed, or do both. *Note: Do not overload the drive.*
Pump Consumes Too Much Power	
13. Head lower than rating, pumps too much water	Shave down impeller's outside diameter in amount advised by pump manufacturer.
14. Stuffing boxes too tight	Release gland pressure. Tighten reasonably. If sealing water does not flow while pump operates, replace packing. If packing is wearing quickly, replace scored shaft sleeves and keep water seeping for lubrication.
15. Casing distorted by excessive strains from suction or discharge piping	Check alignment. Examine pump for friction between impeller and casing, worn wearing rings. Replace damaged parts.
16. Misalignment	Realign pump and motor.

Table 4-5. Pump problems and solutions

- For cast iron radiators, a temperature drop of no more than 30°F.
- Manufacturers indicate the temperature drop in convectors; it may be 10°, 20°, or 30°F.
- Unit heaters, as determined by the manufacturer, may use water with a temperature drop of 50°F or higher.
- Baseboard or commercial finned tube units use a water temperature drop of up to 50°F. Minimum flow rates must be maintained.

Baseboard Units

Baseboard heaters may be cast iron (radiant) or finned tube (radiant-convector[1]). The cast iron type has a solid face and produces radiant heat. The finned tube radiator has two grilles; one grille faces down and the other faces up, producing radiant and convected heat (air moves over the radiator or the fire-tube element). The cast iron type has a lower efficiency than the finned tube type.

Baseboard units are rated in Btuh/linear ft. When heat must be provided to offset cold infiltration from a door or window and a normal baseboard heater cannot be used, a sunken finned-tube radiator is installed below the floor level. This type is commonly called a floor-vector.

The baseboard finned-tube radiator may have a steel pipe or copper tube with fins that are made of aluminum, copper, or steel.

Convectors

A convector depends primarily on gravity convective heat transfer. The heating element is a finned-tube pipe or coil(s), mounted in an enclosure designed to increase the convective effect. The cabinet may be made in many different configurations, including partially or fully recessed into the wall. The usual location is on an exterior wall at or near the floor. Capacity depends on length, depth, height, and heating element design, as well as hot water temperatures or steam pressure.

Convectors are usually taller than baseboard heaters and are open at the bottom and top. The air circulates freely above the heating element, similar to a stack effect. Air enters at the bottom and leaves at the top. Convectors may be floor mounted, free standing, or wall hung right above the floor or windows (valance heating/cooling units). Figure 4-3 shows three different convector types.

Radiators

Radiators are mostly the old-fashioned, cast iron type, consisting of individual sections attached to each other. Modern radiators include radiant panels. A radiant panel is a heating surface designed to transfer heat by radiation. This unit might include a convective component. In the case of radiant floor panels, convective transfer may be predominant.

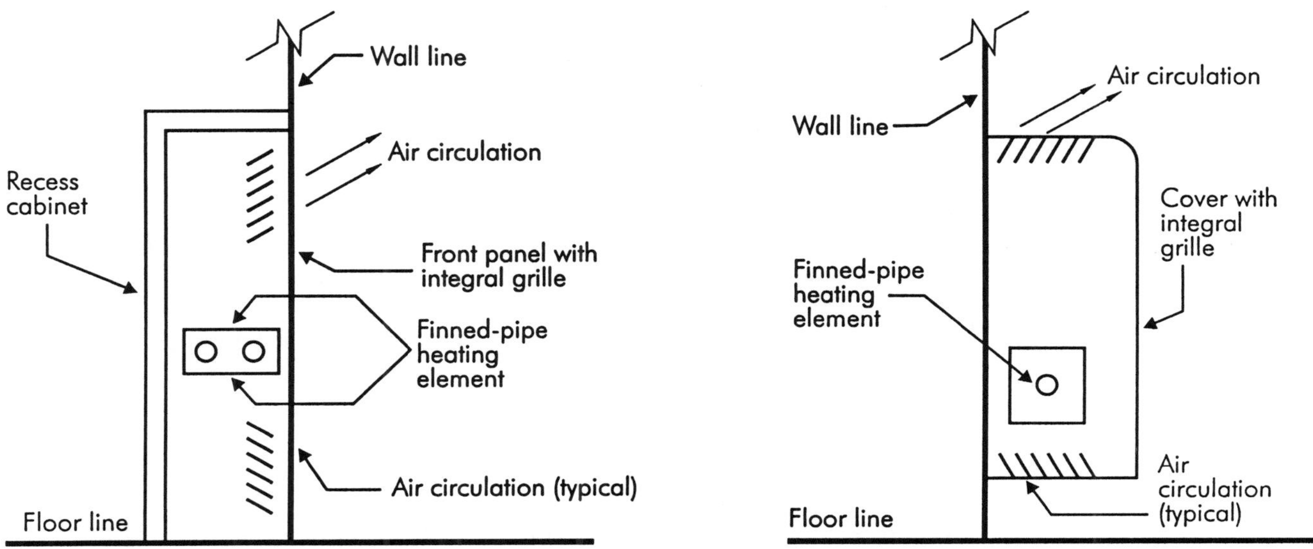

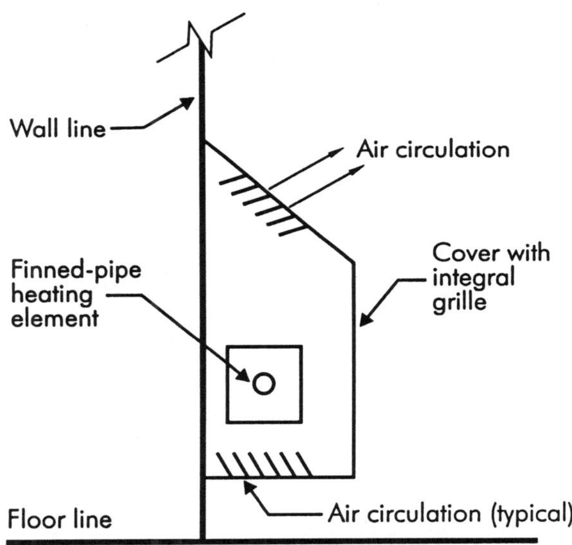

Figure 4-3. Various convector types

Radiant panels may be located in the floor, wall, or ceiling or be free standing. Their surface temperature is limited by demand of building occupants. Typical limitations are 80° to 85°F for floor panels; 100°F for wall panels; and 120° to 130°F for ceiling panels. The heating medium is hot water or electrical resistance heating cable. In general, panels are fabricated in the field using electric heating cable, copper tubing, or steel pipe embedded in the construction. For concrete floor panels, steel pipe is used (3/4" or 1" diameter), because steel has an expansion coefficient similar to that of concrete. Corrosion at the concrete-pipe interface can be severe, and pipe protection must be provided.

Control systems are conventional, because radiant heat sensitive devices are not readily available. Floor panels are very difficult to control due to the relatively large mass providing a slow response.

Fan Coil Units

The fan coil unit consists of a coil (copper tube in a serpentine form in one or more rows) with a fan. The fan pushes the air over the coil. The fan and coil are enclosed in a sheet metal casing. Hot or chilled water may be used within the coil. For cooling purposes, refrigerant might be circulated in the coil. The fan coil unit may be connected inlet-outlet to distribution supply and return ductwork. Unit heaters, cabinet heaters, unit ventilators, and units for make-up air all belong in the fan coil unit category.

The internal flow arrangement in a coil may be one of the four possible types: parallel flow; serpentine coil; coil in series, or split grid, Figure 4-4. Coils of any type may be arranged in one or multiple rows.

Example 4-6. Using Table 4-6, select the length of 1/2 in. (tube size) baseboard required for a room with a heat loss of 12,000 Btuh. The unit has 2 gpm of water flowing through it at an entering temperature of 218°F.

Solution 4-6. The average water temperature in the unit must first be determined. To find this, the following equation is used.

$$\Delta t = \frac{Q}{(500)(gpm)}$$

where: Δt = water temperature difference between entering and leaving the baseboard
Q = heat loss in Btuh
500 = value derived from 60 min/hr multiplied by 8.33 lb/gal

The heating capacity of the unit must equal the room heat loss. Using this equation we obtain:

$$\Delta t = \frac{12{,}000 \text{ Btuh}}{(500)(2 \text{ gpm})} = 12°F$$

The hot water supply temperature (boiler water supply) is 218°F. Therefore, the leaving temperature is 206°F (218° - 12°F), and the average temperature is as follows:

$$\frac{218° + 206°F}{2} = 212°F$$

Using Table 4-6, at a flow rate of 2 gpm, an average water temperature of 210°F (the next lowest temperature rating listed is used to be certain that the unit has adequate capacity), and 1/2-inch pipe diameter, the rated capacity

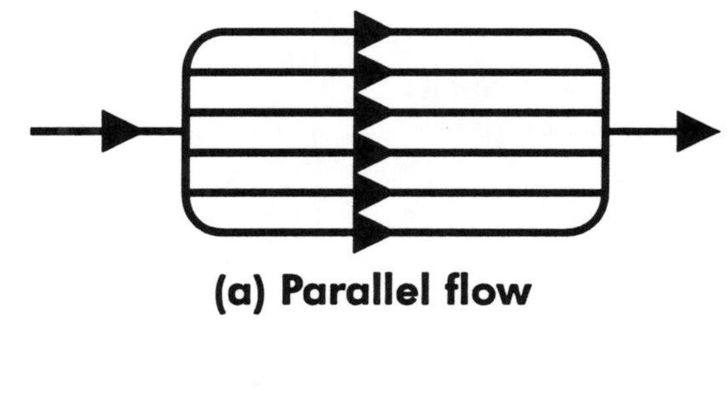

(a) Parallel flow

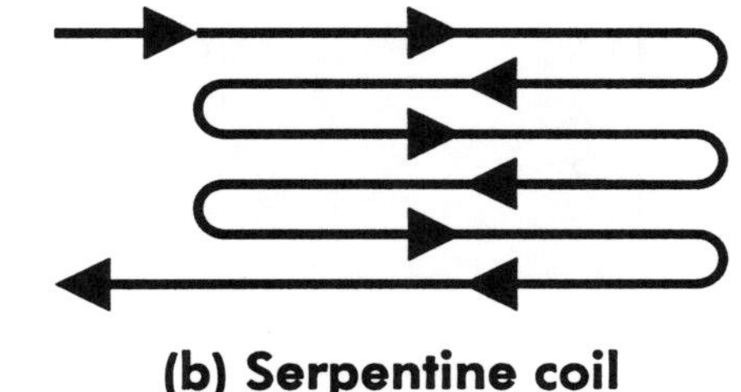

(b) Serpentine coil

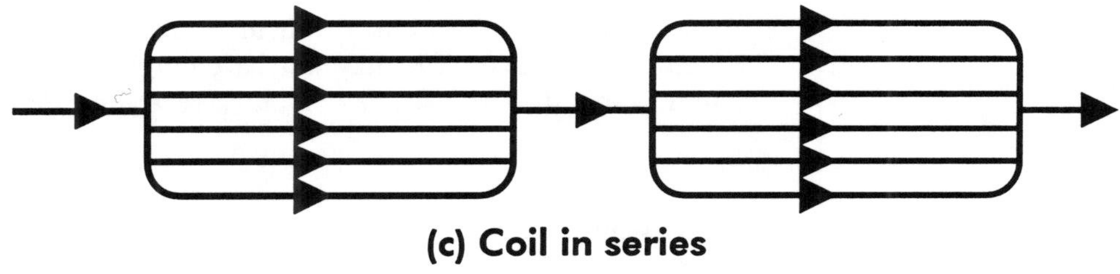

(c) Coil in series

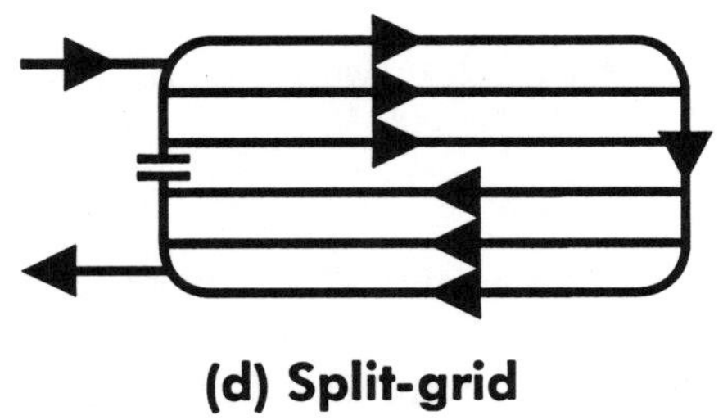

(d) Split-grid

Figure 4-4. Coil flow arrangements

listed is 840 Btuh/ft length. Therefore, the length required for a capacity of 12,000 Btuh is:

$$\text{Length} = \frac{12{,}000 \text{ Btuh}}{840 \text{ Btuh/ft}} = 14.3 \text{ ft}$$

The contractor would order 15 ft, rather than deal with the fractional amount. This also provides a little extra capacity.

Nominal tube size (in.)	Flow (gpm)	Velocity (fps)	Average water temperatures (°F)							
			170	180	190	200	210	220	230	240
3/4	1	0.6	510	580	640	710	770	840	910	970
	2	1.2	520	590	650	730	790	860	930	990
	3	1.8	530	600	670	740	800	870	950	1010
	4	2.4	540	610	680	750	810	890	960	1030
1/2	1	1.2	550	620	680	750	820	880	950	1020
	2	2.4	560	630	700	770	840	900	970	1040
	3	3.6	570	640	710	780	850	920	990	1060
	4	4.8	580	660	720	790	870	930	1000	1080

Table 4-6. Typical baseboard radiation rating (hot water ratings, Btuh/ft length)

The manufacturer should be consulted before using flow rates that vary greatly from the range shown in their tables. A good guideline is to use flow rates between values that result in water velocities between 1 and 5 fps. Velocities below 1 fps may not be enough to carry dirt particles through the unit. Table 4-7 lists water velocities recommended for different flow rates.

Tube diameter (in.)	Flow rate (gpm) 1	2	4	6	8
1/2	1.4	2.8	5.8	8.5	
3/4	0.6	1.5	2.8	4.0	5.5

Table 4-7. Water velocities (fps) for Type L tubing

Example 4-7. A contractor is about to install a hydronic system and notes that the engineer's specifications call for 1/2" diameter tubing with a flow rate of 7 gpm. Based on Table 4-7, this flow rate results in a very high velocity and would probably be very noisy. What should the contractor do?

Solution 4-7. The contractor should call the engineer and discuss possible changes in the design, such as larger radiator pipe.

Rating tables and selection procedures are similar for other types of radiation. Catalogs and manufacturer recommendations can be obtained upon request.

NOTE

[1] This is a pipe or tube with fins attached.

Chapter 5

Hydronic System Configurations

There are various arrangements used to distribute heated or cooled water from the source to consumers. In a hydronic heating system, hot water may be distributed in three basic ways:

- Boiler with simple piping loop, Figure 5-1.
- Diverting flow on a second loop by controlling valve (the valve may not be fully closed, it might just divert some of the flow), Figure 5-2.
- Primary and secondary loops and pumps. The second pump may be programmed to draw water from the main loop when hot water is desired, Figure 5-3. Figure 5-4 shows the diversion of hot water to a radiator.

Piping Systems

There are also a variety of piping systems used to distribute hot (and cold as indicated) water to consumers. These systems are as follows:

- Series loop heating system
- Two-circuit series loop system
- Zone series loop system
- One-pipe heating or cooling system
- Two-pipe direct return system
- Two-pipe reverse return system
- Reverse return system with horizontal one pipe main

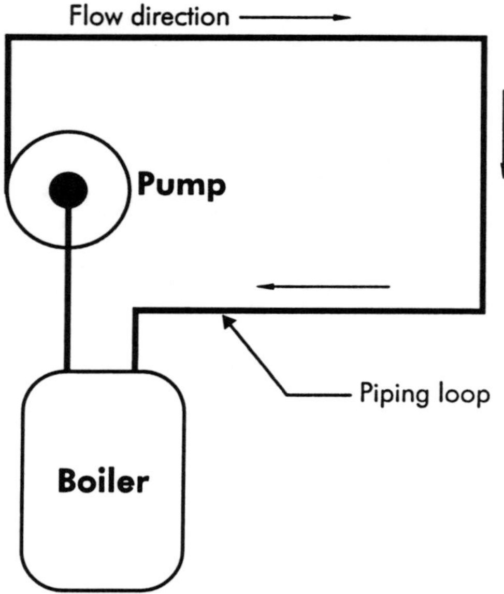

Figure 5-1. Boiler with simple piping loop

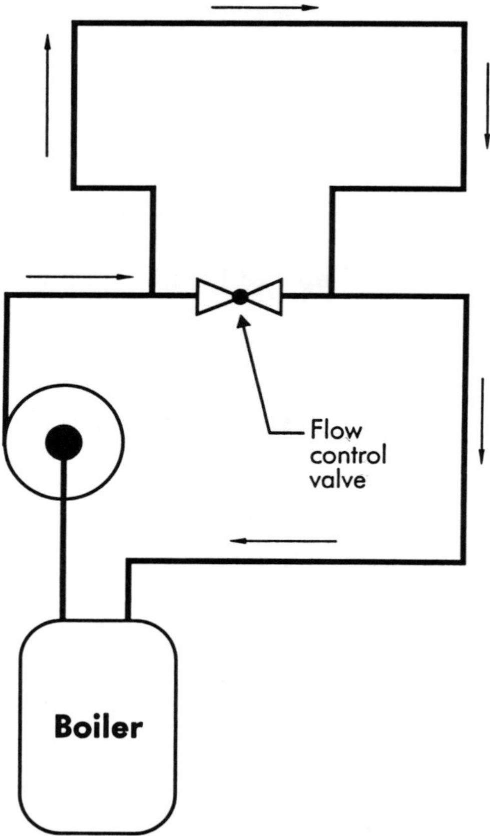

Figure 5-2. Diverting loop with controlling valve

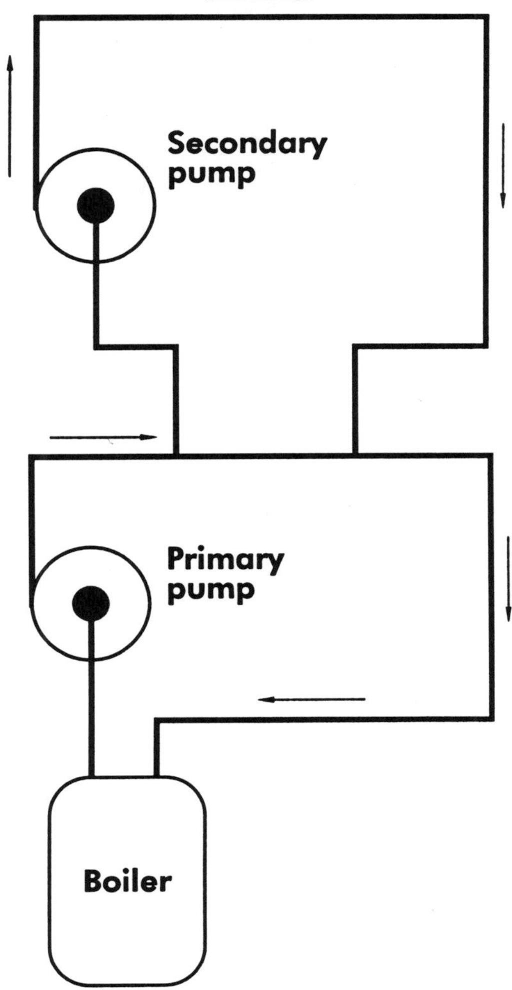

Figure 5-3. Primary and secondary loops and pumps

- Overhead direct return system with downfeed one pipe circuit
- Overhead reverse return system with downfeed one pipe circuit
- Overhead two-pipe direct return with reverse return radiation circuits
- Overhead downfeed reverse return system with reverse return radiation circuits
- Three-pipe, three pump distribution system (include cooling)
- Simple three-pipe water distribution system (include cooling)
- Four-pipe distribution system (include cooling)

The least complicated hydronic heating system used in residential buildings is designated a series loop system. The water heated by the boiler runs in one direction, continuously passing through one radiator or baseboard heater to the next, Figure 5-5. After the last terminal unit, the "cooler" water returns to the boiler to be heated and sent back into the circuit.

Figure 5-5 illustrates that each downstream terminal unit receives slightly cooler water than the preceding one. In order to compensate for the decrease

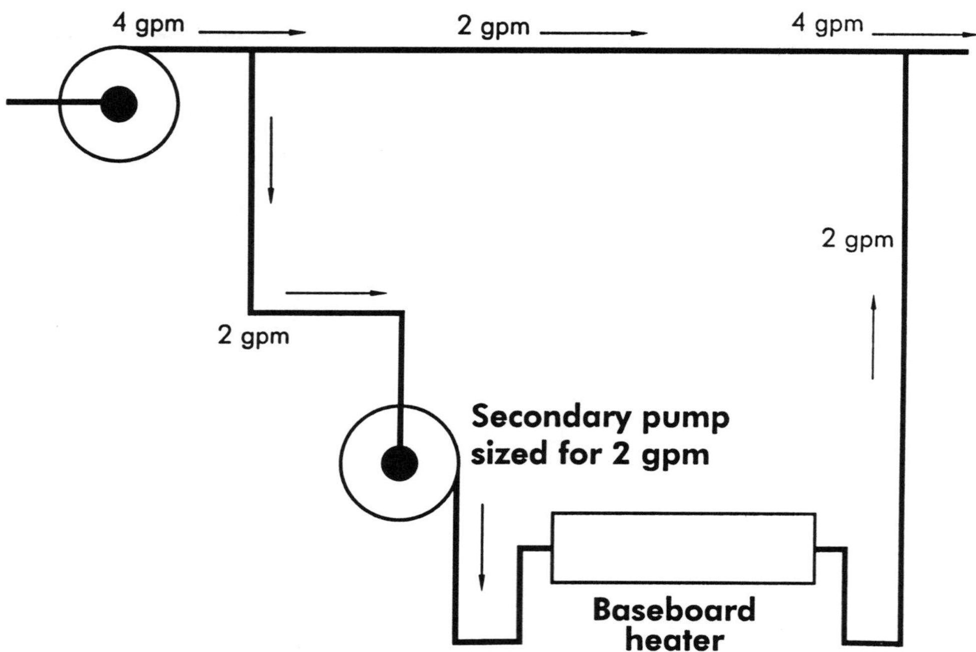

Figure 5-4. Diverting hot water to a radiator

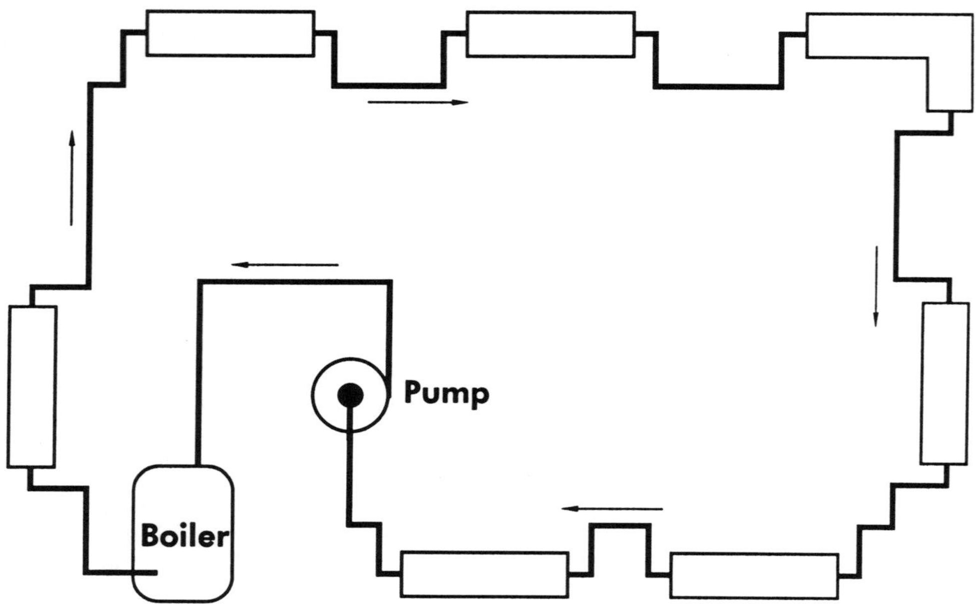

Figure 5-5. Series loop heating system (plan)

in water temperature, the terminal unit must either be physically larger for the same amount of heat released or if the coil or radiator area is the same, the flow must be larger.

Based on practical tests, the temperature drop (reduction) from one terminal unit to the next is considered to be 2°F (1.1°C). When increasing unit size, a rule of thumb is to increase them about 2% over the preceding unit for the first 70% of units downstream from the boiler and 3% for the rest. The formula to determine the size of the pipe surface is:

$$A_n = A_1 \frac{T_s - t - \frac{1}{2}(q_t/q_n)(T_s - T_R)/N}{T_s - t - \frac{1}{2}[2_n - 2 + (q_t/q_n)](T_s - T_R)/N}$$

The temperature leaving coil number (n) is:

$$T_n = T_s - [n - 1 + (q_t/q_n)](T_s - T_R)/N$$

where:

A_n	=	surface of coil number (n)
A_1	=	first coil surface
T_s	=	initial supply temperature
T_R	=	final return temperature
t	=	temperature of space (to be maintained)
N	=	total number of coils
T_n	=	temperature leaving the coil number (n)
q_n	=	flow rate in coil (n)
q_t	=	total flow in system

When the flow is not known, the following formula applies:

$$A_n = A_1 \frac{\frac{1}{2}(T_s - T_1) - t}{\frac{1}{2}(T_s + T_1) - t - (n - 1)(T_s - T_R)/N}$$

A variation of the simple system shown in Figure 5-5 is a two-circuit series system, which distributes the hot water in two directions using two smaller loops, Figure 5-6. A derivation of this is the zoned series loop system, Figure 5-7. Such a system requires careful balancing so each loop receives the correct amount of heated water.

One-Pipe System

The one-pipe system is similar to the series loop system except that the heating element is located off the loop being fed with hot water through a diverting special tee fitting, Figure 5-8. The special tee, which is installed at each heating element, causes a pressure drop. The loop pipe is one size throughout, because the total amount of water circulated is constant. Water always flows from a higher pressure to a lower pressure, which is how some water is diverted through the heating unit. The water that returns to the header loses some heat, so the next heating unit receives slightly cooler water.

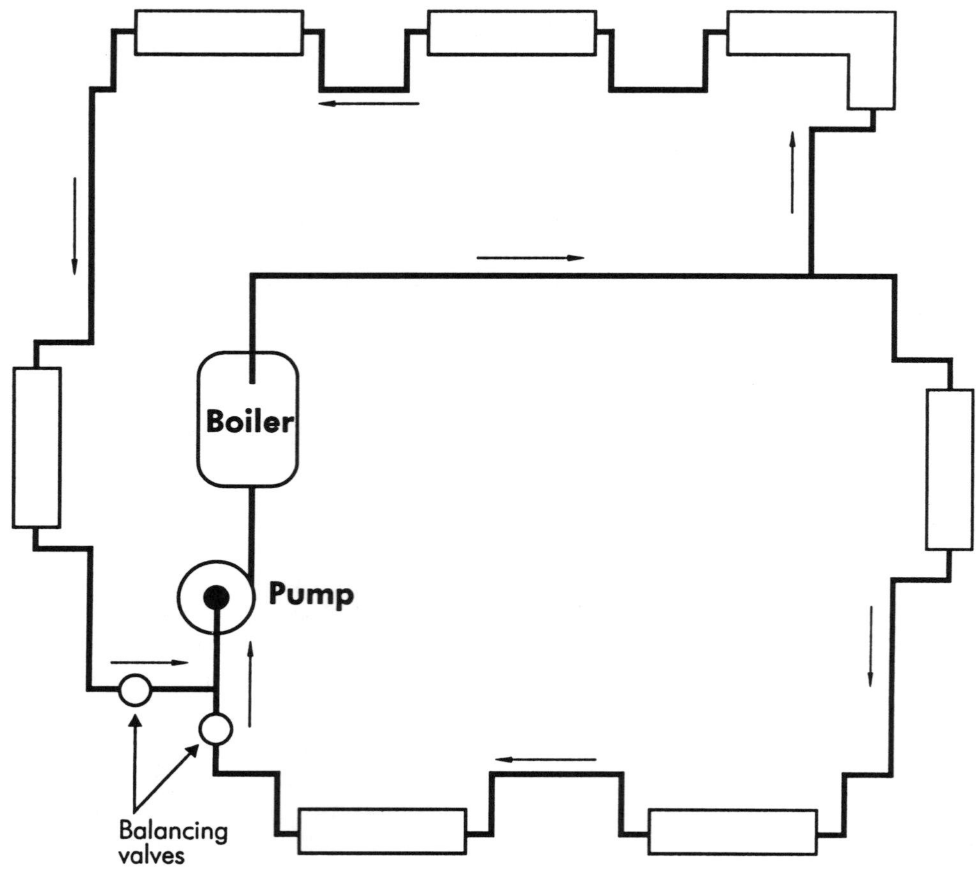

Figure 5-6. Two-circuit series loop heating system (plan)

Such a system is quite simple and is used for residential and small commercial applications.

Example 5-1. Determine the required flow in gpm for a one-pipe series loop system. Assume a heat loss calculation for an area resulted in a heat loss of 36,000 Btuh. The supply water temperature is 200°F, and the water return temperature is 180°F.

Solution 5-1.

$$\text{gpm} = \frac{\text{Btuh}}{(500)(\Delta t)} = \frac{36{,}000 \text{ Btuh}}{(500)(20°\text{F})} = 3.6 \text{ gpm}$$

where:

Btuh = heat loss

500 = value derived from 60 min/hr multiplied by 8.33 lb/gal

Δt = temperature difference between water supply and return temperatures

The required flow for this one-pipe series loop system is 3.6 gpm.

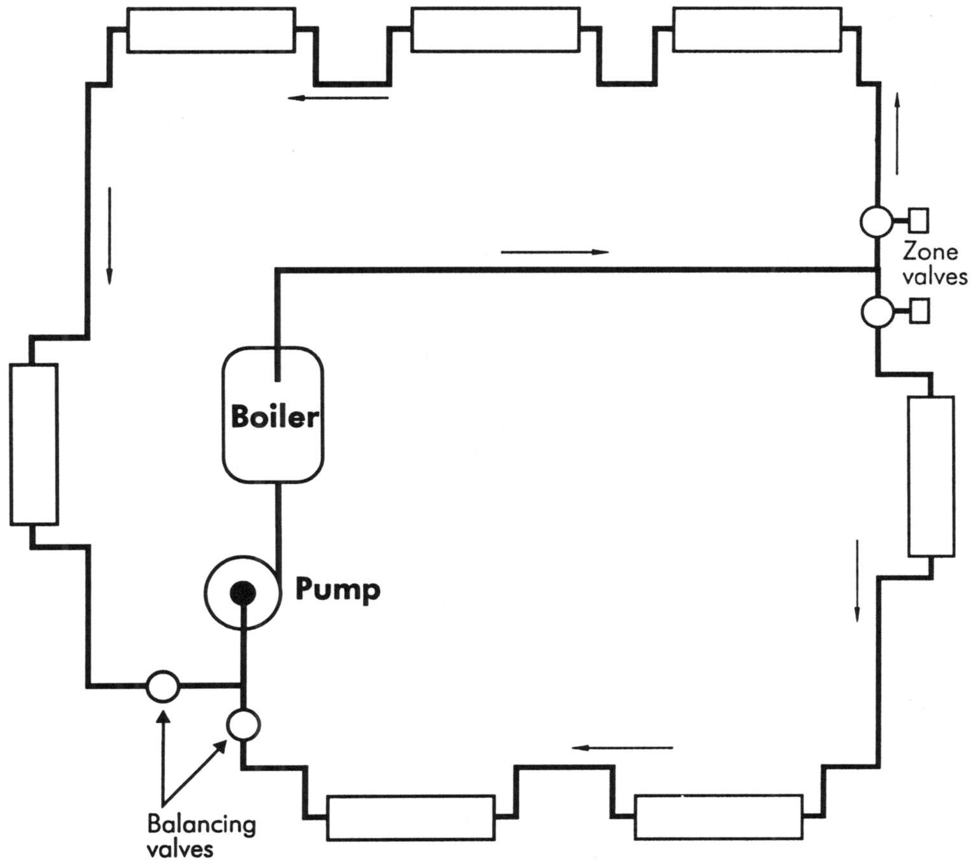

Figure 5-7. Zoned series loop system (plan)

Two-Pipe Direct Return System

The two-pipe direct return system involves conveying water in two parallel and opposite flow directions. There is a supply loop and a parallel return loop, and the loops are interconnected at their ends, Figure 5-9. In order for this system to function well, a balancing valve must be included so that the pressure at each consumer is similar to the next. A different pipe size or an orifice may be substituted for the balancing valve.

A variation of the system is the two-pipe reverse return system. This system includes two parallel pipes, but the supply and return flow directions are the same, Figure 5-10.

The two-pipe system requires more pipe (length), so it is more expensive to install than the one-pipe system. The system pipe diameter varies in the supply pipe (decreases) and in the return pipe (increases). Some people consider such a system to have better efficiency, water distribution, and temperature control.

Figures 5-11, 5-12, and 5-13 show three variations of one- and two-pipe systems.

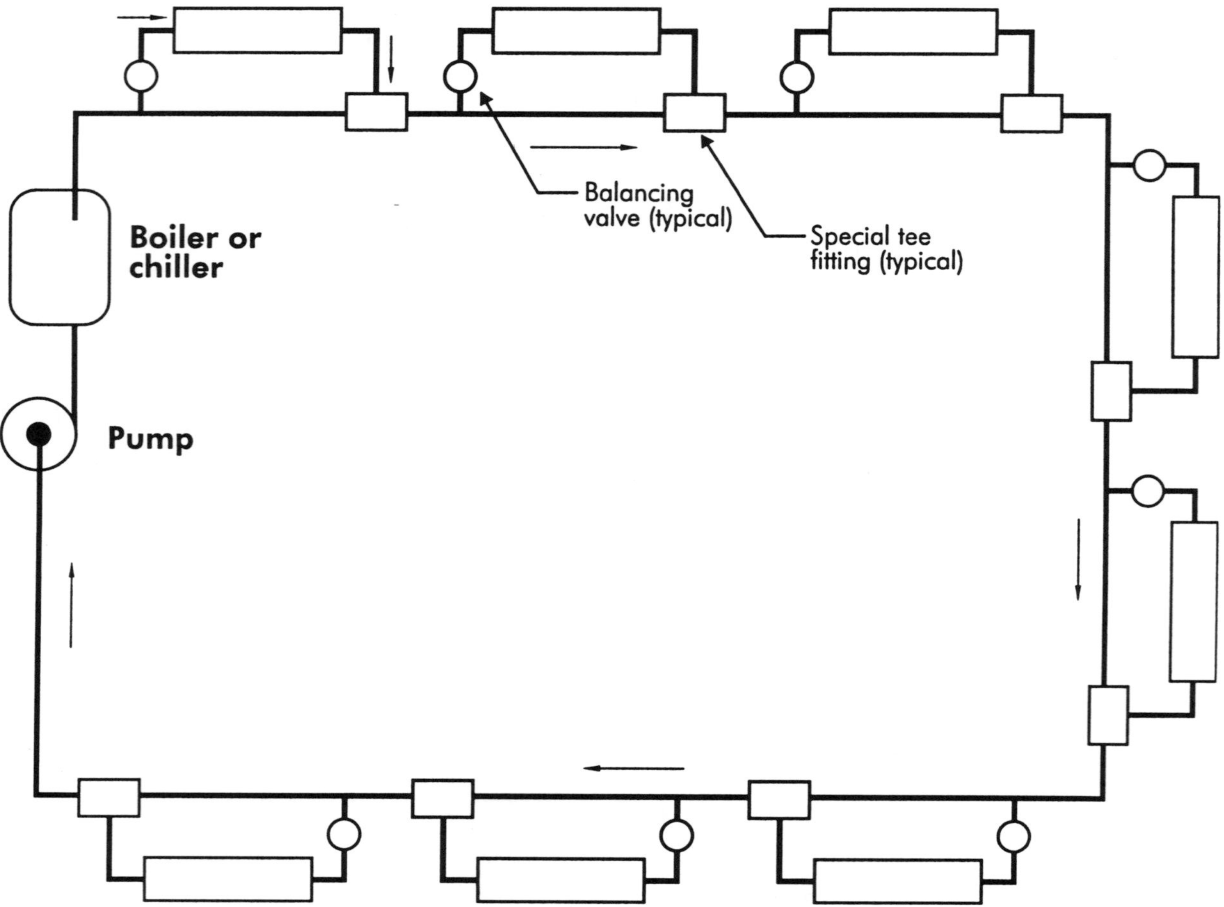

Figure 5-8. One-pipe heating or cooling system single circuit (plan)

Combination One-Pipe and Two-Pipe Systems

Figures 5-14, 5-15, and 5-16 show three possibilities for combination one- and two-pipe systems. These systems are simple and present advantages of both one- and two-pipe arrangements. They are applicable in multi-story buildings.

Three-Pipe System

The three-pipe system simultaneously allows hot water to be fed to some consumers and chilled water to others. This type of situation occurs when certain buildings or areas require continuous cooling even in winter, while other areas require heat. The system shown in Figure 5-17 includes two supply mains and one return main.

One supply pipe may carry hot water and the other chilled water. Each user has a special two inlet valve and one outlet. The hot and cold media do not mix in the supply system, but they do mix in the return pipe. The supply water separation of the two is possible. The system separation is made possible by installing three pumps — one for each circuit. Balancing such a

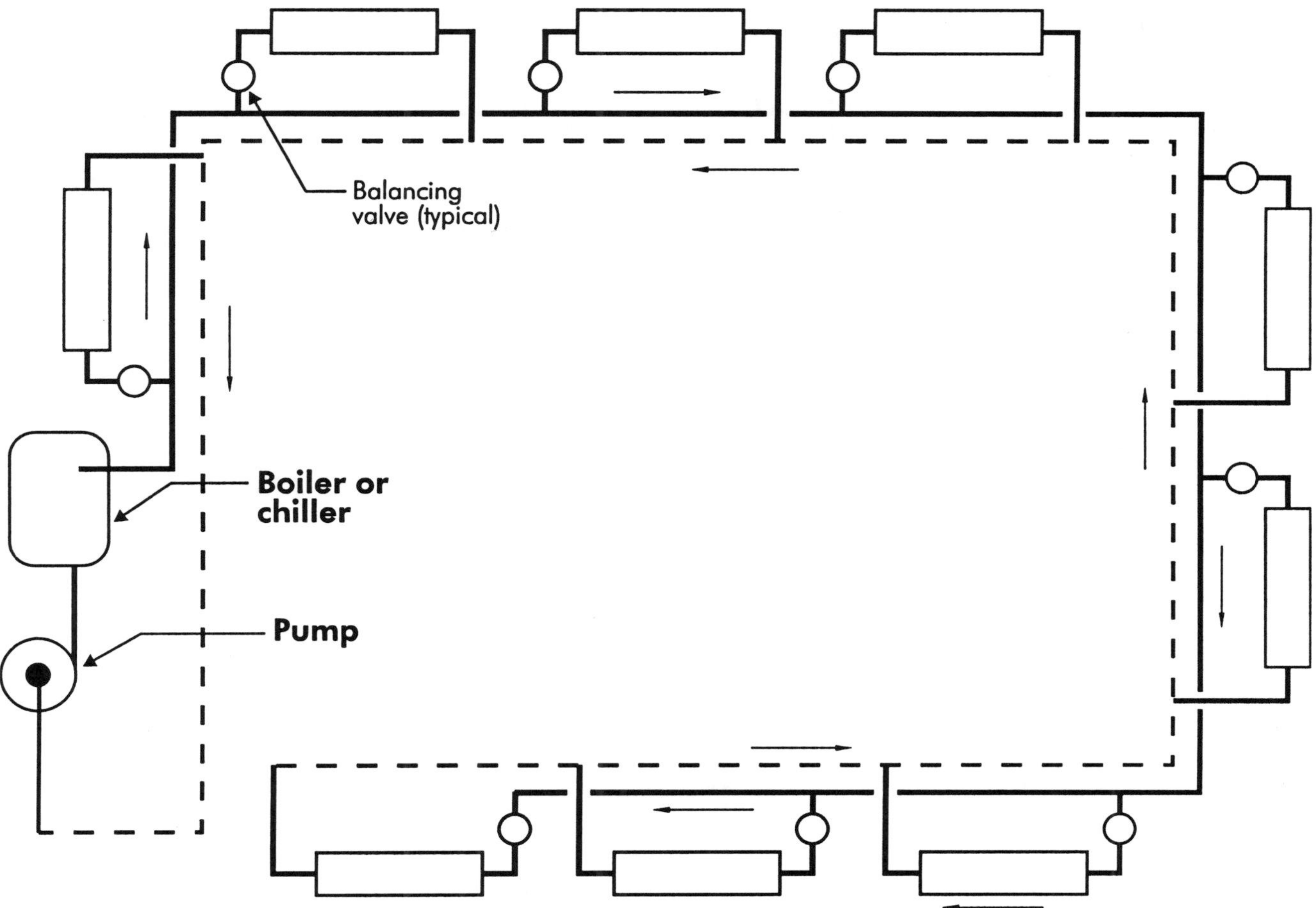

Figure 5-9. Two-pipe direct return system (plan)

system requires skill and accumulated experience. A more simple three-pipe system is shown in Figure 5-18.

Four-Pipe System

The four-pipe system consists of two different two-pipe circuits. It is used for separate and parallel heating and cooling systems. In this system, pipe sizes may be reduced, because water temperature in the hot water system may be raised. The four-pipe system is, in fact, two separate and parallel systems. Such a system was originally developed for a cooling system with the need of a reheat coil, Figure 5-19.

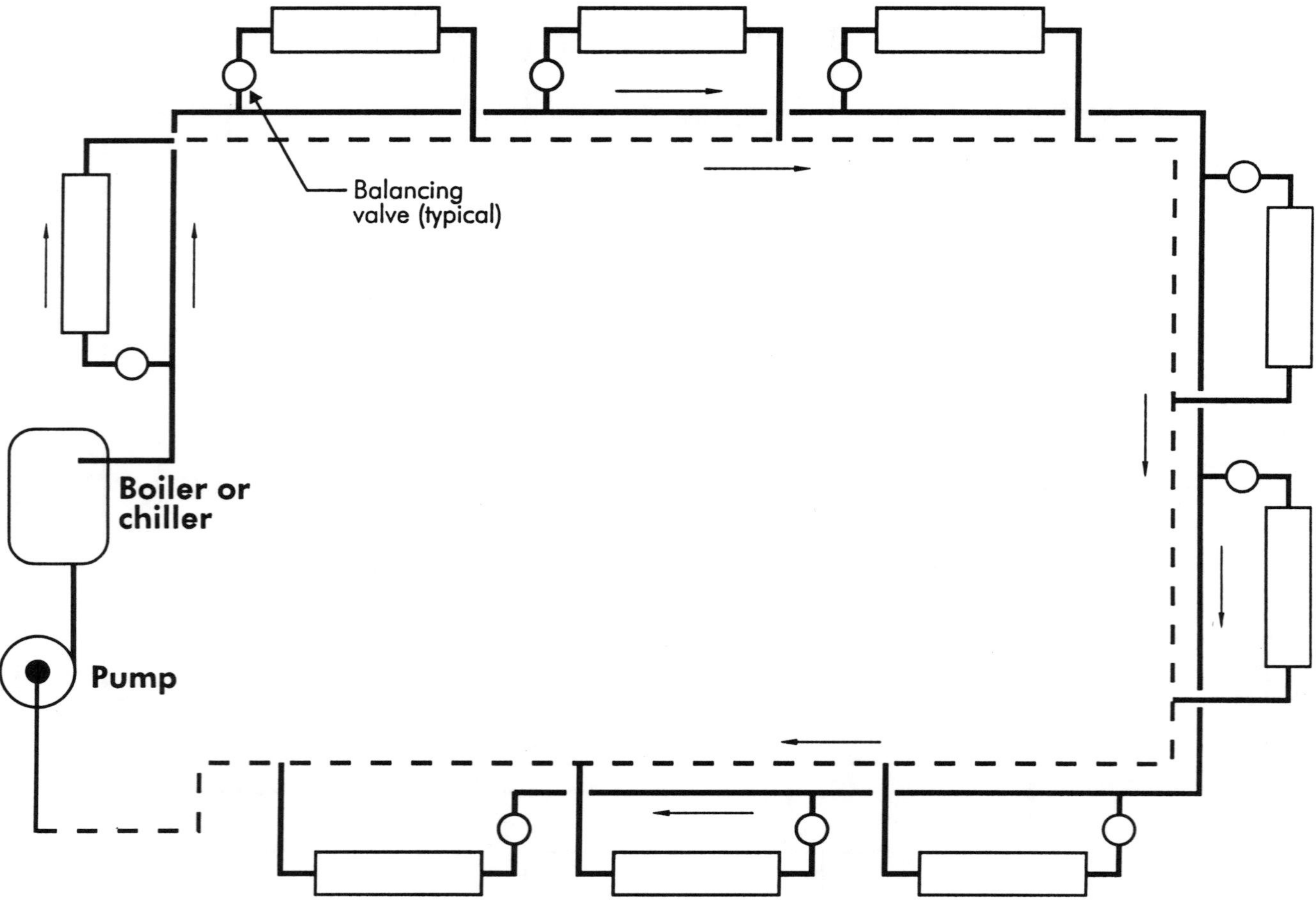

Figure 5-10. Two-pipe reverse return system (plan)

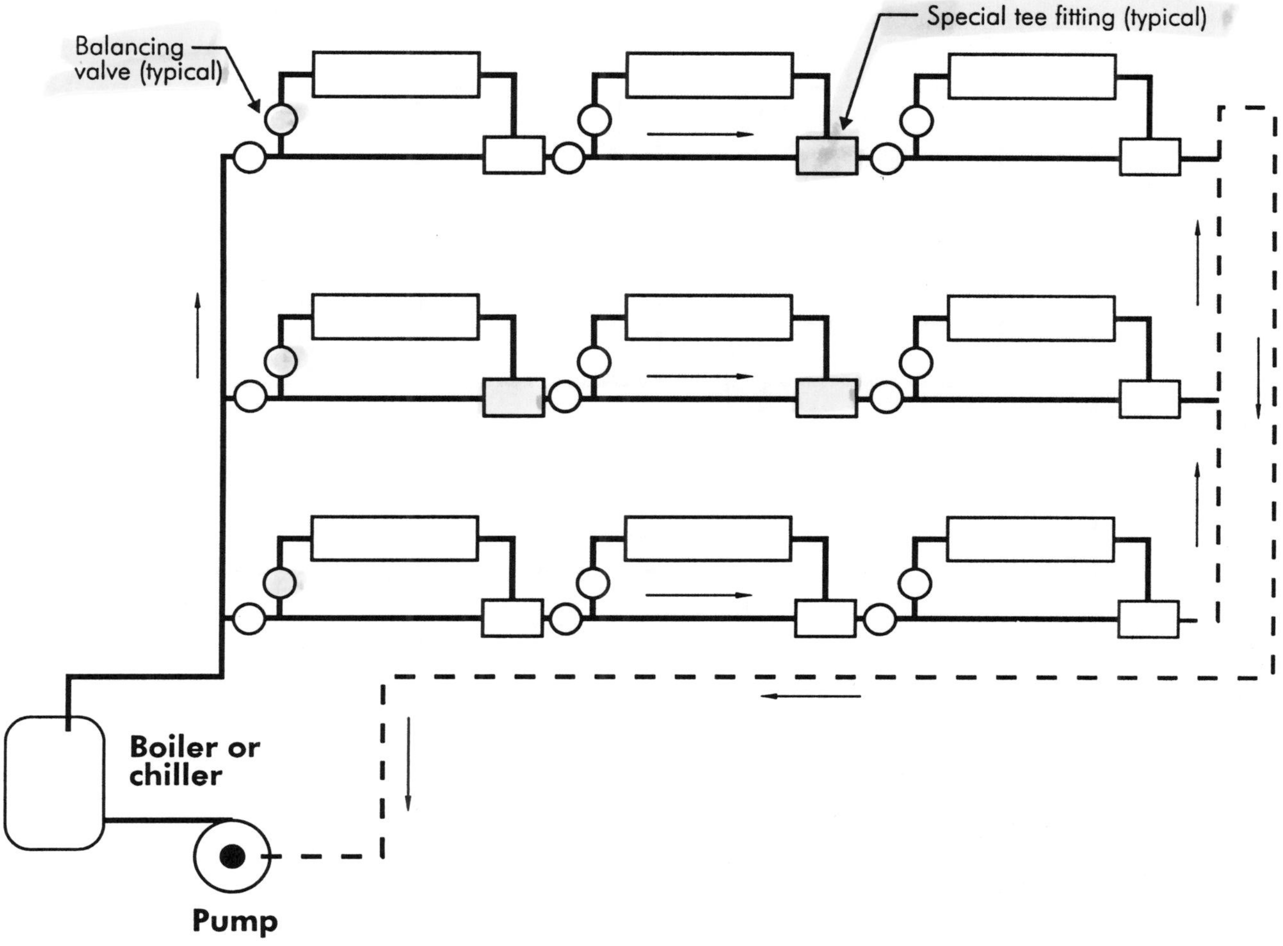

Figure 5-11. Reverse return system with horizontal one-pipe main (elevation)

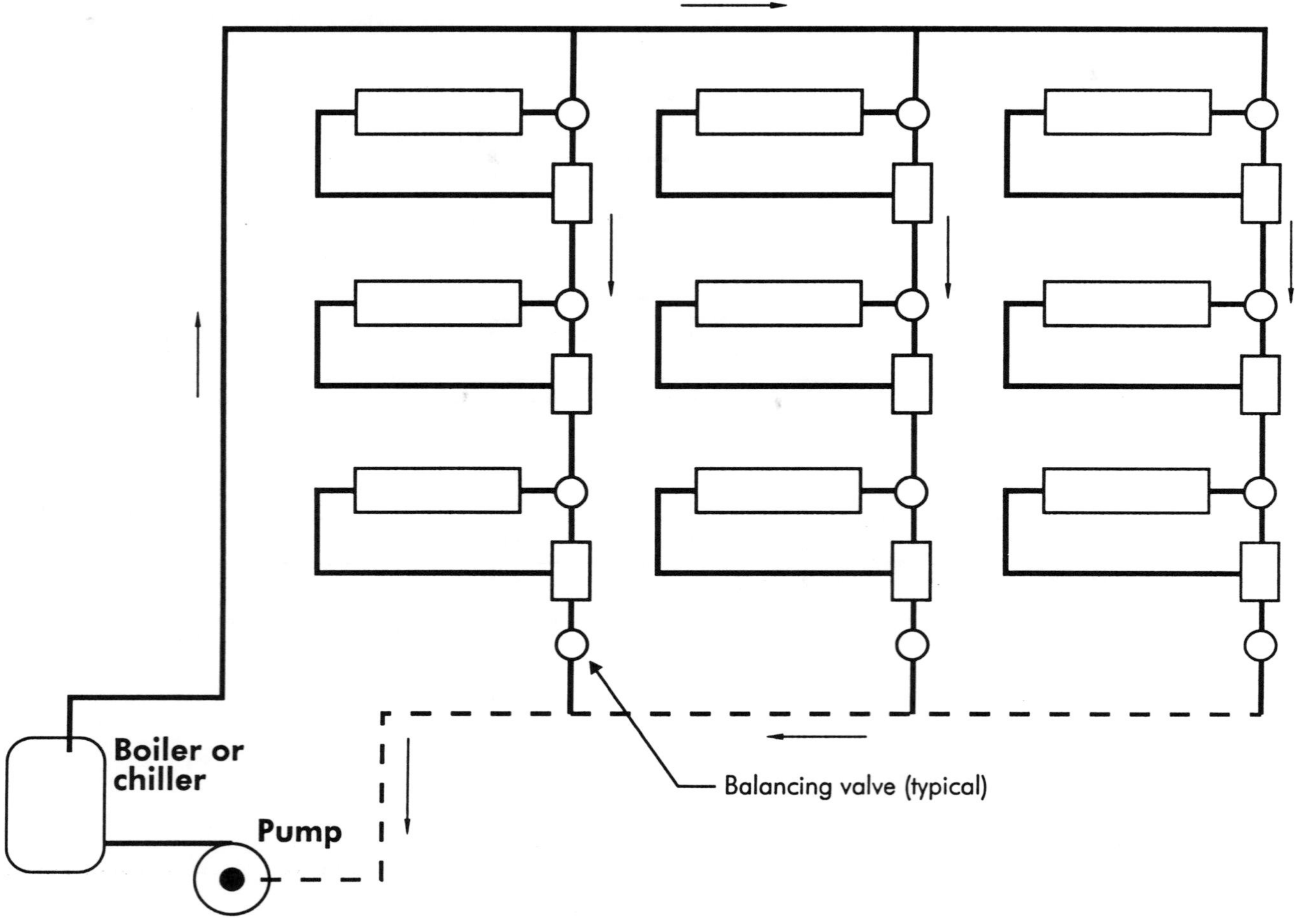

Figure 5-12. Overhead direct return system with downfeed one-pipe circuits (elevation)

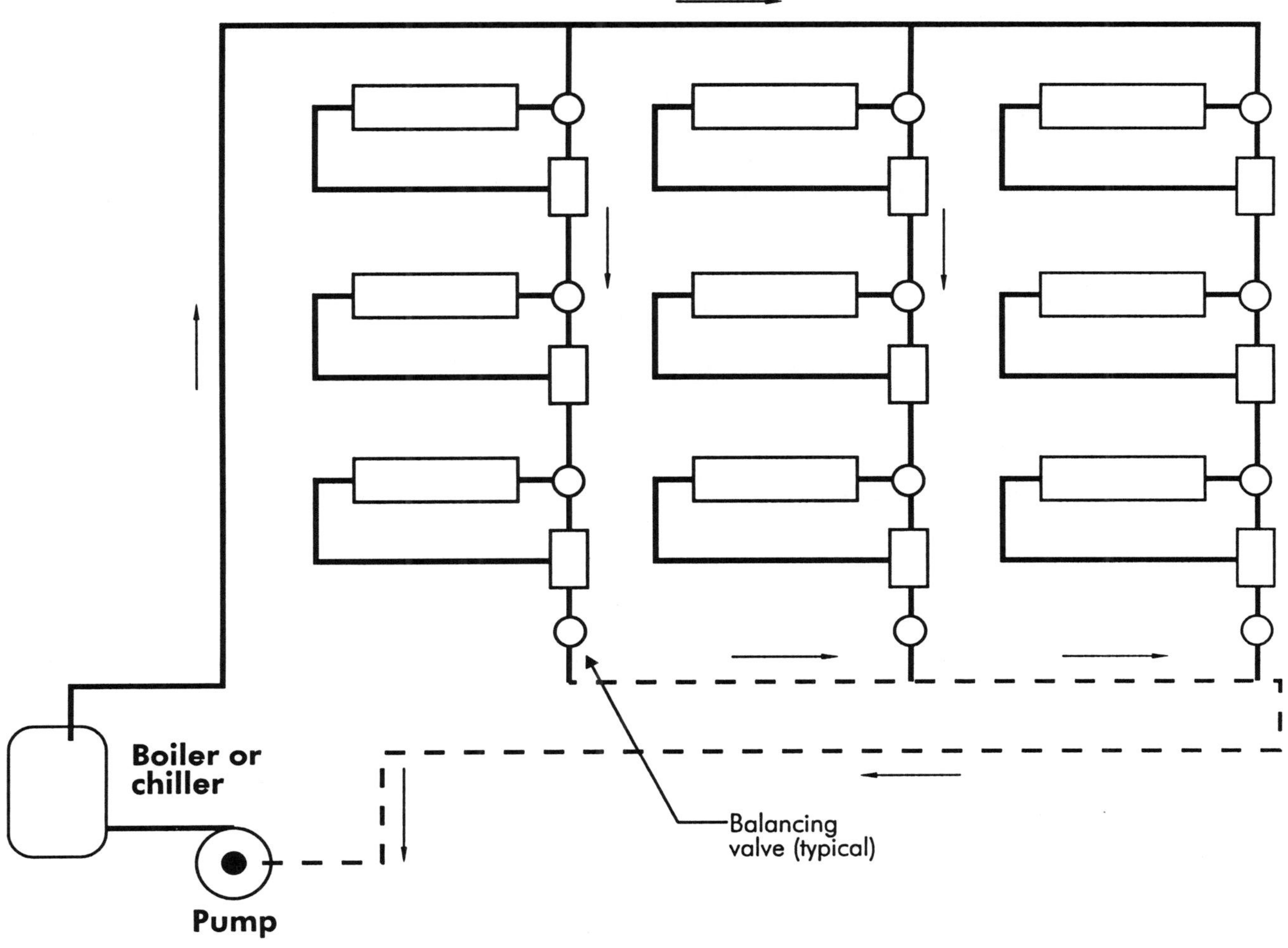

Figure 5-13. Overhead reverse return system with downfeed one-pipe circuits (elevation)

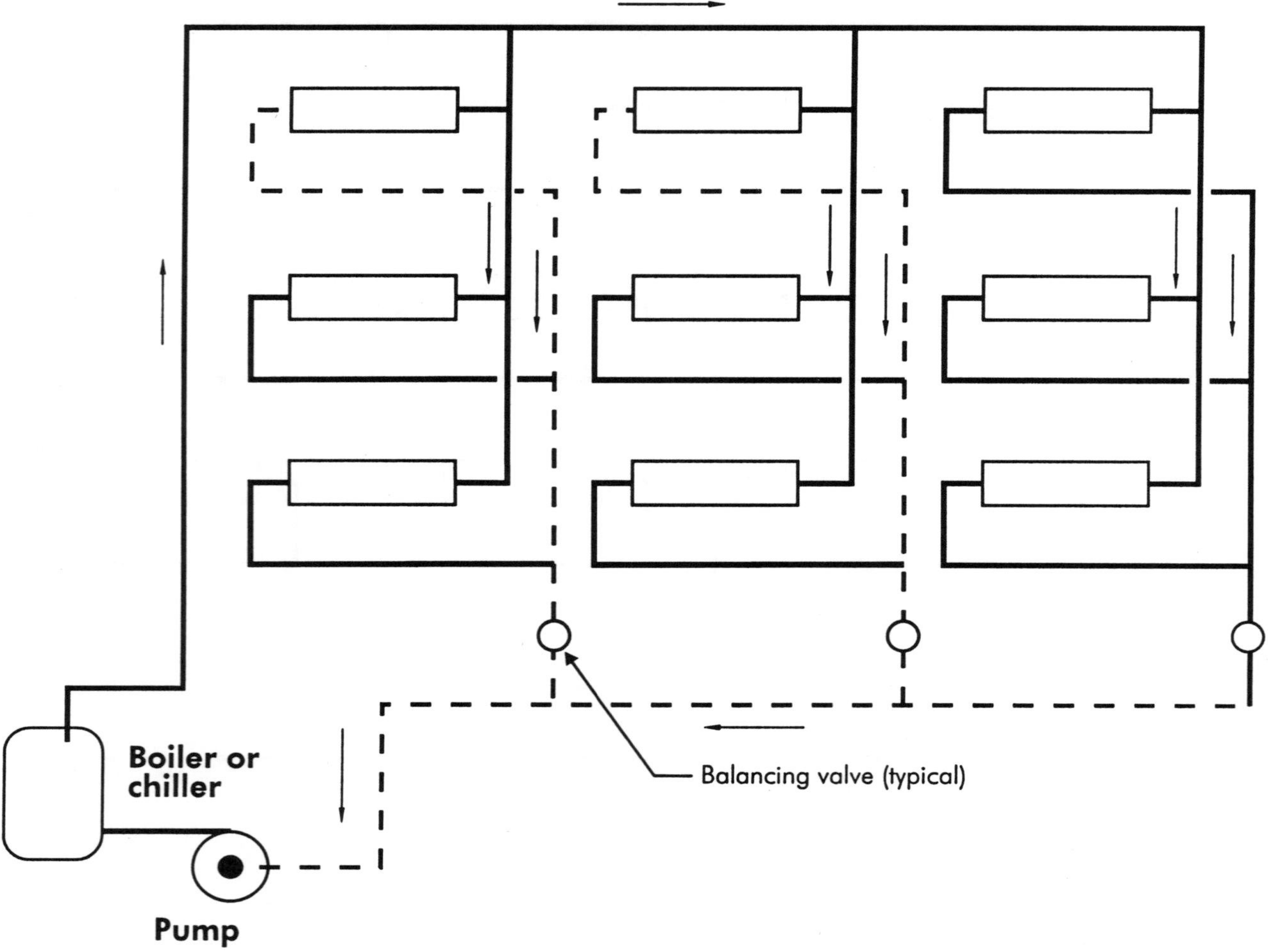

Figure 5-14. Overhead two-pipe direct return with reverse return radiation circuits (elevation)

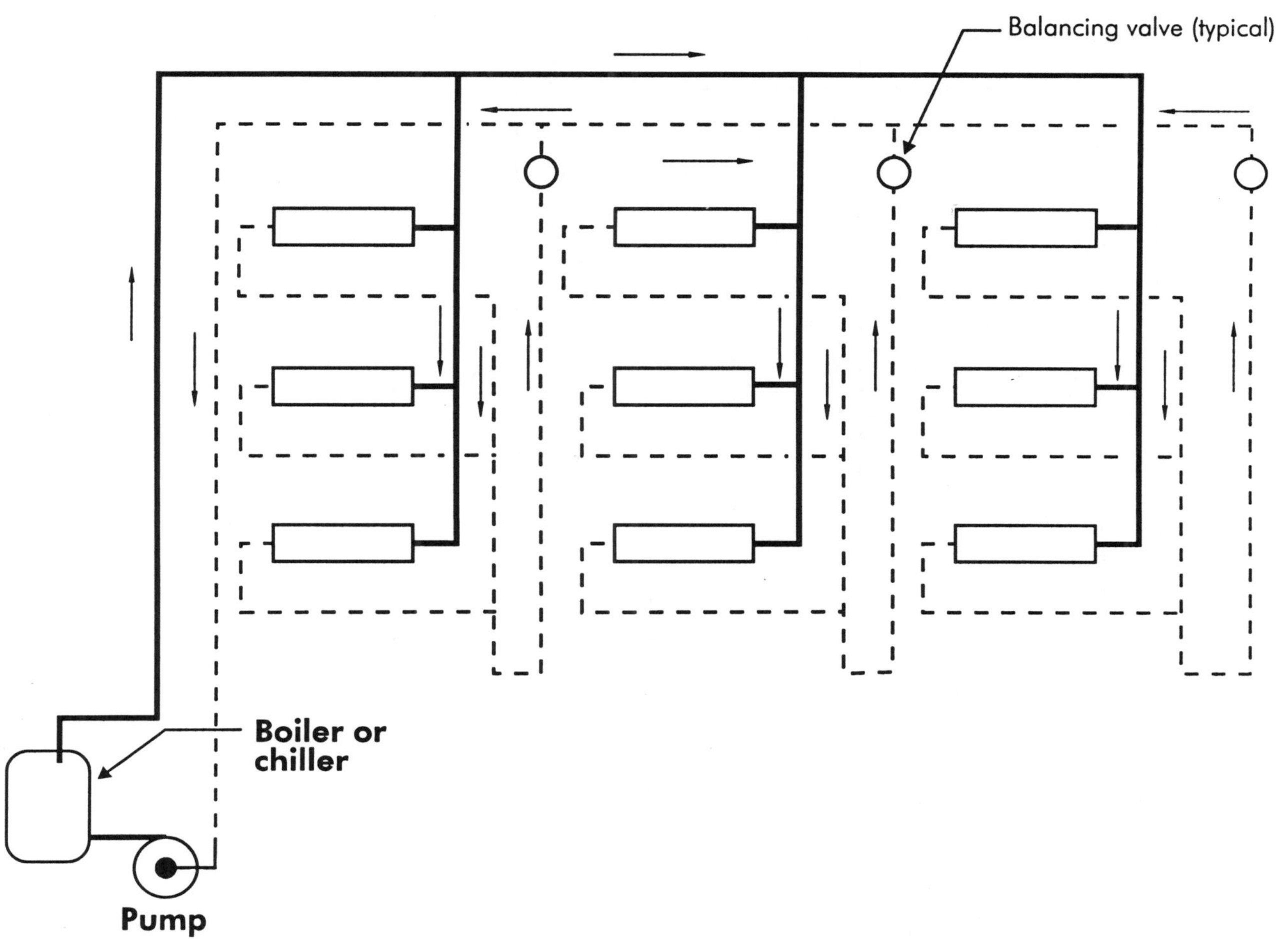

Figure 5-15. Overhead two-pipe downfeed direct return with reverse return radiation circuits (elevation)

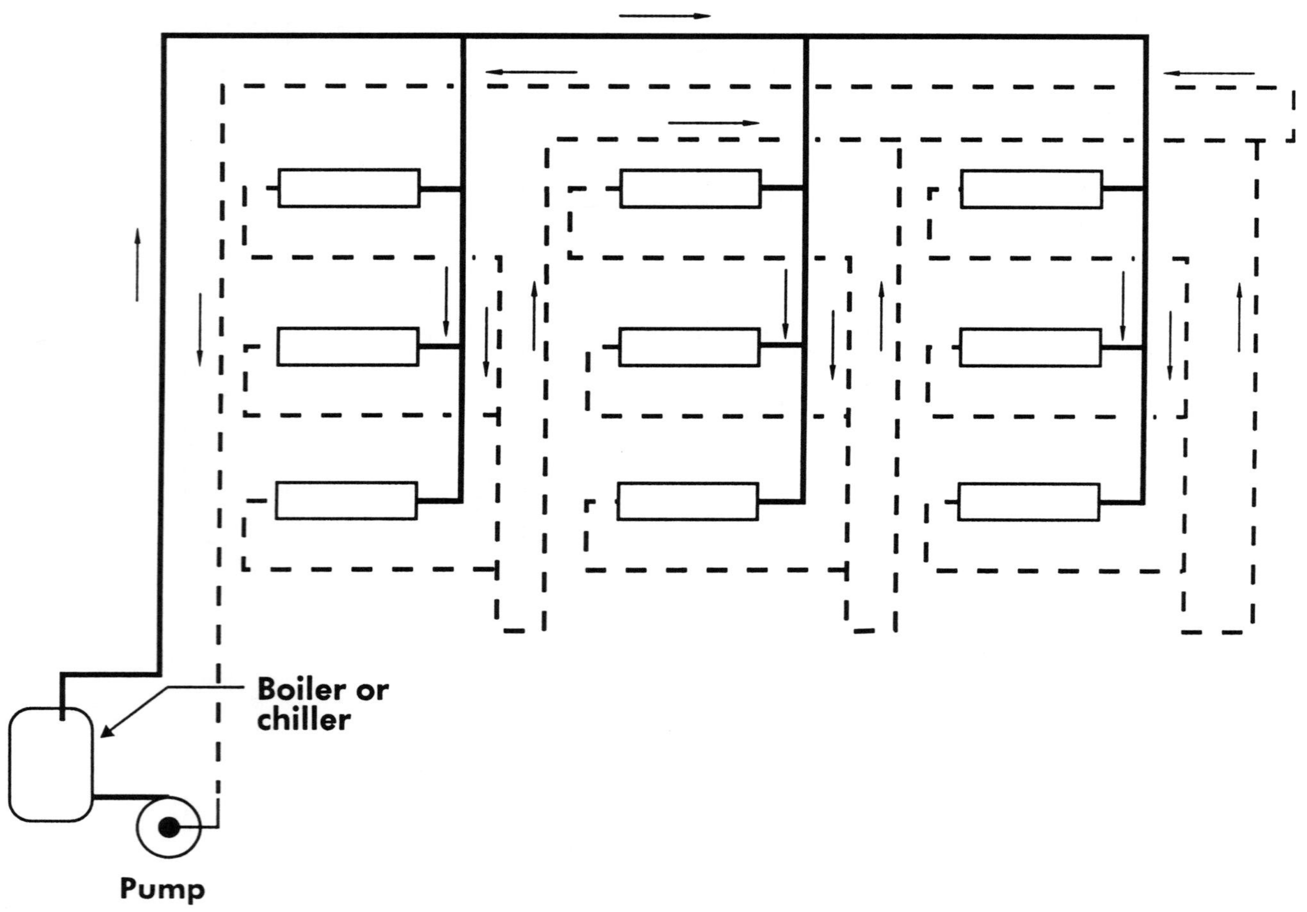

Figure 5-16. Overhead downfeed reverse return system with reverse return circuits (elevation)

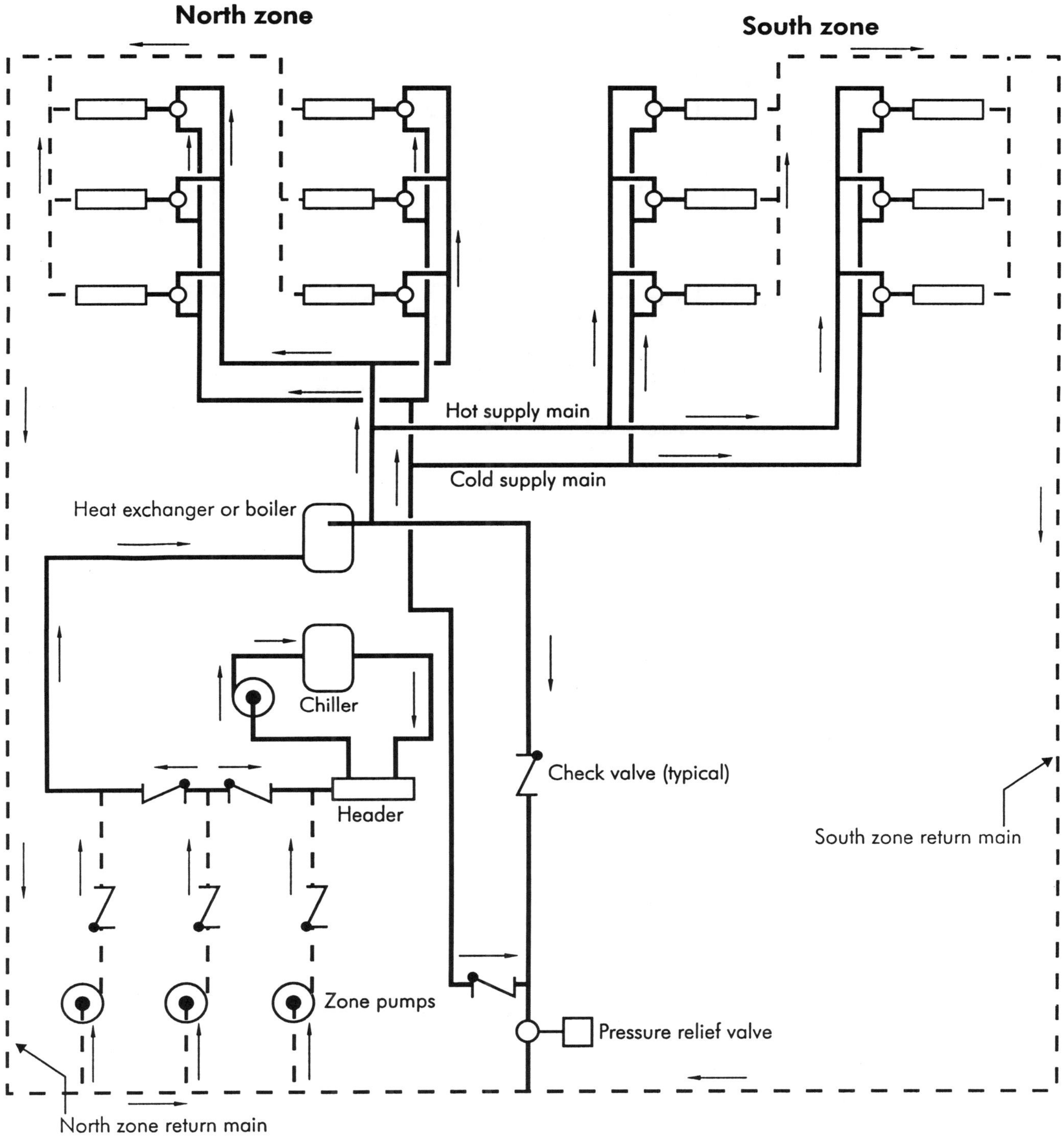

Figure 5-17. Three-pipe, three pump distribution

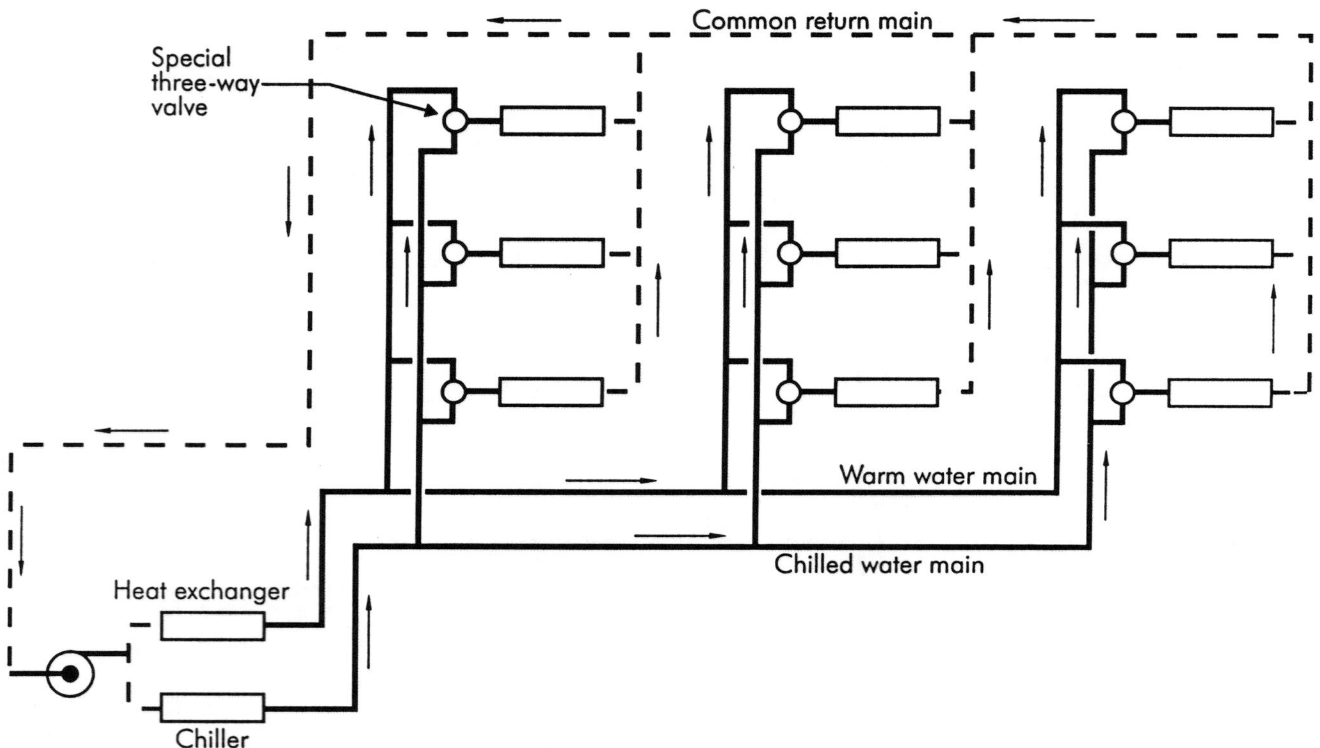

Figure 5-18. Simple three-pipe water distribution system (elevation)

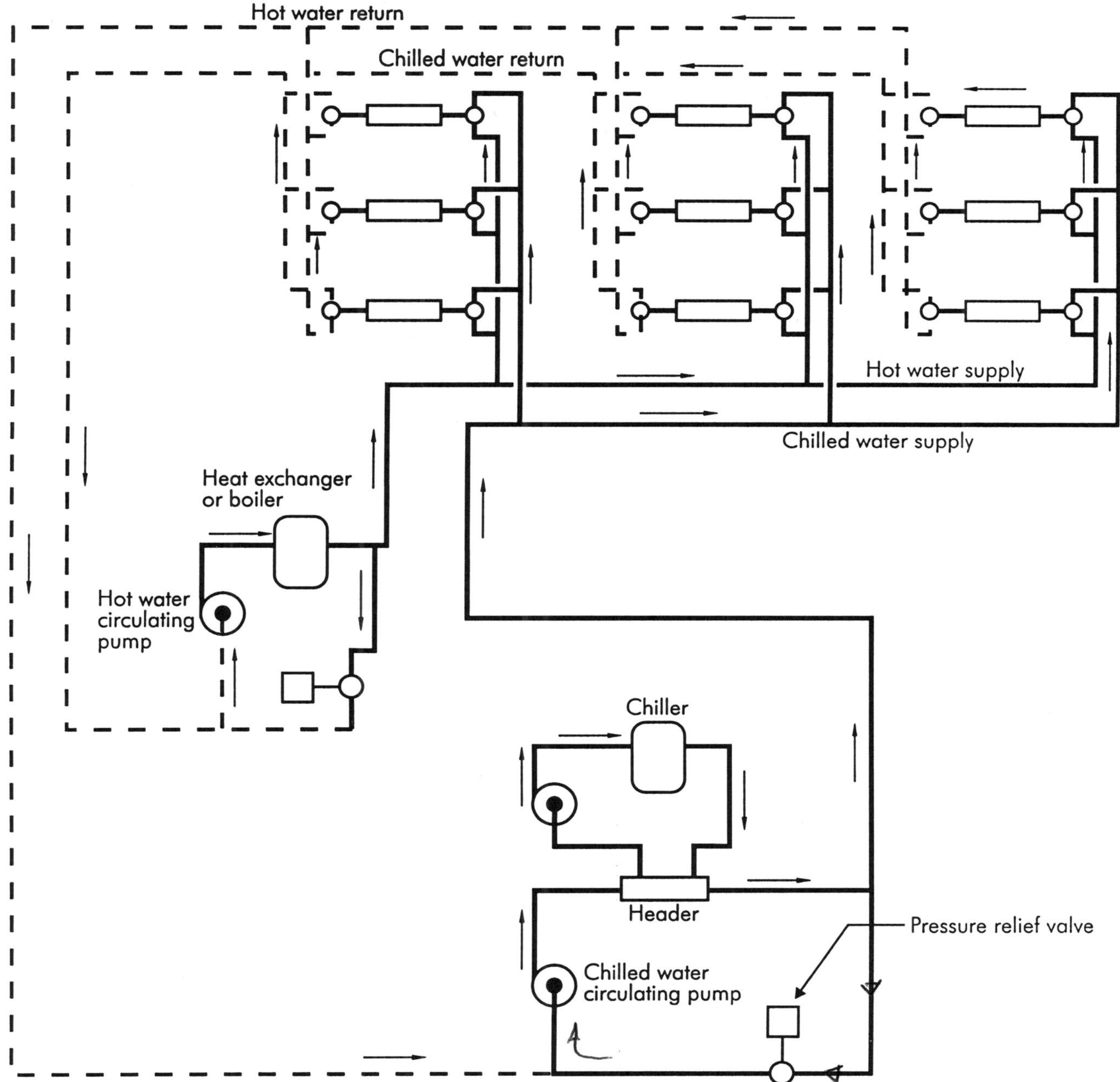

Figure 5-19. Four-pipe distribution system

Chapter 6

Boilers

Before automatic and extended modern boilers and furnaces were developed, space heaters or stoves heated rooms or areas. In parts of the country, many houses (mostly in rural and/or remote areas) are still heated by stoves burning coal, gas, oil, or wood. The heat is furnished by radiation and convection from the stove to the surrounding area. It is not an efficient system, because the location of the source has the highest temperature; cold spots exist close to outside doors, windows, etc. Modern systems transfer the heat produced in a boiler or furnace to water (or steam) or air, respectively.

Heat may produce power or electricity. Electricity may also produce cold water for space or process cooling by way of specialized equipment called a chiller. The resultant hot or chilled circulated water modifies the temperature in the space served. The energy conversion equipment convert the energy stored in fuels into useful heat, which is then transferred to the medium. The energy conversion equipment that produces useful energy in the form of heat are boilers and furnaces. The main components of a boiler are: a combustion chamber where fuel is burned; a heat exchanger for transferring the heat to water; an enclosure; and various apparatus and controls.

TYPES OF BOILERS

Hot water boilers can be fabricated of steel and have pressures between 15 and 160 psi for low pressure applications. For hot water heating applications, the limitations are a temperature of 250°F (121°C) and a working pressure of 30 psi. Steam boilers used for residential and small commercial applications are made of cast iron and have a working pressure between 1 and 15 psi. Residential and commercial boilers usually burn fossil fuels (oil or gas). Besides boilers that burn fossil fuels, there are also electric boilers.

All boilers and their corresponding combustibles are regulated and standardized by the Hydronic Institute. Other standards and rating codes applicable to boilers are issued by the Steel Boiler Institute (SBI), Mechanical Contractors Association (M-C-A), and the American Gas Association (AGA). Certain states and/or cities may also have their own rules and regulations applicable to boilers, furnaces, and chillers. While installing a heating system in most places does not require a permit, a permit is required for the potable water connection. If gas is used as a fuel, a gas permit is also required.

If a large boiler is used in a power plant in which larger boilers with higher working pressures are required, boilers must be tested and approved for the higher pressure. Due to their diversity, there are other classifications applicable to boilers such as:

- type of fuel used (coal, gas, oil, electricity);
- type of application (heating, heating and domestic hot water, power producing);
- type of construction (sectional, round, fire tube, water tube);
- type of combustion function of the gases exhausted (natural draft, induced draft, and forced draft).

Cast iron boilers are constructed of vertical sections that are connected together. Small cast iron boilers are shipped assembled, ready for field connections. These types of boilers are built in a very wide range of capacities.

Steel boilers may either be of the fire tube or water tube type. In the fire tube boiler, combustion gases are within the tubes with the water around them. In a water tube boiler, the water is within the tubes and the hot gas is around them. Both of these types may have the combustion chamber fitted with an integral water-jacket or refractory lined brick. Boilers with a water-jacket are very common and are called *fireboxes*.

Boilers that heat domestic hot water may either be direct or indirect heaters. In direct heaters, the hot water is stored in a storage tank, and the water is supplied at the normal city water pressure. A disadvantage to this type of heater is that it has fairly high water temperature scale deposits in the storage tank (precipitation of minerals from water). Indirect heaters heat domestic water by means of a coil or a jack in which steam or hot water circulates. The heating medium is produced in a boiler.

Definitions

Before beginning a discussion of how to select a boiler, a few definitions must be known:

- *Once-through system* - Water passes through the equipment once (for cooling) and then is wasted. An example is a water-cooled condenser with no cooling tower. This type of system is presently unacceptable, because it wastes a lot of water.
- *Recirculating system* - Water flows in a closed circuit from the heat exchanger (chiller or boiler) to the equipment (air handling unit or radiator) and back.
- *Open system* - Water flows into a tank or reservoir that is open to the atmosphere.
- *Closed system* - Water flows in a closed circuit not exposed to the atmosphere. (For a system to be classified as closed, it must have all its parts (items) *bubble tight.*) Such a system usually contains a compression tank and is not connected to the atmosphere. This tank serves as an air cushion. This type of system presents a few advantages over an open system:
  - — Very little make-up water (if any) is required.
  - — If no make-up water is required, no dissolved oxygen is added.
  - — System is pressurized allowing for higher water temperature supplied into the system and a large temperature drop in radiators (coils, etc.).
- *Gross output* - Total quantity of heat in Btu per hour that the boiler will deliver.
- *Net load ratings* - Quantity of heat available in Btu per hour for the connected load. This rating is obtained by deducting the normal piping and pick-up allowances from the gross output.

SELECTING A BOILER

I=B=R (Institute of Boiler and Radiator Manufacturers) states that when sizing a boiler, no allowance is necessary for domestic hot water service unless there are more than two bathrooms to be served, or if the estimated use of hot water will not exceed 75 gal per day. If the estimated use of domestic hot water exceeds these limits, the following allowances should be made:

- For a storage-type heater, allow 120 Btuh for each gallon of water storage tank capacity.
- For a tankless heater, allow 12,000 Btuh for each bathroom in excess of two.

The maximum gross load (capacity) of a boiler is the sum of the following components:

1. *Radiation load* - Estimated heat emission requirements of the connected radiation equipment (measured in Btuh). The required connected radiation is determined by calculating the heat loss for each room. The sum of these heat losses is the total required heat emission of connected radiation.
2. *Piping heat loss* - Estimated heat loss from all sections of piping and fittings in the system (bare and insulated).

3. *Warm-up or pick-up allowance* - Net boiler load, which is the sum of Items 1 and 2.
4. *Domestic hot water load* - Estimated maximum heat in Btuh required to heat water for domestic use. (If the domestic hot water is supplied by a separate heating device, this (boiler) load does not exist.)

The selection of a boiler involves evaluating design conditions of the cold season and making sure the chosen capacity will cover heat losses plus infiltration and a safety factor.

To calculate boiler capacity, perform the following steps (detailed in Chapter Three pages 41-43):

1. Calculate room area
2. Calculate room volume
3. Establish design temperature outdoor/indoor Δt (temperature difference)
4. Calculate U factor for building shell
5. Tabulate room area, volume, and heat loss (Btuh):
   a. Establish total heat loss and if applicable:
   b. Calculate hot water requirement for process based on the following formula:

$$(\text{gpm})\ (\text{minutes running/day})\ (8.33\ \text{lb/gal})\ (130^\circ - 40^\circ\text{F}) = \text{Btu/day}$$

Note: 130°F is considered temperature of water required and 40°F is considered the cold water supply temperature.

   c. Calculate domestic hot water gpm:

$$(\text{gpm})\left(\frac{(60)(8.33\ \text{lb/gal})}{500}\right)(140^\circ - 40^\circ\text{F}) = \text{Btu/day}$$

6. Add a, b, and c from Step 5 and multiply by 2/3 to establish boiler Btuh. (The reduction is necessary because of non-simultaneous consumption.) A reasonable recommendation is to select two boilers, each representing 2/3 of the total Btuh quantity.

If two boiler controls are required to establish a lead lag system, which will fire them based on demand, the following shall also be determined:

- Oil consumption in gallons per hour (gph) or natural gas consumption
- Oil burner characteristics and type of draft
- Chimney diameter

The size of a boiler for a heating system depends on the heating load calculated for the area considered.

Example 6-1. Assume the total heat load equals 400,000 Btuh. Establish the range of boiler capacity, the type of fuel to be burned, and whether the system is hot water or steam. Select a boiler from one manufacturer.

Solution 6-1. For small systems, residential and small commercial hot water boilers are used. The hot water is then pumped in the system to thermal units.

In large installations, steam boilers including heat exchangers (called converters) may be used to heat water with steam.

The next step is to select the Δt (temperature drop or difference between hot water to and from the system) for the system. The usual Δt is 20°F.

As an example, assume a boiler is selected from the manufacturer's catalog shown in Figure 6-1. In this case, the selection is Model V905.

This boiler will supply hot water at 180°F to the heating system. The amount of hot water flowing in the system (considering a Δt of 20°F) is:

$$\text{gpm} = \frac{400{,}000 \text{ Btuh}}{(8.33 \text{ lb/gal})(60 \text{ min/hr})(1 \text{ Btu/lb - °F})(20\text{°F})} = 40 \text{ gpm}$$

Example 6-2. Size a boiler condensate feed tank capacity (steam system) based on 15 min storage (running time of 25%), and establish the steam capacity (lb). It is known that two boilers are required, and each boiler produces 3013 lb of steam.

Solution 6-2. To establish the total steam capacity, just multiply the amount of steam each boiler produces by two:

$$3013 \text{ lb/boiler} \times 2 = 6026 \text{ lb}$$

Consider the 25% running time:

$$6026 \text{ lb} \left(\frac{15 \text{ min}}{60 \text{ min/hr}}\right) = 1506.5 \text{ lb/hr}$$

The capacity (in gallons) is:

$$\frac{1507 \text{ lb/hr}}{8.33 \text{ lb/gal}} = 181 \text{ gal}$$

Work with a manufacturer to select a boiler that is closest to the 181 gallon capacity (plus a safety factor). After selecting the boiler capacity, it is necessary to select the condensate pump, which includes establishing head and pump discharge.

Example/Solution 6-3. To determine how to size boiler oil tank capacity, follow these steps:

- Take two-thirds of the recommended boiler oil consumption in gph and multiply by two (when two boilers are installed)
- Obtain gallons/day by multiplying the consumption obtained by 24 hours
- Provide a reserve for minimum 15 days

In this case, the oil tank capacity shall be equal to the maximum oil consumption of 15 days.

COMMERCIAL HYDRONIC HEAT PRESSURIZED, WET BASE OIL, GAS OR COMBINATION

Burnham®
AMERICA'S BOILER COMPANY

V9 SERIES

Heating Capacities:
311 to 1445 MBH

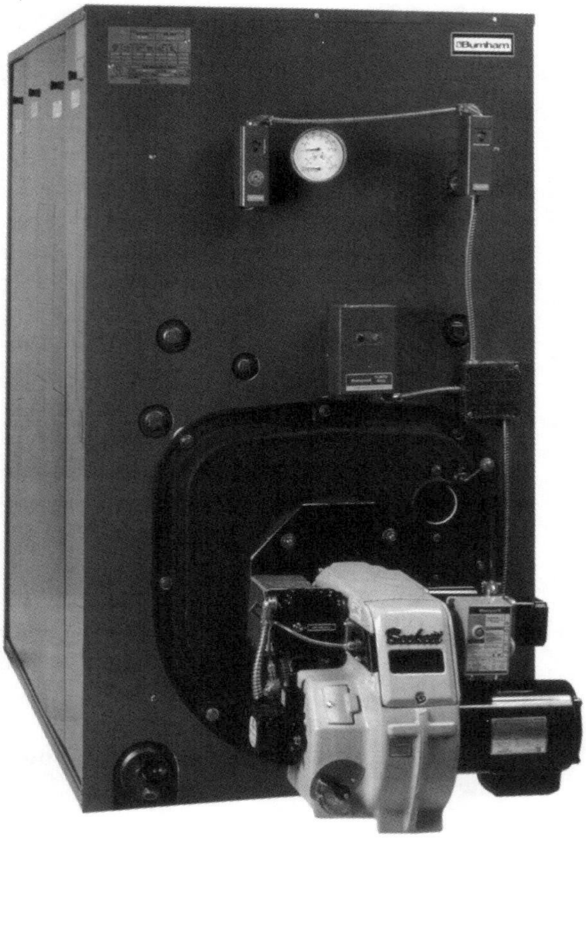

Figure 6-1. Manufacturer's catalog (Courtesy, Burnham), continued

SPECIFICATIONS

V9 RATINGS

BOILER MODEL (1)	BOILER HORSEPOWER	GROSS OUTPUT MBH	NET I=B=R RATING (2)			BURNER INPUT		HEATING SURFACE (SQ. FT.)		NET FIREBOX VOLUME (CU. FT.)	PRESSURE IN FIREBOX (INCHES WTR. COLUMN) (3)
			SQ. FT STEAM	MBH STEAM	MBH WATER	OIL (GPH)	GAS (MBH)	STEAM	WATER		
V-903	9.3	311	971	233	270	2.75	397	34	37	3.2	.28
V-904	12.1	404	1263	303	351	3.5	505	48	54	4.8	.29
V-905	16.0	534	1671	401	464	4.6	668	62	71	6.4	.20
V-906	19.8	664	2075	498	577	5.8	830	77	88	7.9	.29
V-907	23.7	794	2483	596	690	6.9	992	91	105	9.5	.26
V-908	27.6	924	2888	693	803	8.0	1155	105	122	11.0	.29
V-909	31.5	1054	3296	791	917	9.1	1317	119	139	12.6	.28
V-910	35.4	1184	3700	888	1030	10.2	1479	134	156	14.2	.28
V-911	39.3	1314	4125	990	1143	11.4	1642	148	173	15.7	.28
V-912	43.2	1445	4579	1099	1257	12.6	1804	162	190	17.3	.30

(1) Suffix "S" indicates steam boiler, "W" indicates water boiler. Suffix "G" indicates gas-fired, "O" indicates oil-fired. "GO" indicates combination gas-oil fired.

(2) I=B=R net ratings shown are based on piping and pickup allowances which vary from 1.333 to 1.315 for steam and 1.15 for water.
Consult manufacturer for installations having unusual piping and pickup requirements, such as intermittent system operation, extensive piping systems, etc.
The I=B=R burner capacity in GPH is based on oil having a heat value of 140,000 BTU per gallon.

(3) Boiler ratings are based on 12.5% CO^2, + .10" water column pressure at boiler flue outlet.
Ratings shown above apply at all altitudes up to 1000 feet on oil and 2000 feet on gas. For altitudes above those indicated, the ratings should be reduced at the rate of 4% for each 1000 feet above sea level.

NOTE: Maximum Allowable Working Pressure—

Steam	15 PSI
Water (USA)	50 PSI or 70 PSI depending on relief valve setting
Water (Canada)	45 PSI

TANKLESS HEATER RATINGS (Water & Steam)

BOILER MODEL	NUMBER OF V9-2 TANKLESS* HEATERS INSTALLED			
	1	2	3	4
V-903	6.0	–	–	–
V-904	7.5	–	–	–
V-905	7.5	–	–	–
V-906	7.5	13	–	–
V-907	7.5	15	–	–
V-908	7.5	15	–	–
V-909	7.5	15	21	–
V-910	7.5	15	22.5	–
V-911	7.5	15	22.5	–
V-912	7.5	15	22.5	28.5

*Ratings are given in gallons per minute continuous flow of water heated from 40°F to 140°F with 200°F boiler water.

Form No. 4526 C-10/94-20Ma
Printed in U.S.A.

Figure 6-1. concluded

Boiler Efficiency

The variables that influence the operating efficiency of a boiler are as follows:

- Boiler size
- Boiler design
- Boiler operating temperature
- Burner design
- Frequency of cycling

There are several different types of efficiencies that must be calculated before a selection is made. These are as follows:

$$\text{Thermal efficiency} = \frac{\text{Useful boiler output}}{\text{Total boiler input}}$$

$$\text{Combustion efficiency} = \frac{\text{Heat energy in combustion gases}}{\text{Energy within the fuel}}$$

$$\text{Seasonal operating efficiency (SOE)} = \frac{\text{Seasonal output}}{\text{Seasonal input}}$$

RESIDENTIAL/ SMALL COMMERCIAL HEATING SYSTEMS

Hydronic heating systems include the following basic elements:

- Source of energy - gas, oil, electricity
- Heat convecting media - water, steam
- Distribution system - piping and fittings
- Terminal devices - radiators, fan-coil units, etc.

Hot water temperature is based on economic as well as comfort considerations. Higher media temperatures result in smaller distribution units and smaller initial cost; however, the operating costs are higher. Water temperatures usually range between 180° and 200°F (82° and 93°C). The modern control system adjusts the water temperature as a function of outdoor temperature.

Residential water heating systems are designed for a temperature difference or drop (Δt or TD) of about 20°F (11°C) between supply and return. This TD produces satisfactory results and simplifies calculations, because 1 gpm conveys 10,000 Btuh or the equivalent of 2.93 kW (1 kW = 3413 Btuh). The gpm flowing in a system can be calculated by using the following formula:

$$\text{gpm} = \frac{\text{Btuh}}{(8.33\ \text{lb/gal})(60\ \text{min/hr})(\text{TD})}$$

With the TD of 20°F, it can be shown that 1 gpm conveys 10,000 Btuh as follows:

$$\text{Btuh} = (1\ \text{gpm})\ (8.33\ \text{lb/gal})\ (60\ \text{min/hr})\ (20°\text{F}) = 10{,}000\ \text{Btuh}$$

The following pipe sizes are recommended[1] for conveying the listed Btu (other tables may give slightly different figures but the differences are negligible):

Schedule 40 steel pipe size (in.)	Flow (gpm)	Btuh	Water velocity (fps)
1/2	1 to 1-1/2	10,000	1.58
3/4	4	40,000	2.41
1	8	80,000	2.97
1-1/4	14	140,000	3.00

One manufacturer publishes a complete table that contains data comparable with that just shown, Table 6-1.

RECOMMENDED FLOW RATES IN GPM
FOR ALL PIPING EXCEPT BRANCH CONNECTIONS ON TACO VENTURI SYSTEMS

Nominal Pipe or Tubing Size In.	STEEL PIPE				COPPER TUBING			
	Minimum GPM	VELOCITY Ft/sec	Maximum GPM	Velocity Ft/Sec	Minimum GPM	Velocity Ft/Sec	Maximum GPM	Velocity Ft/Sec
1/2"	–	–	1.8	1.9	–	–	1.5	2.1
3/4"	1.8	1.1	4.0	2.4	1.5	1.0	3.5	2.3
1"	4.0	1.5	7.2	2.7	3.5	1.4	7.5	3.0
1 1/4"	7.2	1.6	16	3.5	7	1.9	13	3.5
1 1/2"	14	2.2	23	3.7	12	2.1	20	3.6
2"	23	2.3	45	4.6	20	2.1	40	4.1
2 1/2"	40	2.7	70	4.7	40	2.7	75	5.1
3"	70	3.0	120	5.2	65	3.0	110	5.2
3 1/2"	100	3.2	170	5.4	90	3.1	150	5.2
4"	140	3.5	230	5.8	130	3.5	210	5.8
5"	230	3.7	400	6.4	–	–	–	–
6"	350	3.8	610	6.6	–	–	–	–
8"	600	3.8	1200	7.6	–	–	–	–
10"	1000	4.1	1800	7.4	–	–	–	–
12"	1500	4.3	2800	8.1	–	–	–	–

Table 6-1. Quick pipe sizing table (Courtesy, Taco, Inc.)

The heat transferred from the water to the space can be calculated with the following formula:

$$q_W = 60G_{wc}(t_1 - t_2)$$

In SI units, the equation would be:

$$q_W = G_{wc}(t_1 - t_2)/1000$$

where: q_W = heat transfer rate from water to the space Btuh (w)
G = water flow rate (gpm)
w = water density (lb/gallon)

c = specific heat of water Btu/lb-°F
t_1 = water temperature entering the unit (°F)
t_2 = water temperature leaving the unit (°F)

The value "w" should be obtained from available tables for the temperature of flow. The value "c" should be determined for the average temperature flow.

Selecting Water Temperature[2]

Design water temperature is the maximum (heating) or minimum (cooling) water temperature supplied to the system at operating design conditions. The design water temperature should be determined as a result of an economic analysis of the system requirements unless it is dictated by process load requirements.

The water temperature leaving the boiler depends on the design water temperature and the design water temperature difference. Neither design water temperature nor design water temperature difference have any effect on the boiler size. The temperature of water leaving the boiler is the design water temperature plus one-half of the design water temperature difference. For example, consider a system in which the design water temperature is 200°F and the design water temperature difference is 20°F. The temperature of the water leaving the boiler will be:

$$200°F + \left((20°F)\,(.5)\right) = 210°F$$

In multiple circuit systems, design water temperature differences may be selected for each circuit; however, the water temperature leaving the boiler must be the same for each circuit.

Elements that influence the selection of the operating temperature range include:

- whether the system is for heating or cooling;
- the type and pressure rating of the primary heat source;
- the types of terminal units, pumps, and accessories.

Flow Rate

For large heating systems, a minimum system flow rate should provide the lowest cost while providing the maximum control. The following methods are used to determine minimum flow rate to the individual terminal unit and the total system flow rate:

- Various basic terminal units are capable of full capacity production at specific temperature drops. For example, baseboard and finned-tube

units are selected frequently for a 20° to 50°F (11° to 28°C) drop, unit heaters for a 50°F (28°C) drop, and extended surface coils in air systems for a 100°F (56°C) drop.
- Manufacturers' catalogs list capacity ratings at various flow rates and entering temperatures. Engineers can select units for minimum flow rate (or maximum temperature difference) to produce the desired capacities.

For large, non-residential heating systems, good results depend on the following:

- Higher supply temperatures
- Primary-secondary pumping
- Multiple terminal equipment designed for smaller rates

STEAM HEATING

Steam is the vapor phase of water. It is generated by adding more heat than required to maintain the liquid phase at a given pressure, causing the liquid to change to vapor without any further increase in temperature. Table 6-2 shows some of the properties of saturated steam.

			Specific Volume ft^3/lb		Enthalpy Btu/lb		
Gauge	Absolute psia	Saturation Temp. °F	Liquid V_t	Steam V_t	Liquid h_t	Evap. h_t	Steam h_t
0	14.7	212	.0167	26.8	180	970	1150
2	16.7	218	.0168	23.8	187	966	1153
5	19.7	227	.0168	20.4	195	961	1156
15	29.7	250	.0170	13.9	218	946	1164
50	64.7	298	.0174	6.7	267	912	1179
100	114.7	338	.0179	3.9	309	881	1190
150	164.7	366	.0182	2.8	339	857	1196
200	214.7	388	.0185	2.1	362	837	1179

Table 6-2. Properties of saturated steam (Reprinted by permission of the American Society of Heating, Refrigerating and Air-Conditioning Engineers, Atlanta, Georgia, from the 1980 ASHRAE *Handbook--Systems*)

Steam offers the following advantages:

- It flows through the system unaided by external energy sources such as pumps.
- Because of its low density, it can be used in tall buildings where water systems create excessive pressure.
- Terminal units can be added to or removed from the system without making basic changes to the system design.
- Components can be repaired or replaced by closing the steam supply without the difficulties associated with draining and refilling a water system.
- It is pressure-temperature dependent; therefore, the system temperature can be controlled by varying either steam pressure or temperature.

- It can be distributed throughout a heating system with little loss in temperature along the network of pipes.

In view of these advantages, steam is recommended for the following applications:

- Where heat is required for process and comfort heating, such as industrial plants, hospitals, restaurants, dry cleaning plants, laundries, and commercial buildings.
- Where the heating medium must travel great distances, such as in facilities with scattered building locations, or where the building height results in excessive pressures in water systems.
- Where intermittent changes in heat load occur.

COMBUSTION

To produce heat in a boiler or furnace, fuel is burned in the combustion chamber. Combustion is a chemical reaction between fuel, oxygen, and heat.

There is a theoretical air/fuel ratio that must be satisfied to obtain complete combustion of the fuel. Complete combustion (100%) is difficult to achieve even with the theoretical amount of air. To prevent excess production of carbon dioxide (CO_2), which results from incomplete combustion, excess combustion air is supplied. This excess amount vary from 5% to 50% above the theoretical amount of air required for combustion. The room in which the boiler or furnace is located must have an opening in order to allow this extra air in for combustion.

Table 6-3 shows the decrease of CO_2 produced (in percentage) when excess air (20% and 40%) is supplied.

Heating fuel	Theoretical air/fuel ratio	Theor.	20% excess	40% excess	Value
Natural gas	9.6 ft^3/ft^3	12.1	9.9	8.4	1000 Btu/ft^3
No. 2 fuel oil	1410 ft^3/gal	15.0	12.3	10.5	140,000 Btu/gal
No. 6 fuel oil	1520 ft^3/gal	16.5	13.6	11.6	153,000 Btu/gal
Bituminous coal	940 ft^3/lb	18.2	15.1	12.9	13,000 Btu/lb

Note: Values are approximate as the composition of fuel varies.

Table 6-3. Percent CO_2 in combustion gas quality of air supplied (Reprinted with permission from *Air Conditioning Principles and Systems, Second Edition,* by E. Pita, J. Wiley, 1981)

The following examples illustrate how to calculate the amount of combustion air required.

Example 6-4.[3] A boiler fires 15 gallons per hour (gph) of No. 2 oil and requires 20% excess air for complete combustion. How many cfm of combustion air should be supplied?

Solution 6-4. Using the data from Table 6-3, note that the actual air quantity should be 1.2 times the theoretical (20% excess):

$$\text{cfm} = (1410 \text{ ft}^3/\text{gal})(15 \text{ gal/hr})\left(\frac{1 \text{ hr}}{60 \text{ min}}\right)(1.2) = 423 \text{ cfm}$$

Example 6-5.[4] A technician measures 8% CO_2 in the combustion gas of a natural gas-fired boiler that requires 15% excess air for complete combustion. Is the air/fuel ratio satisfactory?

Solution 6-5. Based on Table 6-3, over 40% excess air is being used, so the answer is no. A great deal of excess hot gas is going up the stack. Adjustments must be made so that excess air does not exceed more than 20% to 25%.

FITTINGS

A hydronic heating system must be equipped with some specific fittings and/ or appurtenances, including:

- air separators;
- venturi fittings;
- safety valves;
- flow regulators;
- venting fitting (for one-pipe system).

Air Separators

All water contains dissolved air. When water is heated, the air is forced out. All air must be vented from a heating system to prevent noisy operation and/ or a complete blockage of water circulation. One way to separate the air from the water is to use a low velocity of water in the system and the tendency of air to collect at the high point of the boiler. When air collects at this point, it may continue up to the compression tank. *Dip tubes*, which are installed in the boiler, prevent air from re-entering the system. A newer system uses a diaphragm or bladder type compression/expansion tank. In this case, the dip tube is eliminated.

Air Eliminators

If air is not eliminated from a system, the air might collect at some point, forming pockets and preventing water flow to radiators. The result is no heat. Air pockets may be eliminated by installing manually operated air vents. An automatic hot water air valve on each heat distributing unit may also be used.

In an automatic hot water air valve (such as the Taco vent, which is a fiber disc air vent), air passes through special fiber discs and vent holes to the atmosphere. Water following the air wets the special fiber discs causing them to swell, completely sealing the valve. As more air accumulates, the special fiber discs dry and shrink, permitting the valve to repeat the cycle. The Taco Hy-Vent™ may be used at all high points in the piping system. When the valve shell is full of water, the valve is closed. When sufficient air accumulates, the float drops and the valve opens. As the air passes out, water again fills the shell, closing the valve. This process repeats itself when air accumulates. Specific types of air valves are manufactured for various locations, i.e., high points, supply lines, etc.

The air scoop, shown in Figure 6-2, operates on the following principles:

- Heated water liberates air; the hotter the water, the more air is liberated.
- Air, being lighter than water, tends to travel along the upper portion of a horizontal pipe at the velocities commonly used in a heating system.
- As the air and water enter the air scoop, the air bubbles are scooped up by the first baffle and moved into the upper chambers. Any air bubbles that get through the first baffle are scooped up by the second or third baffle.
- Air that accumulates in the scoop's chamber is removed from the system by the air valve. Air from the scoop's second chamber passes into the expansion tank to act as an air cushion.
- Should the air completely fill the expansion tank, the excess will be removed by the air valve without disturbing the operation of the system.

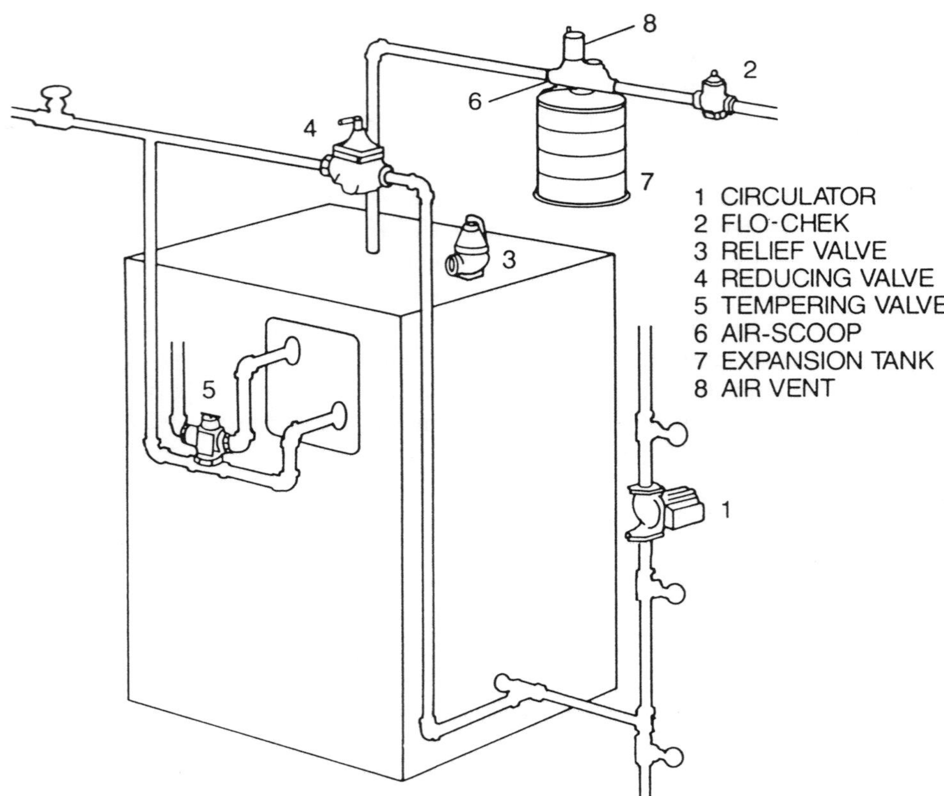

Figure 6-2. Typical boiler layout for residential installation (Reprinted with permission from *Taco Residential Hydronic Heating*)

Another type of air separator has inlet and outlet openings that are installed tangentially, Figure 6-3. Circulation through this separator creates a vortex or whirlpool in the center, where the lighter, entrained air can collect and rise into a compression tank installed above. Instead of relying entirely on low velocity separation, the action of centrifugal force sends heavier air-free water to the outer portion of the tank, allowing lighter air-water mixture to move into the lower velocity center. An air collecting screen located in the vortex aids in developing a low velocity area in the center where air can collect. This type of air separator is advantageous in small tanks.

Figure 6-3. Bell & Gossett Rolairtrol air separator with removable system strainer (Courtesy, ITT Fluid Handling)

Venturi Fittings

A venturi fitting is usually used in a one pipe system (rarely used presently) and/or in kick space heaters:

- when water enters in the fitting from the main pipe;
- when the restriction of the nozzle causes an increase in velocity, resulting in lower system pressure at this point;
- as a differential in pressure, which causes flow from the high pressure to the low pressure region. Water continually enters at this point as long as a flow exists in the main;
- when velocity is converted back to pressure in main with minimum loss.

The venturi fitting results in a constant percentage of water being diverted to the heating unit at any rate of water flow through the main and fitting. The fitting also results in a jet action with its higher efficiency (maximum diversion with minimum pressure drop) within the flow rates.

The venturi fitting acts like a powerful little pump, pulling water through heat distributing units. For one manufacturer (Taco, Inc.), only one venturi fitting is required per heat distributing unit, whether it is located above or below the main. The fitting is located at the return branch connection on the main. The same fitting may be used for upfeed or downfeed heat distributing units.

The Taco super venturi fitting in existing systems has a somewhat higher pressure drop and a substantially higher percentage of diversion and capacity. Super venturis have higher capacities and may be used in combination with a standard venturi leading to reduced branch sizes. In most cases, it is possible to use one branch pipe size throughout, resulting in a definite cost savings. Their greater pull makes them particularly effective on downfeed radiators and radiators located a long distance from the main supply pipe. Super venturis were used extensively with combination heating and cooling units. Taco venturi fittings may be used in conjunction with all types of baseboards, convectors, radiators, and some types of fan coil units (for heating or cooling). Figure 6-4 shows typical water flow pattern through the main and riser connections.

Relief or Safety Valves

The relief or safety valve, Figure 6-5, is designed to protect the boiler from rupturing should the pressure exceed its working pressure. These valves operate on both thermal expansion and/or steam pressures.

Relief valves that carry the ASME symbol must be tested and approved under the ASME *Boiler and Pressure Vessel Code* and rated by their discharge capacity in Btuh. They should be installed close to or directly into the top of the boiler without any shut-off valve between the relief valve and the boiler. The drain line from the discharge of the valve should be full size, and a shut-off valve should never be installed in this drain line. ASME relief valves of lower capacities may be installed in multiple units to total the capacity of the boiler. Relief valves should be installed at the point of no pressure change.

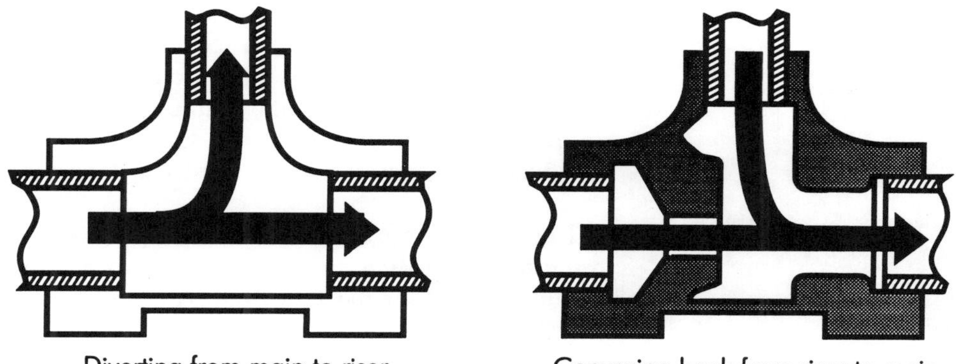

Figure 6-4. Typical water flow through main and riser connection (Courtesy, Taco, Inc.)

Figure 6-5. ASME relief valve

Airtrol™ Tank Fittings

Water that contains previously released air from heating or pressure reduction will attempt to re-absorb any free air that is present. Since the piping to compression tanks should be no smaller than 3/4", a relatively large and direct pipe connection exists between the compression tank and the system. Gravity circulation can easily take place between the cool tank and the hot system. As a result, the relatively air-free system water will attempt to re-absorb the free air that should be confined in the compression tank. Gravity circulation between the compression tank and the system will cause tank temperatures to parallel system temperatures.

While free passage of air rising into the tank is necessary, some method of restricting gravity circulation between the tank and the system is needed. The Airtrol™ tank fittings from Bell and Gossett serve this purpose. They are a form of check valve that allow air to pass freely but reduce the circulation of system water. This fitting reduces tank temperatures and tank size considerably.

Each fitting offers free passage to air rising from the system, but a restriction to water passage exists, which reduces gravity circulation to the tank. As air rises into the tank virtually unrestricted, an equal volume of water must return to the system. While it is impossible to entirely prevent the re-absorption of air from a compression tank, the Airtrol™ tank fittings are particularly helpful in reducing circulation, which results in lower tank temperatures.

Airtrol™ tank fittings also offer the ability to establish the initial water level in the tank on filling the system. Too much air is often trapped when the system is filled for the first time. A built-in or separate manual air vent is available, which consists of a 1/8" or 1/4" tube that extends up into the tank. After filling a system, this manual vent can be opened, allowing excess air in the tank below the top of the vent tube to escape. When the water level reaches the top of the tube, the vent should be closed. After the system is heated and additional air is released from the system water into the compression tank, the air level is lowered further. Mechanically pressurized and attic-installed tanks usually require different initial venting procedures, but vent tubes can be easily adjusted to specific requirements.

Flow Control Valve

This is a specially designed check valve that is installed in the main supply pipe close to the boiler. This valve closes when the circulator (pump) is not in operation. Its function is to prevent uncontrolled gravity hot water circulation; in other words, to prevent unwanted distributed heat. Such valves may be installed in each zone of a system that is not controlled by a water-operated valve.

Pressure Reducing Valve

This valve is installed to feed the make-up water to the boiler automatically when at the right pressure. The valve operates (allows water into the system) when the pressure in the hot water system drops below a set point. The location of this valve is at a point where there is no pressure change (at the compression tank connection). Since the hot water is treated with chemicals ahead of the pressure reducing valve, a reduced pressure backflow preventer must be installed to prevent potable water contamination.

MULTI-BOILER INSTALLATIONS

A boiler is sized for the worst-case condition, which is the coldest day of the year for the location. During the coldest day of the year, a boiler would run continuously in order to keep the occupant comfortable. The heat loss calculation is produced based on the coldest day, and the boiler is sized for the same. Consider, however, that just about every other medium cold day of the year that boiler is capable of doing a lot more than is needed. It does not make any sense for a boiler to generate its maximum heat on a 30° or 40°F degree day.

A simple way to reduce fuel consumption is to install two or more smaller boilers rather than one large boiler. Multiple boilers are a better match for outdoor temperature variation. For example, if a building requires a total net heat of 600,000 Btuh, it would be possible to size a single boiler for that load. It would heat the building during any day of the year, but on most medium cold days it would be oversized and would operate on and off with reduced efficiency. By splitting the total load between two smaller boilers, each capable of generating approximately 350,000 Btuh, there are two main advantages. First, on an average winter day, just one boiler may be in line to produce the heat required for the building. Since this smaller boiler is sized more closely to the actual heat loss of the building on that day, chances are it will run longer than a single larger boiler would. The fuel consumption is less, and as the weather gets colder, the second boiler will come on in parallel to help the first maintain the required temperature. Second, when two boilers are installed, stand-by capacity is automatically built in.

One large single boiler failure means no heat, whereas one multiple boiler out of service means slightly lower capacity. Another advantage of using multiple boilers is environmental protection. Small modular boilers burn gas or No. 2 fuel oil, which are cleaner and produce fewer pollutants than heavier oils. In addition, multiple smaller boilers operate longer at full capacity, are more efficient, and consequently reduce pollutant emissions. One large boiler cannot operate at low heat demand at maximum efficiency. Incomplete combustion flue losses contribute at lower efficiency. Multiple boilers are usually combined with heating zone control.

According to Dan Holohan, Dan Holohan and Associates, Inc., Bethpage, NY, as quoted in the magazine *Fuel Oil & Oil Heat with Air Conditioning*:[5]

> Traditionally, we size boilers for the coldest day of the year (and then some!) because we don't want to have trouble with our customers. On that frigid January day, we have to be sure there will be enough heat going into all the rooms. If we undersize, we know we're going to have callbacks.
>
> The trouble with this, though, is the "coldest day" may not even last a full day. There are times when the design condition may happen for just a few hours, at most. Sure, the boiler will rise to the occasion on that day, but every other day of the year, it's oversized. And it's wasting fuel oil.
>
> Since the yearly cost of operation plays a large part in the marketing of fuel oil and natural gas, I think it makes sense for the building owner to take advantage of the reality that on most days, a "smaller" boiler will get the job done as well or better than a full-load-sized boiler. In other words, it ain't always that cold!
>
> And this is where multiple-boiler systems come in. We can split the worst-case load in half, thirds or even quarters, and then fire only the load we need on any given day. At a certain job, the total heating and domestic hot water load was 313,926 Btuh. We decided to use two boilers, each rated for 175,000 Btu. On all but the coldest days, only one of the boilers would be fired. And since it would be firing on longer cycles than a full-load-sized boiler, we'd have a better rating. In short, we'd save fuel without sacrificing comfort.

It may be more beneficial to install a number of small boilers instead of one or two large ones. Small boilers are called *modules* if they are installed without separate shut-off valves. If they are provided with shut-off valves, they are just called multiple boilers. They work in sequence or all together following the heat demand. A small boiler usually has a capacity of up to 500,000 Btuh.

Considering all the facts, including the first cost of equipment and installation, there is usually a very small difference for a multiple boiler installation. Initial higher costs will probably be canceled by savings over a longer term. With multiple boilers and higher efficiencies operating costs are lower.

COMPRESSION TANKS

As water heats up, it expands and air is eliminated. The expansion and contraction of water is accommodated in the compression tank. When the water expands, it compresses the air that is present in the system. The compression tank collects this compressed air. The compression tank in a closed system does not vent this air to the atmosphere. Once the system cools down, the water returns to its normal state, causing the compressed air in the compression tank to also return to its normal pressure. The compression tank may be called an expansion tank when it is open to the atmosphere; however, these terms are occasionally interchangeable.

The size of the compression tank depends on the volume of water in the system. Other elements that influence tank size include system operating temperature, location (height) of the expansion tank, and the relation between the boiler and the height of the terminal user (radiation unit). The location of the circulator (pump) with respect to the expansion tank and the boiler must also be considered. The volume of water required in the system may be determined after the boiler and other system components are sized. Table 6-4 lists the water volume contained in various pipe sizes.

Nominal Pipe Size		Standard Steel Pipe					Type L Copper Tube			
		Schedule	Inside Diameter		Water Content				Water Content	
In.	(mm)	No.	In.	(mm)	gal/ft	f/m	In.	(mm)	gal/ft	f/m
3/8	(9)	–	–	–	–	–	4.430	(1.09)	0.0075	(0.09)
1/2	(12)	40	0.622	(1.58)	0.0157	(0.19)	0.545	(1.38)	0.0121	(0.15)
5/8	(15)	–	–	–	–	–	0.666	(1.69)	0.0181	(0.22)
3/4	(19)	40	0.824	(2.09)	0.0277	(0.34)	0.785	(1.99)	0.0251	(0.31)
1	(25)	40	1.049	(2.66)	0.0449	(0.56)	1.025	(2.60)	0.0429	(0.53)
1 1/4	(30)	40	1.380	(3.50)	0.0779	(0.97)	1.265	(3.21)	0.643	(0.81)
1 1/2	(38)	40	1.610	(4.09)	0.106	(1.32)	1.505	(3.82)	0.0924	(1.15)
2	(50)	40	2.067	(5.25)	0.174	(2.16)	1.985	(5.04)	0.161	(2.00)
2 1/2	(62)	40	2.469	(6.27)	0.249	(3.09)	2.465	(6.26)	0.248	(3.08)
3	(75)	40	3.068	(7.79)	0.384	(4.77)	2.945	(7.48)	0.354	(4.40)
3 1/2	(88)	40	3.548	(9.01)	0.514	(6.38)	3.425	(8.70)	0.479	(5.95)
4	(100)	40	4.026	(10.23)	0.661	(8.21)	3.905	(9.92)	0.622	(7.73)
5	(125)	40	5.047	(12.82)	1.04	(12.92)	4.875	(12.38)	0.970	(12.05)
6	(150)	40	6.065	(15.41)	1.50	(18.63)	5.845	(14.85)	1.39	(17.26)
8	(200)	30	8.071	(20.50)	2.66	(33.03)	7.725	(19.62)	2.43	(30.18)
10	(250)	30	10.136	(25.75)	4.19	(52.04)	9.625	(24.45)	3.78	(46.95)
12	(300)	30	12.090	(30.71)	5.96	(74.02)	11.565	(29.38)	5.46	(67.81)

Table 6-4. Volume of water in standard steel pipe and copper tubes (Reprinted by permission of the American Society of Heating, Refrigerating and Air-Conditioning Engineers, Atlanta, Georgia, from the 1980 ASHRAE *Handbook--Systems*)

The size of the compression tank must be adequate to receive the increased volume of water from expansion due to heating, while keeping the pressure within minimum and maximum limits. Compression tank size depends on the following sources of pressure:[6]

- *Static pressure* - The pressure due to the height of water above any point. Usually the critical point is the boiler, which is often at the bottom of the system.
- *Initial fill pressure* - If the system were initially filled without pressure, the pressure at the highest point in the system would be atmospheric. In order to provide a safety margin to prevent the pressure from going below atmospheric and thus allowing air in, the contractor should fill the system under pressure. A pressure of 4 to 5 psig at the top of the system is adequate for hydronic systems.
- *Pressure-temperature increase* - The pressure that occurs after the system is filled with cold water and pressurized and the system temperature is raised. The pressure will increase further due to expansion of the water compressing air in the tank.
- *Pump pressure* - When the pump is operated, the pressures change in the system by the value of the pump head. As explained before, this depends on where the compression tank is located. If the tank is connected at the pump suction, the head is added at every point. If the tank is connected at the pump discharge, the pump head is subtracted from the pressure at every point.

For large systems, the size of the compression tank and the pressure control can be critical. To reduce compression tank size and produce satisfactory pressure control, consider the following points:

- A compressed gas, if available, can be admitted to the compression tank to obtain the desired pressure.
- The additive pressure of the pump may be reduced by using larger pipe sizes or greater design temperature drop.
- The boiler may be constructed for pressures higher than 30 psig (207 kPa).
- The steam boiler may be used with a steam-to-water heat exchanger to heat the water circulated.

Compression Tank Piping

Efficient air separating equipment and correctly sized and positioned compression tanks require only simple piping connections for proper performance. In its simplest form, piping need only be a direct-connected riser pitching up to the tank. However, a number of variations exist that can have a considerable effect on the overall efficiency of a system, as well as on the required compression tank size.

As air bubbles are separated from a system, either with a boiler dip tube or an external air separator, they must be allowed to rise into the compression tank. As free air rises into the tank, an equal volume of water must be displaced and returned to the system. This means that the piping connecting the air separation equipment and the compression tank must always pitch upward and be large enough in diameter to allow water and air to pass each other at the same time. Tests have shown that a 3/4" pipe is the smallest diameter pipe that should be used to permit simultaneous water and air passage. When horizontal piping is used, a reduced gravity head results, making it necessary to use piping larger than 3/4".

Water that has previously released air from solution through heating or pressure reduction will attempt to re-absorb any free air that is present. A relatively large and direct pipe connection exists between the compression tank and the system, and gravity circulation can easily take place between the cool tank and the hot system. As a result, the relatively air-free system water will attempt to re-absorb the free air that should be confined in the compression tank. Gravity circulation between the compression tank and the system will cause tank temperatures to parallel system temperatures.

While free passage of air rising into the tank is necessary, some method of restricting gravity circulation between the tank and system is needed. A special fitting will serve this purpose, allowing air to pass freely but reducing the circulation of system water. This fitting helps reduce tank temperatures and tank size considerably. As air rises into the tank virtually unrestricted, an equal volume of water must return to the system. Although it is impossible to entirely prevent the re-absorption of air from a compression tank, these fittings are particularly helpful in reducing circulation, which results in lowered tank temperatures.

This type of fitting also offers the ability to establish the initial water level in the tank on filling the system. When the system is filled for the first time, there is often too much air trapped. A built-in or separate manual air vent is available and consists of a small 1/8" or 1/4" tube, which extends up into the tank. After filling a system, this manual vent can be opened, allowing excess air in the tank below the top of the vent tube to escape. When the water level reaches the top of the tube, the vent should be closed. After the system is heated and additional air is released from the system water into the compression tank, the air level is lowered even more. Mechanically pressurized and attic installed tanks usually require different initial venting procedures, but vent tubes can be easily adjusted to specific requirements.

The point of connection between the expansion tank and the system should be carefully selected so that automatic, check, or manual valves will not isolate the tank from a hot boiler or any other part of the system. The point where the compression tank is connected to the system is called the point of *no pressure change.*

In a closed hydronic heating system, the compression tank contains a certain volume of air. The air is compressible while the water is not. In order to raise the pressure in the compression tank, the air must be compressed, which requires more water to enter the compression tank. However, the system is closed and water cannot be added, so the pressure in the tank remains the same, independent of the pump operation. For this reason, it is called the point of no pressure change.

Pump Location

The location of the pump in relation to the compression tank connection determines whether the pump pressure is added or subtracted from the system static pressure. This is because the junction of the tank with the system is a point of no pressure change whether the pump operates or not. The pump location is also important in deciding whether or not the hydronic system will operate as a closed, sealed, and corrosion-free system, or as an open system subject to corrosive attack. Mechanical designs that place the compression tank at the pump discharge (pump discharging into boiler) can also establish below-atmospheric-pressure at air vents, etc., so that air can be sucked into the piping system. Such mechanical designs change the closed system to an open design and can quickly destroy boilers and piping by corrosion. A proper mechanical design, usually established by pumping away from the compression tank, maintains closed system characteristics because pump pressure is added to system pressure, and pressures are maintained at high levels.

Figure 6-6a shows the pump discharging away from the boiler and expansion tank. Full pump pressure will appear as an increase at the pump discharge, and all points downstream will show a pressure equal to the pump pressure minus the friction loss from the pump to that point. The fill pressure must be only slightly higher than the system static pressure.

When the pump discharges into the boiler and expansion tank, Figure 6-6b, full pump pressure appears as a decrease below system fill pressure. Unless the fill pressure is higher than the pump pressure, a vacuum can be created within the system. This arrangement is not generally used except in small systems or systems in low-rise buildings where pumps have a low total head capability.

Tanks located in the attic, Figure 6-6c, are often used to reduce expansion tank size. Points between the pump discharge and the tank will show a pressure increase when the pump operates, and points between the tank and the pump suction will show a pressure decrease. The required fill pressure no longer needs to include system static pressure.

Sizing the Compression Tank

The sizing of the expansion tank is usually based on the following ASME formula for temperatures of 160° to 280°F (71.1° to 137.8°C):

$$V_t = \frac{(0.00041\ t - 0.0466)\ V_s}{(Pa/Pf) - (Pa/Po)}$$

where:
- V_t = minimum volume of the expansion tank (gal)
- V_s = volume of water in the system (gal)
- t = average operating temperature (°F)
- Pa = pressure in expansion tank when water enters (ft of H_2O absolute or atmospheric pressure ft of H_2O absolute)
- Pf = initial fill pressure plus static pressure or minimum pressure at tank (ft of H_2O absolute)

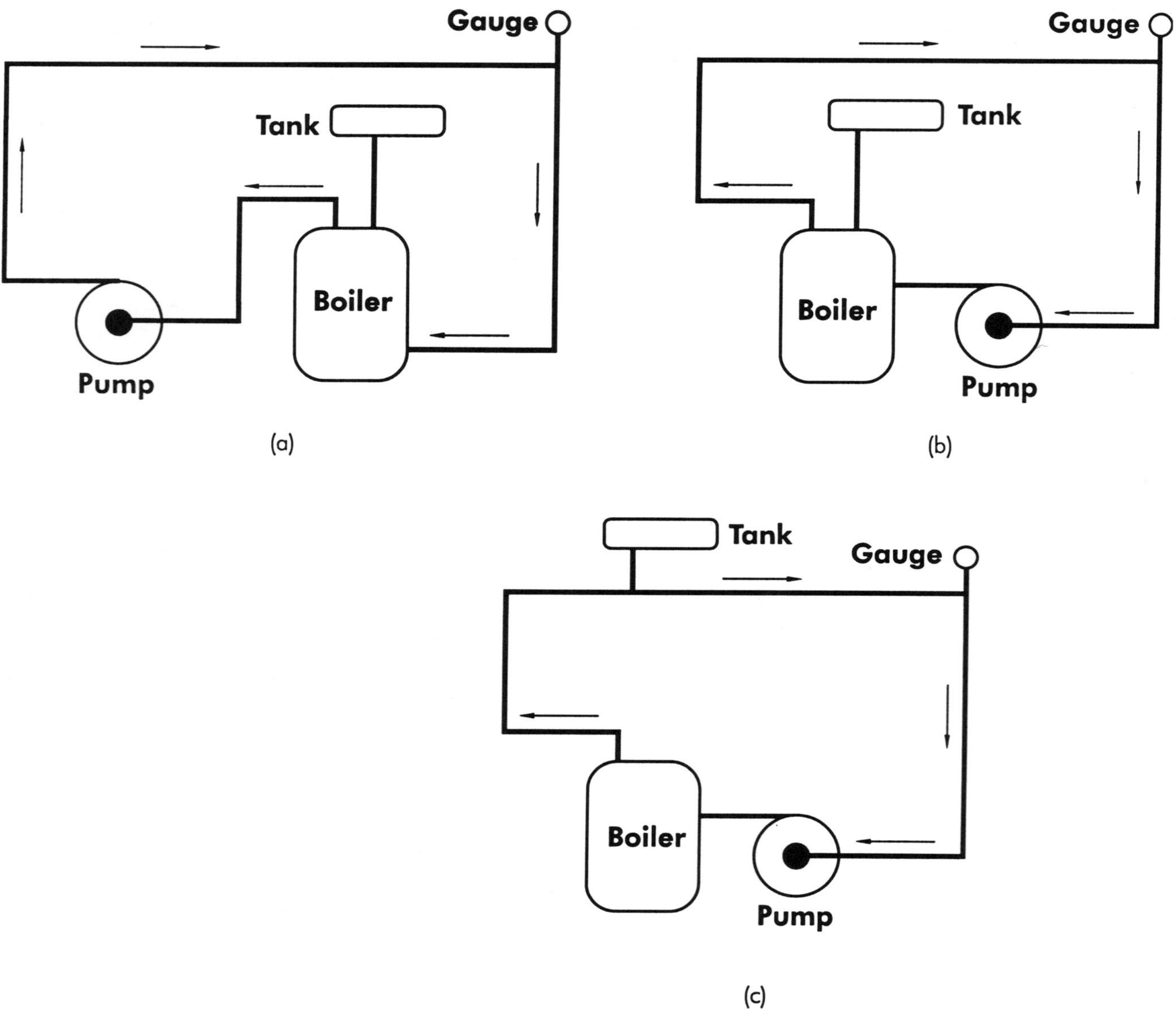

Figure 6-6. Effect of pump location and expansion tank location

Po = maximum operating pressure at tank (ft of H_2O absolute)

0.0466 = a coefficient (no dimensions)

0.00041 = a coefficient (no dimensions)

A more simple formula to use when water temperature does not exceed 160°F (71.1°C) is:

$$V_t = \frac{E}{(Pa/Pf) - (Pa/Po)}$$

where: E = expansion of water in the system from minimum to maximum temperature (gal)

The other items are the same as in the preceding formula.

Figure 6-7 is a diagram showing the water expansion function of temperature.

Example 6-6. What is the compression tank capacity in the system that was shown in Figure 6-6a. The data for this system is as follows:

Water volume = 500 gal
High point of the return = 20 ft
Pump head = 20 ft H_2O
Boiler pressure = 30 psi
Atmospheric pressure absolute = 34 ft H_2O
Water temperature average = 200°F
High point above the boiler = 25 ft

Solution 6-6. Calculate the first part of the formula given earlier:

$$\left((0.00041)\ (200°F) - 0.0466\right)(500\ gal) = 17.70\ gal$$

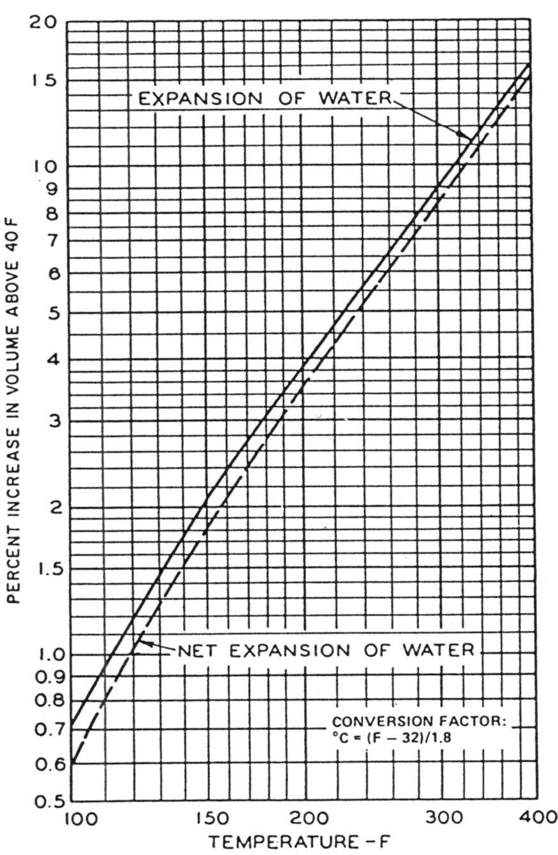

Figure 6-7. Expansion of water above 40°F (Reprinted by permission of the American Society of Heating, Refrigerating and Air-Conditioning Engineers, Atlanta, Georgia, from the 1980 ASHRAE *Handbook--Systems)*

Then calculate the second part of the formula:

$$Pf = 25 \text{ ft } H_2O + 4 \text{ ft } H_2O \text{ (added for positive venting)} = 29 \text{ ft } H_2O \text{ gauge pressure}$$

$$Pf = 29 \text{ ft } H_2O + 34 \text{ ft } H_2O = 63 \text{ ft absolute (abs)}$$

The value Po is the maximum boiler operating pressure minus 5 psi or 10% (whichever is larger):

$$Po = 30 \text{ psi} - 5 \text{ psi} = (25 \text{ psi})(2.31) = 57.8 \text{ ft } H_2O \text{ gauge pressure}$$

$$Po = 57.8 \text{ ft } H_2O + 34 \text{ ft } H_2O = 91.8 \text{ ft absolute}$$

$$V_t = \frac{17.70 \text{ gal}}{(34 \text{ ft } H_2O/63 \text{ ft abs}) - (34 \text{ ft } H_2O/91.8 \text{ ft abs})} = 104 \text{ gal}$$

The calculated tank capacity is 104 gal.

Example 6-7. The difference between this example and Example 6-6 is that the circulator is located at the intake of the expansion tank, Figure 6-6b. The problem data is the same as in Example 6-6. What is the compression tank capacity?

Solution 6-7. When the pump operates, the pressure at the top of the return pipe is reduced by an amount equal to the pump head. To maintain a positive pressure at the system high point, the minimum pressure must be increased by an amount equal to the pump head:

$$Pf = 29 \text{ ft } H_2O + 20 \text{ ft } H_2O = 49 \text{ ft } H_2O \text{ gauge pressure} + 34 \text{ ft } H_2O = 83 \text{ ft } H_2O \text{ absolute}$$

$$V_t = \frac{17.70 \text{ gal}}{(34 \text{ ft } H_2O/83 \text{ ft abs}) - (34 \text{ ft } H_2O/91.8 \text{ ft abs})} = 442.5 \text{ gal}$$

The calculated tank capacity is 442.5 gal.

Example 6-8. The difference between this example and the two previous examples is that the circulator is located on the return and the expansion tank at the high point of the system, Figure 6-6c. The problem data is the same as for the previous examples. What is the compression tank capacity?

Solution 6-8. The tank is far enough from the boiler so that friction in piping must be added. For this example, the friction loss is equal to 8 ft of H_2O. The minimum pressure (Pf) in the tank is 4 ft H_2O for positive venting plus the 8 ft H_2O friction. Add the atmospheric pressure of 34 ft H_2O, and the total obtained is 46 ft H_2O absolute.

Since the pump is located between the expansion tank connection and the boiler, the maximum pressure (Po) in the tank must be less than the operating pressure of the relief valve by an equal amount to the effect of pump operation on the boiler pressure. In this case, the effect of the pump is reduced by the friction loss between the tank and the boiler:

$$\begin{aligned} Pf &= 20 \text{ ft } H_2O - 8 \text{ ft } H_2O = 12 \text{ ft } H_2O \text{ gauge} + 34 \text{ ft } H_2O \\ &= 46 \text{ ft } H_2O \text{ absolute} \end{aligned}$$

Calculate the difference in height:

$$25 \text{ ft} + 12 \text{ ft} = 37 \text{ ft}$$

$$\begin{aligned} Po &= 57.8 \text{ ft } H_2O - 37 \text{ ft } H_2O = 20.8 \text{ ft } H_2O \text{ gauge} + 34 \text{ ft } H_2O \\ &= 54.8 \text{ ft of } H_2O \text{ absolute} \end{aligned}$$

Substitute into the equation:

$$V_t = \frac{17.70 \text{ gal}}{(34 \text{ ft } H_2O/46 \text{ ft abs}) - (34 \text{ ft } H_2O/54.8 \text{ ft abs})} = 147.5 \text{ gal}$$

The calculated tank capacity is 147.5 gal.

Examples 6-6, 6-7, and 6-8 show that the compression tank volume is based on its position in the system as well as the position of the pump. The tank is not always located in the same position.

Example 6-9. Calculate the compression tank capacity for a system in which the compression tank is located at the boiler elevation. The data is as follows:

Water volume = 600 gal
High point above the boiler = 25 ft above the boiler
Pump head = 20 ft H_2O
Boiler pressure = 30 psig
Average water temperature = 200°F
Fill pressure = 5 psig
Atmospheric pressure absolute = 34 ft H_2O

Solution 6-9. From the above data, the terms in the equation are:

$t = 200°F$
$V_s = 600$ gallons
$Pa = 34$ ft H_2O

$Pf = (5.6 \text{ psig})(2.31) + 25 \text{ ft} = 38 \text{ ft } H_2O \text{ gauge} + 34 \text{ ft } H_2O = 72 \text{ ft } H_2O \text{ absolute}$

$Po = 30 \text{ psig} + 15 \text{ psig safety} = (45 \text{ psia})(2.31) = 104 \text{ ft } H_2O \text{ absolute}$

Calculate the first part of the formula:

$$\left[(0.00041)\ (200°F)\ -\ 0.0466\right](600\ gal)\ =\ 21.2\ gal$$

Substitute all the numbers into the equation:

$$V_t = \frac{21.2\ gal}{(34\ ft\ H_2O/72\ ft\ abs)\ -\ (34\ ft\ H_2O/104\ ft\ abs)} = 146\ gal$$

A compression type expansion tank of approximately 150 gal capacity would be used on this system, if connected at the suction side of the pump. If the tank is connected to the discharge side of the pump, the minimum pressure at the tank must be increased by the amount of pump head:

$$Pf\ =\ 72\ ft\ H_2O\ absolute\ +\ 20\ ft\ H_2O\ =\ 92\ ft\ H_2O\ absolute$$

The required tank size in this case becomes:

$$V_t = \frac{21.2\ gal}{(34\ ft\ H_2O/92\ ft\ abs)\ -\ (34\ ft\ H_2O/104\ ft\ abs)} = 505\ gal$$

The tank size is larger due to its location.

One recommendation (rarely used) for compression tank size involves measuring the radiation (using manufacturer's information if possible) and installing the tank size as follows:[7]

Connected radiation in sq ft EDR (equivalent direct radiation)	Tank size (gal)
350	18
450	21
650	24
900	30
1100	35
1400	40

These sizes are recommended for a gravity system, and they are larger than for a forced circulation system. Commercial sizes of compression tanks are limited to 80 gallons. If a larger capacity is necessary, then multiple tanks must be used.

Chilled Water System-Closed Tank Sizing

Closed tanks are often used on chilled water and dual temperature systems to maintain the closed, corrosion-free characteristics of the hydronic system. Tank sizing by equation results in tanks of very small size because of low coefficients of expansion. For this reason, closed compression tanks in chilled water systems are sized to 50% of the equivalent hot water size, where piping systems are considered to operate from 70° to 200°F (21.1° to 93.3°C).

ITT Fluid Technology Corporation suggests the following formula be used when the system is closed and has a compression tank that is equipped with an Airtrol™ tank fitting:

$$V_t = \frac{(Ew - Ep)\ V_s}{(Pa/Pf) - (Pa/Po)} - 0.02\ V_s$$

where:
- V_t = compression tank size in gallons
- V_s = volume of system in gallons
- Ew = unit expansion of water
- Ew-Ep = unit expansion of system
- Pa = atmospheric pressure, psi absolute
- Pf = initial pressure in tank, psi absolute
- Po = final pressure (relief valve setting) psi absolute
- $.02\ V_s$ = approximate amount of air released from new system water upon heat-up, which is 2% of system water volume. Omitting this results in a slightly larger tank selection.

Another possibility for selecting the compression tank capacity is to use the information in Table 6-5.

Tank volume in gallons
Tank equipped with Airtol Tank Fitting
12 psig Initial fill pressure,—30 psi Relief Valve Setting

SYSTEM WATER VOL IN GALS	MEAN DESIGN WATER TEMPERATURE						
	150°	160°	180°	200°	220°	240°	250°
10	0.6	0.8	1.0	1.3	1.6	1.9	2.0
20	1.2	1.7	2.0	2.6	3.2	3.8	4.1
30	1.8	2.5	3.0	4.0	4.8	5.7	6.1
40	2.4	3.3	4.0	5.3	6.4	7.6	8.2
50	3.0	4.2	5.0	6.6	8.0	9.5	10
60	3.6	5.0	6.0	7.9	9.7	11	12
70	4.2	5.8	7.0	9.2	11	13	14
80	4.7	6.7	8.0	11	13	15	16
90	5.3	7.5	9.0	12	14	17	18
100	5.9	8.0	10	13	15	19	20
200	12	17	20	26	32	38	41
300	18	25	30	40	48	57	61
400	24	33	40	53	64	76	82
500	30	42	50	66	80	95	102
600	36	50	60	79	97	114	122
700	42	58	70	92	113	133	143
800	47	67	80	110	129	150	163
900	53	75	90	120	145	170	184
1000	59	80	100	130	161	190	200
2000	120	170	200	260	320	380	410
3000	180	250	300	400	480	570	610
4000	240	330	400	530	640	760	820
5000	300	420	500	660	800	950	1020
6000	360	500	600	790	970	1140	1220
8000	470	670	800	1100	1290	1500	1630
10000	590	800	1000	1300	1610	1900	2000

This table is based on tank and relief valve installation at same level.

Table 6-5. Compression tank capacity (Courtesy, ITT Fluid Handling)

STEAM SYSTEMS

The steam pressure used in low pressure heating systems is between 1 and 15 psi. The steam velocity in pipes depends on the pressure difference in the system. The pipe diameter must be carefully selected to establish a balance between the difference in pressure and the steam velocity. A low pressure drop along the pipe means larger pipes must be selected. Better efficiency may be obtained by using the right type of radiator.

Hot water boilers have an entire network of pipes full of water. The steam boiler contains only a certain amount of water that is recommended by the manufacturer. The steam boiler also contains air, and its distribution pipes are full of air before the boiler starts.

When the steam boiler starts, the burner fires, water starts to pick up heat, and water temperature increases to the point of boiling. The water expands somewhat when heated, but the expansion is much greater when the water changes state from liquid to vapor (steam).

As explained previously, liquid or gas flows from high pressure to low pressure. Before the steam boiler starts, the distribution pipes, terminal heaters, and coils are under little or no pressure. Steam that is generated travels from the boiler to the low pressure areas in the system. Consequently, a steam boiler does not require a circulator (pump).

Steam penetrating the piping system replaces air in the heating units. As in a hot water system, if the air is not evacuated it might prevent system operation. The air is evacuated or vented through special vents. Steam heating systems also may use a vacuum pump, which helps eliminate air in the pipes and radiators. As steam accumulates in the piping, it gives up heat and condenses into water. The condensate returns to the boiler, where it starts the whole cycle over again. Steam traps help condense the water so that only condensate flows back to the boiler.

A safety valve must be installed in a steam boiler, as well as a low water fuel cut-off. If the minimum amount of water in a boiler is not present, the burner will not fire. Boilers are provided with a make-up water connection. A flooded boiler also cannot operate properly, so it is important to make sure the boiler contains the proper amount of water.

Steam may be distributed to the heating elements by a one-pipe or two-pipe system. The one-pipe steam heating system has one pipe connected to each terminal unit. The same pipe supplies the steam and carries the condensate back to the boiler. The pipes used in one-pipe systems are larger than the ones used in two-pipe systems. The condensate return system usually includes a reservoir (condensate tank) and two transfer pumps. The condensate level in the tank is closely monitored.

The two-pipe system includes a return pipe for each terminal unit, which means more pipe is used for the job. However, some problems connected to the one-pipe system (e.g., noise, condensate accumulation, etc.) are eliminated.

FUEL

One way to produce energy or power is to use the heat stored in fossil fuels. Fossil fuels are called non-renewable fuels. Solar wind and tide waves are renewable but are not practical due to unstable weather conditions, expense, etc. For this reason, fossil fuels are normally used for hydronic applications.

The basic components of any combustible material are carbon, hydrogen, and oxygen. Fossil fuels used in boilers or furnaces may be divided into three categories: solid (wood, coal, and their derivatives), liquid (oil), and gas (natural or man-made byproducts of refineries).

The heating value of a fuel refers to the amount of heat generated when the fuel is burned completely. It is measured in Btuh. The heat values for various fuels are listed in the following paragraphs.

Wood

The heating value of dry wood is between 8560 and 9130 Btu/lb. With a high moisture content, the heating value for wood varies between 1750 and 8700 Btu/lb.

Wood may be burned as a fuel by itself or mixed with other low moisture combustibles to improve its efficiency. One type of derived fuel is called wood charcoal, which is obtained by heating wood at a high temperature in an airless chamber. The wood charcoal loses up to 75% of its weight and 50% of its volume during this process. Consequently, wood charcoal has a higher heating value per pound than raw wood. Besides being used as a combustible, wood charcoal may be used as a filter.

Coal

Coal is a natural product that is usually mined from underground deposits. The present natural reserves of coal in the U.S. are tremendous. Coal formed over millions of years through the decomposition of vegetation in the absence of air. Naturally, there were other elements that contributed to this process, such as moisture, pressure, heat, and biochemical action.

Coal may be classified in different ways; for example, ASTM classifies the coal by rank or class. Based on its caloric value, the coal classification is as follows:

Type of coal	Caloric value (Btu/lb)
Anthracitic	9310 to 13,880
Bituminous	10,810 to 14,400
Sub-bituminous	8560 to 10,650
Lignitic	7000

Each of these types of coal used as a combustible has different physical and chemical characteristics that make it better than others.

Liquid Fuels

Liquid fuels (petroleum or oils) in their natural form are found in large and usually very deep underground deposits. Petroleum is detected in its underground location with a variety of sophisticated instruments and then extracted through oil wells. Petroleum, or crude oil, consists of hydrocarbons, which are combustible. It also contains a multitude of other chemicals in very small amounts, such as sulfur, oxygen, nitrogen, metals, etc.

The physical and chemical characteristics of crude oil are very different. The viscosity varies between 2.3 and 23 centistokes. The carbon content is around 85% and hydrogen around 12%.

After extraction, crude oil is refined or distilled. This operation separates crude products based on the boiling range. Refined products include gasoline, kerosene, heating oil, lubricating oil, and residuals.

The specific gravity of oil is given on a scale designated by the American Petroleum Institute (API). The units are measured in degrees API. The heat value is summarized in Table 6-6.

	Per pound		Per gallon	
Typical degree API at 60°F	**At constant volume**	**At constant pressure**	**At constant volume**	**At constant pressure**
10	18,540	17,540	154,600	146,200
20	19,020	17,930	148,100	139,600
30	19,420	18,250	141,800	123,300
40	19,750	18,510	135,800	127,300
50	20,020	18,720	130,100	121,700
60	20,260	18,900	124,800	116,400
70	20,460	19,020	119,800	112,500
80	20,630	19,180	115,100	107,000

Table 6-6. Heat value range in Btu

For better understanding, Table 6-6 should be correlated to the oil grade numbers, which are as follows:

Grade No.	Gravity API
1	38 to 45
2	30 to 38
4	20 to 28
5 (light)	17 to 22
5 (heavy)	14 to 10
6	8 to 15

A well known product is diesel oil, which is Grade No. 2. However, there are a few subdivision grades of this type of oil depending on its particular use. ASTM D396 limits the sulfur content of fuels heavier than No. 2, and they must conform to the legal requirements of the locality in which they are to be used. The physical characteristics of fuel oil are detailed in ASTM D396.

The ASTM oil grades are defined as follows:

- No. 1: A distillate oil intended for vaporizing pot-type burners and other burners requiring this grade of fuel.
- No. 2: A distillate oil for general-purpose domestic heating or for use in burners not requiring No. 1 fuel oil.
- No. 4: Preheating not usually required for handling or burning.
- No. 5 (Light): Preheating may be required depending upon climate and equipment.
- No. 6. (Heavy): Preheating may be required for burning and in cold climates, may be required for handling.
- No. 7: Preheating required for burning and handling.

Gaseous Fuels

Combustible gaseous fuels are found in deep underground deposits and usually accompany liquid petroleum products. Natural gas is the best known product and consists mainly of methane. It is odorless and colorless. Most gas distributing regulations require a specific odor be mixed into the gas, so leaks can be detected early.

When natural gas is not available from a utility piping network, a gas company may provide bottled LPG (liquefied petroleum gas), which is another combustible gas. It consists of propane, butane, isobutane, propylene, and butylene. Propane and butane are natural products extracted from natural gas or petroleum.

Fuel gas is measured in cubic feet. Table 6-7 shows the comparative thermal value of coal, natural gas, oil, and electricity.

Fuel	1.0 million Btu	24.0 million Btu	0.0916 million Btu	0.125 million Btu	0.139 million Btu	0.15 million Btu	0.003412 million Btu
Natural gas (1000 Btu/ cu ft)	1.000 cu ft	24.000 cu ft	91.600 cu ft	125.000 cu ft	139.000 cu ft	150.000 cu ft	3.412 cu ft
Coal (12,000 Btu/lb)	83.333 lb	2.000 lb	7.633 lb	10.417 lb	11.583 lb	12.500 lb	0.2843 lb
Propane (91,600 Btu/gal)	10.917 gal	262.009 gal	1 gal	1.265 gal	1.517 gal	1.638 gal	0.0373 gal
Gasoline (125,000 Btu/gal)	8.000 gal	192.000 gal	0.733 gal	1 gal	1.112 gal	1.200 gal	0.0273 gal
Fuel oil #2 (139,000 Btu/gal)	7.194 gal	172.662 gal	0.659 gal	0.899 gal	1 gal	1.079 gal	0.0245 gal
Fuel oil #6 (150,000 Btu/gal)	6.666 gal	160.000 gal	0.611 gal	0.833 gal	0.927 gal	1 gal	0.0227 gal
Electricity (3412 Btu/ kWh)	293.083 kWh	7033.998 kWh	26.846 kWh	36.635 kWh	40.739 kWh	43.962 kWh	1 kWh

Table 6-7. Comparative thermal values

NOTES

[1] Dan Holohan and Associates, Inc., 63 N. Oakdale Ave., Bethpage, NY 11714.

[2] Information in this section reprinted by permission of the American Society of Heating, Refrigerating and Air-Conditioning Engineers, Atlanta, Georgia, from the 1980 ASHRAE *Handbook-- Systems.*

[3] Reprinted with permission from *Air Conditioning Principles and Systems, Second Edition*, by E. Pita, J. Wiley, 1981.

[4] Ibid

[5] Holohan, Dan, *Fuel Oil and Oil Heat with Air Conditioning*, January, 1993, reproduced with permission.

[6] Reprinted with permission from *Air Conditioning Principles and Systems, Second Edition*, by E. Pita, J. Wiley, 1981.

[7] Holohan, Dan, "How to Survive Gravity Hot Water Systems," *Fuel Oil and Oil Heat with Air Conditioning*, July, 1991, reproduced with permission.

Chapter 7

Chillers and Cooling Towers

A chiller is a machine that lowers the temperature of a liquid (usually water or brine), which is then circulated by pumps through a coil. Air passes over the coil and conditions a space for comfort or specialized manufacturing requirements. The cooling media may be used to keep a low temperature for various applications that require lower than ambient temperatures.

Chillers operate based on the refrigeration cycle. The following pieces of equipment or basic components work together to produce the refrigeration effect:

- Evaporator: a coil where the refrigerant evaporates, absorbing heat from the surroundings
- Throttling valve: expansion valve
- Condenser: where the gaseous refrigerant is condensed into a liquid
- Compressor: where the refrigerant is compressed into a high pressure gas at a higher temperature

Figure 7-1 shows a water-cooled condenser. This type of machine is able to condense refrigerant because the high pressure refrigerant condenses at a high temperature. A source of water must be available at a temperature lower than the condensing temperature of the refrigerant. When this cooler water is circulated through the water-cooled condenser, heat flows from the refrigerant into the water.

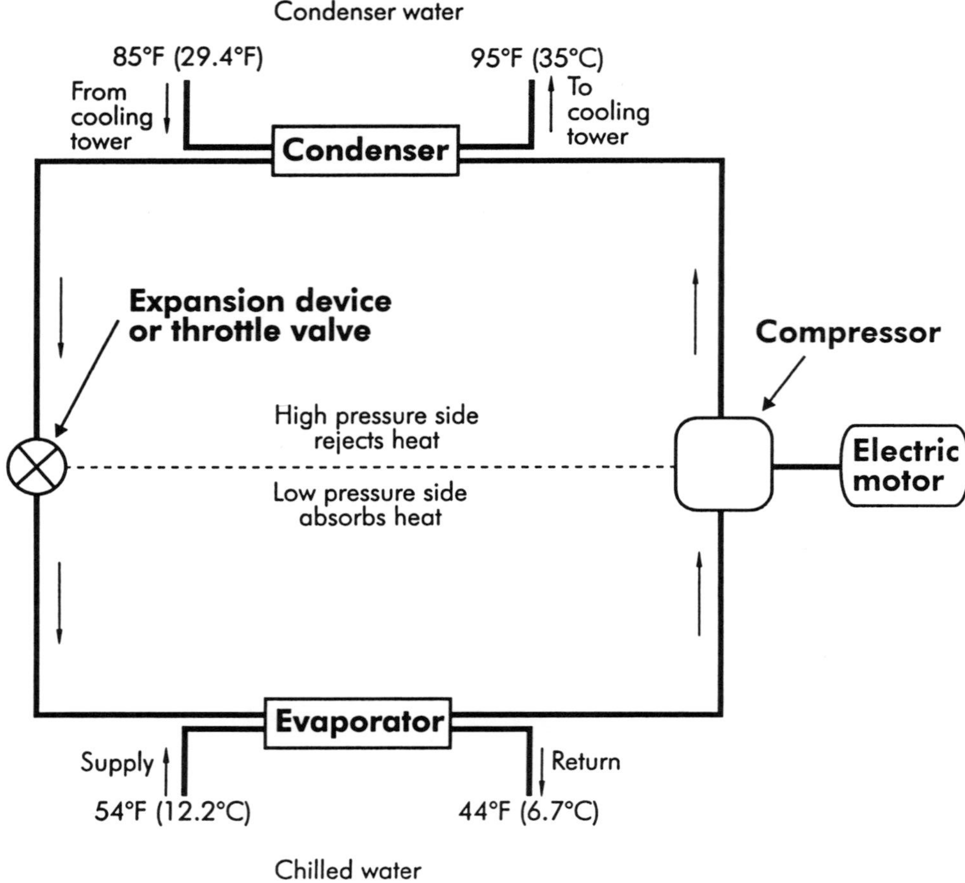

Figure 7-1. Water-cooled condenser (the temperatures shown are for exemplification only)

In a chiller, the cooling media (usually water or air) enters the evaporator, which acts as a heat exchanger. The media temperature is lowered with the help of a liquid refrigerant, which evaporates and becomes a gas. The refrigerant circulates in a separate closed loop. The refrigerant gas then enters the compressor, where its pressure and temperature are increased. This high pressure, high temperature gas moves into the condenser, where it condenses into a liquid. The refrigerant is cooled there by condensing water or air. Heat is ultimately rejected to the atmosphere either by a cooling tower or a dry air-cooled condenser. The heat may also be dissipated in a body of natural water. The throttling valve is located close to the evaporator and controls the liquid refrigerant flow into the evaporator. Another function of the expansion device is to maintain the high pressure in the condenser. To illustrate the end result of a chiller, cold water may enter the evaporator at 54°F (12.2°C) and leave at 44°F (6.7°C), while cooling tower water may enter the condenser at 85°F (29.4°C) and leave at 95°F (35°C). Chillers produce a lower temperature of the refrigeration media, which in turn lowers the temperature of the water circulated in the coils or air supply.

CHILLER TYPES

There are two basic types of chillers: *hermetic* and *external drive*. The external drive type uses a compressor, which is driven by an external electric motor or an engine or turbine. The hermetic unit also has a compressor but is driven by an electric motor totally enclosed in the refrigerant atmosphere. These chillers are quieter, smaller in size, and less expensive.

Chillers are selected based on a certain load range. ASHRAE suggests the following:

- Up to 80 tons (280 kW): reciprocating
- 80 to 120 tons (280 to 420 kW): reciprocating or centrifugal
- 120 to 200 tons (420 to 700 kW): screw, reciprocating, or centrifugal
- 200 to 800 tons (700 to 2800 kW): screw or centrifugal
- Above 800 tons (2800 kW): centrifugal

For chillers ranging between 80 and 200 tons, reciprocating and screw liquid chillers are more frequently employed than centrifugals. Centrifugal liquid chillers (particularly multistage machines) may be applied quite satisfactorily at high head conditions. Factory packages are available to about 2400 tons (8400 kW) and field-assembled machines to about 10,000 tons (35,000 kW).

As with boilers, it is possible to have a multi-chiller system; however, the initial cost is higher. When two or more chillers are arranged to work in parallel, the chilled water amount (at full demand) is divided by the number of machines. If the cooling load demand decreases, one chiller stops and the one(s) remaining furnishes the amount of chilled water necessary at the required temperature.

Various chilled water system arrangements are possible. Hydronic contractors should contact one of the reputable and recognized chiller manufacturers to discuss the various alternatives for a particular application.

Sizing a Chiller

Chillers are expensive machines that are sized to offset a calculated heat gain. These calculations must be accurate, precise, and include a small safety coefficient (5% to 10%). Any unjustified or artificial oversizing increases the price of installation and does not ensure better operation. The size of a chiller depends on the total heat load released and the area to be cooled. This includes a heat gain calculation, which is based on the design outdoor temperature. Outdoor maximum temperature occurs 1% of the total 2928 summer hours. Therefore, "normal" design (maximum) temperatures are exceeded only 30 hours each year. For energy conservation purposes, select a design outdoor temperature that is normally recommended and represents 2.5% of summer maximum temperatures.

There are four costs that must be considered when selecting a chiller: equipment (including all necessary appurtenances), installation, energy (power), and maintenance.

Example 7-1. Determine how much chilled water is required for a chiller based on the following conditions:

Cooling load: 350,000 Btuh
Temperature difference for media: 10°F
Supply chilled water temperature: 45°F

Solution 7-1. The flow is easily calculated using the following formula:

$$\text{gpm} = \frac{350{,}000 \text{ Btuh}}{(8.33 \text{ lb/gal})(60 \text{ min/hr})(1 \text{ Btu/lb°F})(10°\text{F})} = 70 \text{ gpm}$$

The chiller capacity is determined as follows:

$$\frac{350{,}000 \text{ Btuh}}{12{,}000 \text{ Btu/ton of refrigerant}} = 29.16 \text{ tons of refrigerant}$$

Note: One ton of refrigeration equals 12,000 Btu.

Now it is possible to determine the chilled water per refrigeration ton:

$$\frac{70 \text{ gpm}}{29.16 \text{ tons of refrigerant}} = 2.4 \text{ gpm chilled water/refrigerant ton}$$

Instructions for the installation of chillers are furnished by the manufacturer.

COOLING TOWERS

The cooling tower is an integral part of a chiller application, which uses a combination of heat and mass to cool the water. The cooling tower rejects heat accumulated by cooling water from the condenser to the atmosphere by way of a natural or mechanical process, Figure 7-2.

Before water conservation took effect in the 1970s, some systems cooled the condenser with water obtained either from a natural body of water or from the city. The disadvantages of these two possibilities were that water from a lake or river created water temperature pollution, and once-through city water used for cooling is a terrible waste of treated water. Cooling towers recirculate about 95% of the water used for cooling the refrigerant (5% of the water circulated is lost through evaporation, drift, etc.)

The temperature difference between water entering and water leaving the cooling tower is called the *range*. The range is determined by the heat load and water flow rate in the condenser. The difference between the leaving water temperature and the entering air wet bulb temperature is called the *approach*. The approach depends on the type and size of the cooling tower. A closer approach means colder leaving water.

The amount of heat rejected to the atmosphere by the cooling tower is equal to the heat load rejected by the condenser. The temperature at which the heat is transferred by the cooling tower is determined by its thermal capability.

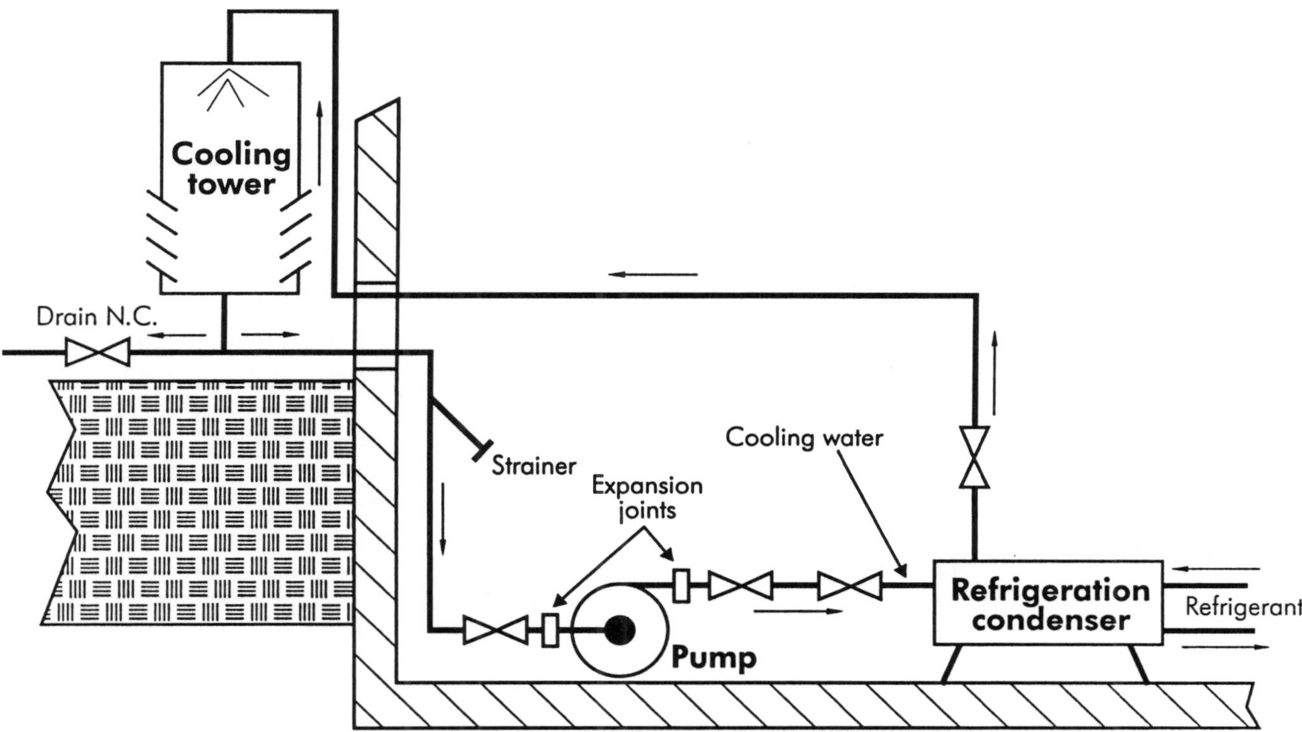

Figure 7-2. Open loop water system

The thermal performance of any cooling tower is strongly influenced by the entering air wet bulb temperature. The dry bulb temperature and the relative humidity of the air influence the water evaporation.

In order to determine the heat load to be dissipated by a cooling tower, use the following formula:

$$\text{Heat load} = Q \times 8.33 \times R = \text{Btu/min}$$

where: Q = water circulated (gpm)
8.33 = lbs per gallon of water
R = range

Cooling tower operation depends on the following items:

- Entering water temperature
- Leaving water temperature
- Entering air wet bulb and dry bulb temperatures
- Water flow rate

There are different types of cooling towers that are made from various construction materials. Such materials of construction include treated wood (decay preservation), metal, plastic, and reinforced concrete. Cooling tower types depend on how much contact there is between the water and the surrounding atmosphere. Direct contact cooling towers have direct contact between air and water. This type of cooling tower includes *fill*, which increases the contact and exposure of air and water. Indirect contact cooling towers, also called closed circuit cooling towers, have water that circulates within a coil. Figure 7-3 illustrates the direct and indirect types of cooling towers.

Cooling towers also depend on air movement. The three basic types of cooling towers are *atmospheric*, *hyperbolic natural draft*, and *mechanical draft*. The atmospheric type of cooling tower has no moving parts and is generally used for small installations. The heated water, which originates at the condenser, enters at the top of the tower. It is sprayed down in the form of droplets, collected in a bottom reservoir, and returned back to the condenser. As the spray travels downward, it is cooled when it comes into contact with air drawn into the tower through louvers. Hyperbolic natural draft type cooling towers are similar to atmospheric cooling towers, except that they are quite large, expensive, and mostly used for large power generation plants. In mechanical draft type cooling towers, the air movement is produced with the help of mechanical fans. In this manner, the air flow is known and controlled.

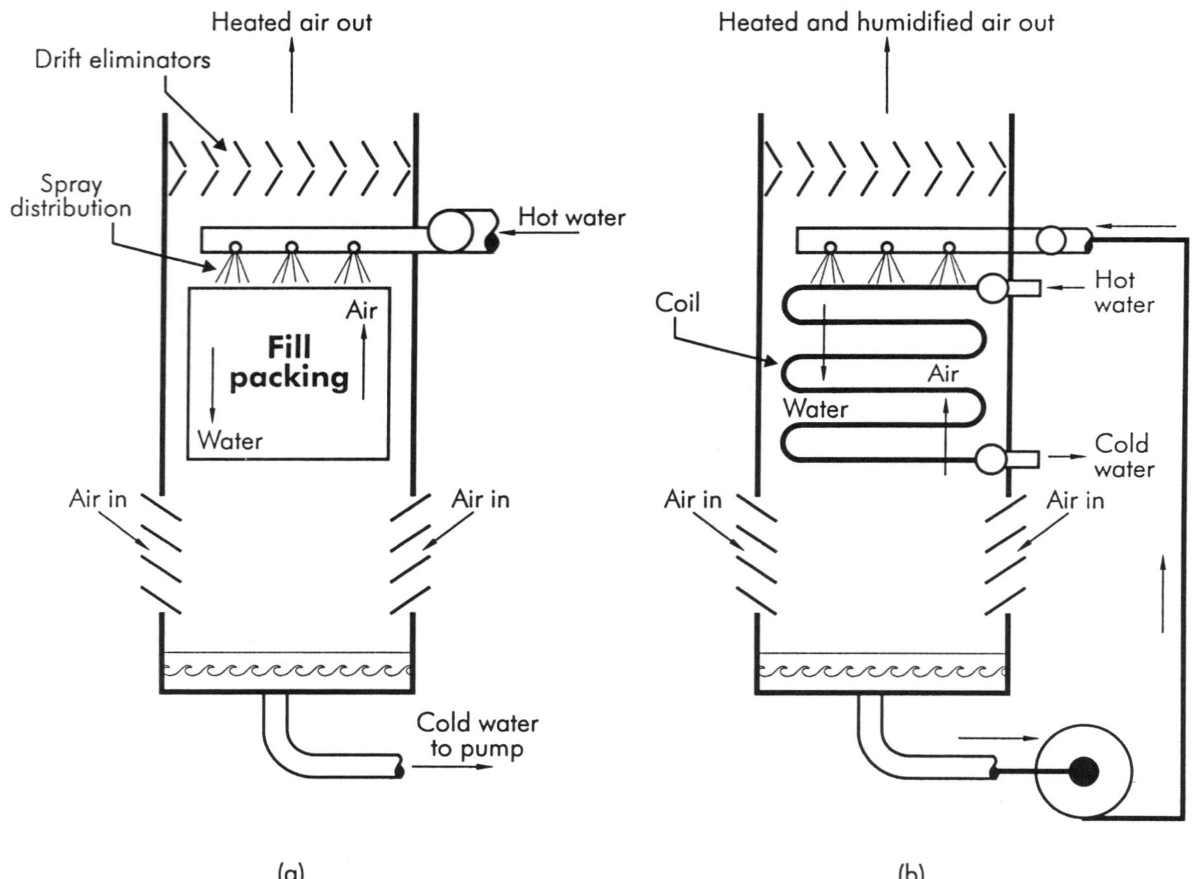

Figure 7-3. a) Direct contact evaporative cooling tower; b) indirect contact evaporative cooling tower

Mechanical draft towers are divided into two groups: forced draft and induced draft. In forced draft towers, the fan blows the air at the bottom of the tower and the air exits at the open top of the tower. In induced draft towers, the fans are located at the top of the tower and draw the air through the tower's side louvers. The fans are located in the air stream.

Cooling towers may be either counterflow or crossflow. In counterflow towers, the air moves vertically upward through the fill, which is counter to the downward movement of the water. In crossflow towers, the air flows horizontally (perpendicular) to the downward flow of the water.

Each type of cooling tower has advantages that make them better than others for certain applications. Elements that influence selection include specific weather conditions, size of job, etc.; however, the final factor is usually economics.

Aside from carefully sizing and selecting the equipment, piping to and from the cooling tower (including fittings) and the circulating pump(s) must be carefully sized for good results. If the cooling tower is used in colder climates, it must be protected from freezing. In addition, if the cooling tower is fabricated of combustible materials, a fire suppression system must be installed.

When the job requires the installation of a cooling tower, it is always wise to contact several reputable cooling tower manufacturers to discuss all conditions and requirements before making a selection. The equipment may be different, but the hydronic system is always the same and requires the skills and experience of a hydronic specialist to ensure proper work.

DEFINITIONS

The following is a list of important terms and definitions relating to cooling towers.

Approach: Difference between the leaving cold water temperature and either the ambient or entering air wet bulb temperature.

Blowdown: Water discharged from the system to control the concentration of salts or other impurities in the circulating water. Unit of measurement expressed in percent of circulating water or gpm.

Btu (British thermal unit): The amount of heat gain (or loss) required to raise (or lower) the temperature of one pound of water 1°F.

Counterflow: Air flow direction through the fill counter to that of the falling water.

Crossflow: Air flow direction through the fill essentially perpendicular to that of the falling water.

Drift: Circulating water lost from the tower as liquid droplets entrained in the exhaust air stream. Unit of measurement expressed in percent of circulating water or gpm.

Dry bulb temperature: The temperature of the entering or ambient air adjacent to the cooling tower as measured with a dry bulb thermometer. Unit of measurement expressed in °F.

Evaporation loss: Water evaporated from the circulating water into the air stream in the cooling process. Unit of measurement expressed in percent of circulating water or gpm.

Fill: That portion of a cooling tower which constitutes its primary heat transfer surface. The cooling tower fill is a large, open heat transfer mass. It is formed by a series of baffles and is designed to break up the water into very fine droplets for maximum air/water contact. Sometimes called "packing."

Forced draft: The movement of air under pressure through a cooling tower. Fans of forced draft towers are located at the air inlets to push the air through the tower.

Hot water temperature: Temperature of circulating water entering the cooling tower's distribution system. Unit of measurement expressed in °F.

Make-up water: Water added to the circulating water system to replace water lost by evaporation, drift, wind, blowdown, and leakage. Unit of measurement expressed in percent of circulating water or gpm.

Mechanical draft: The movement of air through a cooling tower by means of a fan.

Natural draft: The movement of air through a cooling tower purely by natural means; typically, by the driving force of air density differential.

Range: The difference in temperature (°F) between entering and leaving water into the cooling tower, Figure 7-4.

Wet bulb temperature: The temperature of the entering or ambient air adjacent to the cooling tower as measured with a wet bulb thermometer. Unit of measurement expressed in °F.

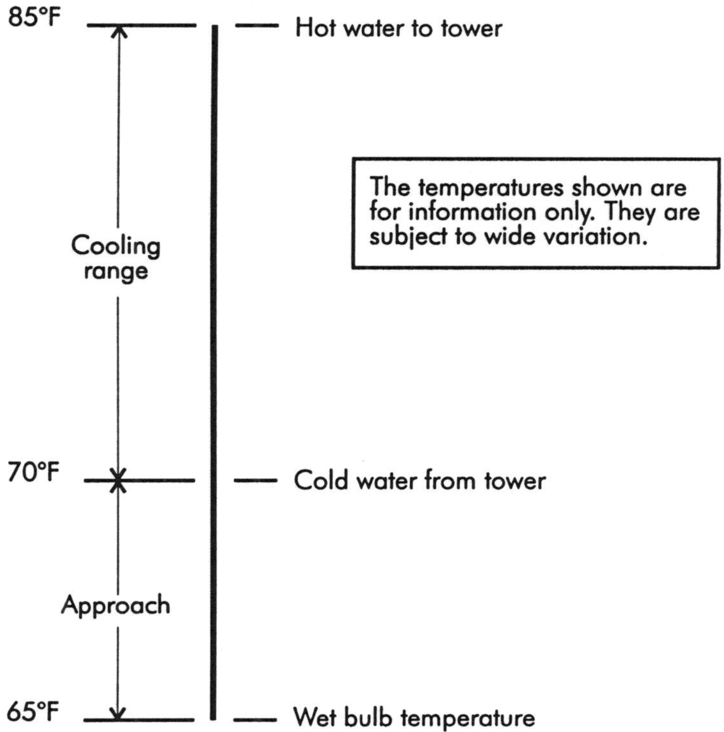

Figure 7-4. Example of the cooling range and approach in a cooling tower

Chapter 8

Radiant Heating

This chapter reprinted with permission by the Hydronics Institute, Inc.

Hydronic radiant heating (HRH) has many applications. This chapter is primarily concerned with residential and light commercial construction, although the fundamentals can be applied to a variety of situations. This chapter will cover typical applications, such as single and multi-family residences, shops, warehouses, garages, small commercial buildings, and manufacturing facilities. Buildings that are built on a concrete slab are generally suited to HRH. New construction techniques have made suspended floors and wood floors candidates as well.

History

Radiant floor heating has been utilized for centuries by many cultures. The Romans used elaborate fire trenches and under-the-floor ducting systems to heat the stone floors of bath houses. The Koreans have used a similar system of fire pits filled with stones under their homes for hundreds of years. In 1909, an Englishman filed a patent on a radiant heating system using tubing embedded in concrete or plaster.

While Europeans and far eastern cultures adopted radiant floor heating as a viable heating system, it was not until the end of World War II that much interest was shown in the United States. During the 1940s and 1950s, the floor heating industry began to grow. There was a considerable amount of interest generated primarily by those soldiers who had been in other countries and had experienced this type of heating system first hand. The famous architect, Frank Lloyd Wright, used radiant floor heating extensively. Unfortunately, the high cost of materials and labor to install copper or black iron pipe, coupled with some performance problems, caused interest to wane in the 1960s. Lack of insulation in buildings often required floor surface temperatures above comfortable levels and also produced wide temperature swings within the space.

The 1970s brought higher standards of construction and insulation, as well as the acceptance of new materials for piping, such as plastic and rubber compounds. Germany, Switzerland, and other European countries reported an astounding increase in radiant floor activity. In the United States, floor heating was rediscovered by the solar industry as a perfect match for solar collection systems. Hydronic radiant heating required water temperatures far less than that of more conventional fin tube convectors or fan coils. This, combined with the outstanding comfort floor heating provides, has made radiant floor heating grow into an important part of the hydronics industry.

Radiant Panel Heating

Unlike convection heaters, which heat air that in turns fills the room, radiant panels heat by radiation. Heat will leave a warmer object and travel to a cooler object if one is present. The greater the difference in temperature between the two objects, the faster the heat will transfer. The sun is warmer than the earth; therefore, it radiates its heat to the earth in the form of short wave energy. Empty space between the sun and earth is not heated; it is only when the short wave energy strikes the earth that it converts into heat and long wave radiant heat energy.

A wood stove is another example of radiant energy. If a person stands in front of a hot wood stove, the part of the body facing the stove will be warm while the part facing away will be cool. This is because the radiation given off by the stove only strikes the front of the person, where it converts to heat. It does not heat the air, so the back of the person senses the cooler air temperature. The only air that is heated by the stove is the air that comes in direct contact with the stove surface and has been heated by the stove radiant energy.

Surface area is an important factor in radiant heat transfer. A large surface area at a warm temperature can radiate as much heat as a small surface area at a high temperature. Three factors determine radiant energy transfer:

- Size of the radiating surface area
- Size of the receiving surface area
- Surface temperature difference between the radiating surface and the receiving surface

Heat does not rise, hot air does. Radiant heat travels in all directions. Although some convection does occur when cool air comes in contact with a warm floor, virtually all but 10% of the heat generated by a typical floor heating system is radiant. Air is warmed by contact with the surfaces within a structure.

INSULATION

A well-insulated building is the best assurance to designing a successful hot water heating system, or any heating system for that matter. The poorer the insulation, the higher the heat loss. High heat loss requires warmer floor surface temperatures. Poorly insulated buildings may produce floor temperatures beyond human comfort levels and cause large temperature swings because of excessive heat build up in the floor.

A concrete HRH slab constructed on or below grade will have increased heat loss to the ground due to its higher than normal temperature. Insulation placed around the perimeter and/or under the slab will reduce that heat loss. Perimeter losses are more significant than downward losses; therefore, it is recommended that insulation be applied to or below the frost line. Buildings built in moderate climates on dry soil generally do not require insulation under the slab; the earth itself acts as an insulative barrier. Structures built in cold climates, on damp soil, or in areas with high water tables should have under-slab insulation to prevent excess downward heat loss. Response time can also be increased in most applications by installing under-slab installation.

Insulation under a suspended floor that is over a heated space is seldom required, although it can be used to improve thermal control. A highly insulative carpet and pad may create an upward thermal resistance greater than the downward resistance, causing an excess downward heat flow. Under-floor insulation can help counteract this situation. A radiant barrier may be installed under the floor as an inexpensive but effective means to reduce downward heat travel.

It is highly recommended that insulation be applied to suspended floors over unheated spaces such as vented crawl spaces, cantilevered floors, unheated garages, or basements. A radiant barrier in conjunction with stem wall foundation wall insulation may be used for a suspended floor over a controlled ventilation crawl space.

FLOOR COVERING

The temperature of the floor surface is very important. Hard surface floors such as concrete, tile, or linoleum are ideally suited to HRH. As thicker floor coverings are added, the thermal mass under the floor covering must increase in temperature to maintain the proper floor surface temperature. Although wood flooring and carpeting can be used quite successfully with HRH, calculations must be made to ensure that the proper surface temperatures are reached without exceeding acceptable temperatures in the thermal mass.

Insulative floor coverings such as carpet and pad will adversely affect the response time. The more insulative the floor covering, the higher the temperature required for the thermal mass. The R-value of a floor covering is its resistance to the flow of heat. The higher the R-value, the more thermal resistance. High R-value floor coverings slow the heat transfer from the thermal mass to the floor surface and thus reduce the recovery time as well as limit the output of the floor.

Hard surface floors are more desirable in front of large picture windows or frequently opened outside doors. A hard surface has more heat output capacity and can react quicker to counteract cold drafts or brief influxes of cold air. Table 8-1 shows the R-values of various floor coverings.

Floor covering	Thickness (in.)	R-value
Carpet*	1/8	0.6
	1/4	1.0
	1/2	1.4
	3/4	1.8
	1	2.2
Rubber pad	1/4	0.3
	1/2	0.8
Urethane pad	1/4	1.0
	1/2	2.0
Vinyl, tile	-	0.2
Hardwood	3/8	0.5
	3/4	1.0

*For wool carpets multiply R-value by 1.5.

Table 8-1. R-values of various floor coverings

Room Temperature

Since HRH systems heat by radiation and not by heating air, the room air temperature is not a direct result of floor heating. The surfaces in the space are warmed by radiation from the floor. Air is warmed by coming into contact with room surfaces. Air temperatures remain fairly constant throughout the space. In a properly designed HRH system, the floor surface temperature will be only a few degrees warmer than the air temperature.

Most conventional heating systems use convective means to transfer heat into the space. Forced air systems use a combination of forced and natural convection to achieve heat distribution. Hot air is forced by means of a blower into the space. This hot air rises to the ceiling, cools, and then falls, where it is returned to the furnace for reheating.

The effect of an HRH system on air temperature is one of minimal drafts and temperature stratification. Structures with high ceilings can benefit greatly from an HRH system. Air temperatures at the ceiling remain cool while the air nearer the floor at the occupied level maintains the desired temperature. The need for ceiling fans to force down wasted hot air from the ceiling is eliminated. The actual temperature stratification will vary, but typically a structure with a 20 foot ceiling will have ceiling air temperatures 2°F cooler than at the five foot level.

Comfort

Indoor climate comfort is a difficult concept to describe. What is considered comfortable to one person can be uncomfortable to another. The same person can experience various degrees of comfort or discomfort at different times in the same environment depending on levels of activity or biological states. A person at rest after eating or exercising may feel warmer than at other times. It is not possible to maintain a constant comfort level for everyone all the time, therefore, compromises must be made.

The comfort of the human body is affected externally by three items: contact with the ambient air, evaporation of skin moisture, and radiation. Depending on the ambient air temperature, the body will either absorb heat from the air or give up heat to the air by conduction. Air movement across the skin (drafts) causes increased evaporation of skin moisture, reducing skin temperature. This is desirable if a person is overheated but detrimental to comfort if one is trying to keep warm.

The human body either radiates its heat to cooler surrounding surfaces or absorbs heat from warmer surfaces. A person standing near a cold window on a winter day can feel heat leaving his or her body. The perception is that the window is cold or that there is cold air coming from the window. In reality, it is the warm surface of the body radiating its heat to the cold surface of the window. The larger the temperature difference between the body and the surrounding surfaces, the faster heat is lost or gained. If the heat gain is rapid, the perception is one of a hot surface; if body heat loss is rapid, the perception is one of a cold surface.

The *mean radiant temperature* (MRT) is a term used to describe the collective effect on an occupant of all the surrounding surfaces in an indoor environment. It is subject to surface area, surface temperature, and proximity. HRH systems help in counteracting a low MRT produced by cold outside walls, ceilings, and windows by raising the MRT, thereby lowering body heat loss.

HRH systems provide a high level of comfort by reducing drafts, creating even air temperature distribution, and raising the MRT. HRH is particularly useful in counteracting large windows and high ceilings where falling drafts may be a problem.

Applying HRH under a bathtub or shower, or even set in surrounding tile can be very pleasing. Radiant heat hangers for towels and house robes are used in bathrooms and are the new trend in modern houses in Europe. Besides looking attractive, they are practical.

Floor Space

Of particular interest to the architect, interior designer, and hydronic specialist is the increased use of floor space provided by the HRH system. Allowances do not have to be made for baseboard units or floor registers, because the entire floor space is usable. Furniture and appliances can be set up to the wall, and floor registers are not required under the windows and in front of sliding glass doors. Furniture or objects set on the floor will absorb heat from the floor and, in turn, their surface temperatures will be raised, thus raising the MRT.

The use of HRH under food cabinets is not recommended due to the excess heat buildup, which can cause food to deteriorate at an accelerated rate.

RESPONSE TIME

Once the HRH control calls for heat in a space, time is required for the system to respond. Typically, HRH systems react slowly due to the fact that they incorporate a large mass at low temperatures. There is an inertia that must be considered both in heating up and cooling down. HRH systems operate best when designed to maintain a constant temperature under continuous operation. Modern construction techniques and improved insulation have greatly reduced the adverse effects of a slow response time.

Although it is true that an HRH system will take considerably longer than a forced air system to bring the air temperature of a space up to design temperature, it will remain much more constant once the desired air temperature is achieved. As discussed earlier, comfort does not necessarily relate to air temperature. An HRH system will provide radiant comfort well before the desired air temperature is reached. The result is that a person may feel comfortable before the air temperature reaches the desired setting.

Because of the flywheel effect and slow response time of the thermal mass, HRH systems do not perform well with night setback thermostats. Rooms that are set back do not cool quickly, and rooms that are recovered must begin reheating early to be back at the set point at the desired time. Little savings result and temperature overshoot can create a comfort problem. A steady state with gradual change is the most desirable mode for HRH systems.

A major advantage of the flywheel effect is experienced when sudden brief inrushes of cold air are experienced, such as when a door is opened. When the cold air travels along the floor, it is rapidly heated by contact with the warm floor. Then, once the door is closed, the floor instantly radiates heat to the occupants unlike a forced air system, which must re-heat all of the air within the space to achieve comfort. In this case, response time is dramatically improved due to the thermal mass.

ZONING

A distinct advantage of HRH systems is their zoning flexibility. Each tube loop can be designed as an individual heating zone, or an entire system can be operated from a single thermostat. Although a thermostat in every room may be the ultimate in control, it is seldom necessary. It is more common to group spaces with similar use, size, or location on one control.

The HRH system is unique in its ability to naturally self-regulate heat output. The rate of heat output is determined by the difference in the floor surface temperature and the MRT of the surrounding surfaces. As a result, the output of the floor near a sliding glass window will be greater than that of a floor in a interior room, even though the floor surface temperatures are the same. The design challenge is to ensure that proper water temperatures and flow are allowed for each location.

As a general rule, it is best to group spaces on the same side of the building with similar floor coverings and heat losses. The self-regulating ability of an HRH system along with water temperature and flow design lend flexibility to zone planning. Zone size is only limited by manufacturer rated manifold and pump capacity.

Excessive zoning increases system cost, complexity, owner interaction, and may be of little benefit. Items to consider in zoning are as follows:

- Occupant wishes
- Physical separation of spaces
- Floor levels
- Frequency of use
- Type of use
- Heat loss characteristics
- Solar gain
- Internal gain
- Types of floor coverings
- Types of floor construction

TUBING

Tubing for HRH systems can be divided into three categories: metallic, synthetic, and composite. Each has advantages and disadvantages, but all can be suitable for radiant floor heating when used with the proper components. As a general rule, all pipe used for floor heating should be flexible enough to be laid out in continuous loops with no joints in the floor.

When selecting tubing, compare pressure and temperature ratings as well as the wall thickness for heat transfer and durability. Although most systems operate between 15 and 20 psi, some codes require a hydrostatic test of 100 psi for 30 minutes. Others require that the tube withstand 180°F water at 100 psi. Keep in mind weight and flexibility for installation. Test the system before being embedded, and check compatibility with embedding material.

All hydronic heating systems are susceptible to oxygen entering the system through numerous sources such as threaded fittings, air vents, and gas permeable materials. Excess amounts of oxygen in a system can lead to premature failure of ferrous metal components due to corrosion. While copper tube is, for all practical purposes, impervious to oxygen migration through its walls, all synthetic tubes display a degree of permeation. Whether this characteristic can lead to problems for ferrous metals in radiant floor heating systems is dependent on a variety of factors. The temperature must be given the strongest consideration as a determining factor. The internal system pressure and the speed of flow are less important. Water quality can also play an important role.

Metallic

The most common metal pipe used today is soft copper. Copper is highly conductive, meets all plumbing standards, stays put when bent, and is readily available. Although it has been used successfully, it has several disadvantages. It is heavy and clumsy to work with, it kinks easily, it is susceptible to internal corrosion, and it can be attacked externally by acids in cementitious thermal mass.

Synthetic

The vast majority of radiant floor heating systems installed in the last 20 years have used synthetic tubing. It is lightweight, flexible, inert to corrosion, resistant to scaling, and not easily crushed, crimped, or kinked. The internal smoothness of synthetic tubing creates less friction, so there is less resistance to flow than in metallic tubing.

Synthetic tubing is available in coils of relatively long lengths, which allow for a variety of tubing layouts without joints. Synthetic pipe can be divided into two distinct categories: plastics and rubber compounds. Plastics are more rigid and generally have thinner tube walls and better heat transfer characteristics than rubber tubes. On the other hand, rubber tubes can be extremely flexible and durable.

The base material for most rubber tubing is EPDM (ethylene propylene diene monomer). This material has a long and proven history of use in industry for such items as automotive hose, O-rings, window channeling, and other products requiring resistance to weathering. It can be extruded as a simple tube, or it can be reinforced with fiber webbing and jacketed with other materials. Temperature requirements for radiant floors are well within the range of EPDM. Unreinforced tubing has a pressure limit of approximately 40 psi while reinforced tubing can withstand high pressures dependent upon the degree of reinforcement. EPDM has a broad resistance to chemicals but not to oils and other hydrocarbon fluids. An oil-resistant rubber jacket may be extruded over the base EPDM to make it more compatible for installations where oil or asphalt is present.

Polyethylene tubing for floor heating is extruded from high density polyethylene (HDPE) or cross-linked polyethylene (PEX). HDPE is more flexible than PEX but is not as tolerant to temperature and pressure. The cross linking process of PEX links molecular chains into a three-dimensional network, which makes the tube durable within a wide range of temperatures and pressures. Polyethylene tubing has been the most widely used tubing for floor heating in Europe. PEX is the most rigid of the synthetic tubing, and it will meet or exceed any of the most stringent pressure and temperature tests required by the codes. It also returns to its original shape when the proper amount of heat is applied. This is particularly useful if the tube is accidentally kinked or crushed during installation. Polyethylene is highly resistant to chemicals including oils.

Polybutylene (PB) is a plastic developed in the United States and has been used almost exclusively in hot and cold water piping since the late 1960s. It has had widespread use in radiant floor heating for over 20 years. PB tubing has a high strength to wall thickness ratio, which results in the thinnest walls and greatest flexibility within the plastic tube group. The temperature and pressure properties of PB tubing fall in the range of code requirements and surpass the 100 psi, 180°F water test. It also retains its flexibility over the wide range of temperatures, which makes it easy to work with in extremely cold conditions. If kinked or damaged, PB may be fusion welded in the field using a special fitting and heat tool. PB has excellent chemical resistance but is susceptible to oils and other hydrocarbon fluids.

Composite

To address the issue of oxygen permeation, recoil memory, and bending, a new generation of tubing is entering the marketplace. This is a tube that uses the benefits of both metal and synthetic tubing. This tube incorporates a thin metal layer, usually aluminum, sandwiched between layers of plastic, which gives the tube many good properties. It has the advantage of being inert to corrosion and impervious to oxygen permeation, while being semi-rigid. It bends much easier than soft copper, retains the bent shape, and stays put. It can meet or exceed all radiant floor temperature and pressure requirements. Damage in the field by kinking or crushing would require tube replacement or a mechanical joint.

Table 8-2 is a comparison between the different types of tubing available for floor heating The values shown in Table 8-2 are 1 to 10, where 1 = least and 10 = most. The values represented are typical and do not represent individual products.

SYSTEM COMPONENTS

While tubing is the most obvious component in an HRH system, other components are required. These components include manifolds, a heat transfer medium, circulating pumps, and a supplementary heating system.

Manifolds

The length of any particular tube circuit in the floor is limited by flow and pressure drop; therefore, many circuits are generally used to cover the floor area. The device used to distribute water to each individual tube from a single supply and return is a manifold (sometimes referred to as a header). The manifold can be as simple as a series of Ts, or it can incorporate sophisticated controls, Figure 8-1.

Heat Transfer Medium

Unlike radiators or fin tube convectors, which transfer heat directly into the space, radiant floor tubing must have an additional, field-installed medium to do that job. Heat in the tubes must be transferred to the broad surface area of

	Metal	Plastic			Rubber		Other
Features	**Copper**	**PE**	**PEX**	**PB**	**EPDM**	**Reinforced**	**Composite**
Safe working temperature	180°F		180°F	180°F	285°F	180°F	180°F
Safe working pressure	100 psi		100 psi	100 psi	30 psi	100 psi	100 psi
Inside diameter restriction	1	5	4	2	9	8	3
ASTM Standard	B88	None	F876	D3309	None	None	None
Thermal conductivity	10	6	6	7	5	4	6
Suitable for tight bends	2	6	5	7	10	9	7
Flexibility @70°F	1	5	4	5	10	9	4
Flexibility @32°F	1	4	3	6	10	9	3
Kink resistance	2	5	6	5	8	9	4
Recoil memory	5	4	4	6	1	2	3
Corrosion inert	1	10	10	9	9	9	10
Scale resistant	1	10	10	10	9	9	10
Oil resistant	10	9	9	4	3	6	10
Field repair	Solder	heat/ mech.	heat/ mech.	weld/ mech.	mech.	mech.	mech.

Table 8-2. Floor heating tube comparison chart

the floor to effectively heat the space. This means that the heat must travel out horizontally between the tubes and rise to the surface of the floor. The more evenly this heat can be distributed between the tubes, the more consistent the floor surface temperature will be across the floor. System design calculations are based on the assumption that the floor surface temperature is consistent. A poor heat transfer medium will not properly conduct the heat laterally and may cause hot spots directly above the tube and cold spots between the tubes.

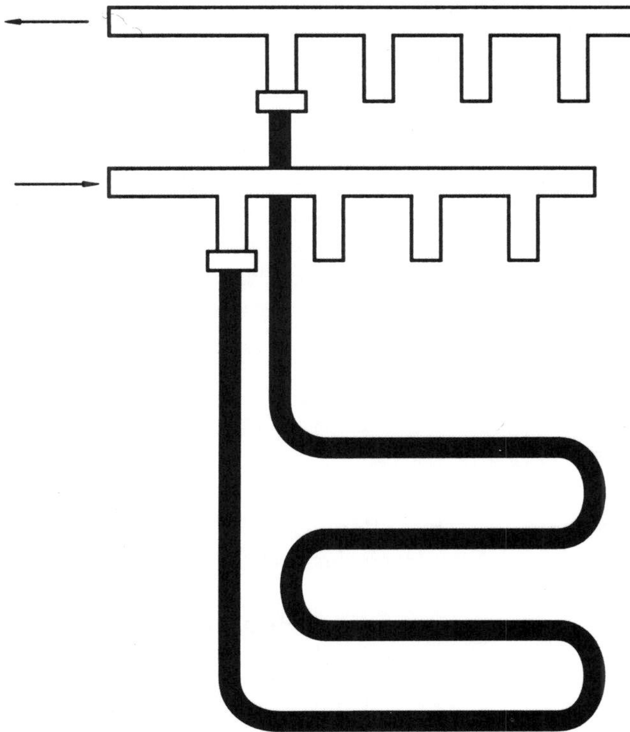

Figure 8-1. Example of a manifold

Building construction has a bearing on what medium is used. The most common types of construction where floor heating is used include slab-on-grade, slab-below-grade, wood suspended floors, and concrete suspended floors. The hydronic system designer together with the architect and structural engineer must select the heat transfer medium that best suits the building construction and the particular application. This medium is usually poured concrete. Keep in mind that the greater the thermal resistance between the water in the tube and the surface of the floor, the higher the required water temperature and the longer the response time.

Circulating Pumps

Movement of water through the HRH system is accomplished with one or more circulating pumps. Although there are several types of pumps, centrifugal pumps are the most commonly used in HRH systems. These pumps use an impeller with curved vanes turning at high rpm to create a low pressure, which draws the water through the pump. If a large quantity of air enters the pump along with the water, the pump will cavitate and not circulate the water. Centrifugal pumps require a certain amount of water pressure with a minimum amount of air trapped in the water in order to function properly. A common problem with HRH pumps is trapped air. The system should be designed to vent any air in the system before it reaches the pump.

Supplementary Heat

There are times when heating apparatus other than HRH is either required or requested as part of the system. It may be that the owner or designer is utilizing the HRH to do only a portion of the building heating, such as a basement or main floor, and using conventional devices, such as convectors or fan coils, to provide the balance of the heating requirements. There may also be cases in which HRH is not capable of delivering all the heat required in a space without exceeding generally accepted maximum surface temperatures. As a result, additional heating may be provided by convectors or wall radiators. In either case, special allowances must be made for this equipment when performing the system design calculations.

DEFINITIONS

The following definitions will help clarify some of the terms often used in hydronic heating.

Convection: Heat transfer by movement of fluid (i.e., water, air). Natural convection is due to differences in density from temperature differences; warm air rises and cool air falls, causing a circular flow. Forced convection is produced by mechanical means.

Degree Day: A unit based on temperature difference and time, which is used in estimating heating system energy consumption. For any one day whose mean temperature is below 65°F, the degree days for that day is the difference between 65°F and the mean for that day. Degree days for any period is the sum of the degree days for each day in that period.

Design Temperature: The temperature a system is designed to maintain (inside design temperature) or operate against (outside design temperature).

Infiltration: Air flowing inward, as through a crack between window and frame.

Radiation: Energy radiated in the form of waves or particles.

Radiant Barrier: A membrane with a polished aluminum surface, which reflects long wave radiant energy. It is typically a composite of aluminum laminated to a plastic film, which is available in rolls or as the face of fiberglass batt insulation.

Thermal Conductance (C): The number of heat units (Btu) that will pass through 1 square foot of non-uniform material in 1 hour for each degree F difference in temperature between the two bonding surfaces of the material.

Thermal Conductivity (k): The number of heat units (Btu) that will pass through 1 square foot of uniform material 1-inch thick in 1 hour for each degree F difference in temperature between the two surfaces of the material.

Thermal Mass: A dense material used to store and transfer heat. Generally in the form of a concrete-like material poured as a finished floor or sub-floor in which hot water tubes are embedded.

Thermal Resistance (R): The ability of a material or combination of materials to retard or resist the flow of heat. It is the reciprocal of U.

Thermal Resistivity (r): The ability of unit thickness of a uniform material to retard or resist the flow of heat. It is the reciprocal of thermal conductivity (1/k).

Transmission: In thermal load calculations, a general term for heat travel (by conduction, convection, radiation, or any combination thereof).

Chapter 9

PROJECT BIDDING

The previous chapters covered the components and details of a hydronic system. The right selection of equipment is very important, because it will be used to generate comfort for building occupants.

A contractor who installs these systems and subsystems is aware that a job must pay for expenses, time consumed, materials used, skill/expertise, and include some profit. In order to work on a particular hydronic installation or project, the contractor must first be awarded the job. Obtaining a job is a challenging process at times. Larger projects are usually advertised, but that is not the only way to find out about a project. Former satisfied customers and word-of-mouth recommendations are often the best way to learn about a new project.

The contractor should bid for projects within his/her scope, capabilities, and current workload. Contractors should not tackle projects if they are not fully qualified. If quality work is completed in a timely manner, the contractor will continue to receive work and stay in business. Competition is fierce in all types of trades, but nothing will replace the following basic characteristics:

- Timely completion of the job
- Quality workmanship
- Job completion within budget

The goal is to have a project awarded. Coincidence or luck should not be counted on; instead, a very careful and thorough estimate is the key to a bid price low enough and competitive enough to be awarded a job. In soliciting a bid, the design architect-engineer will prepare the design documents containing the drawings and specifications. Hydronic drawings include plans and diagrams as well as specific details about various types of connections and schematic arrangements. Appendix G contains two small sample hydronic projects for sizing a hot water heating system and a chilled water system.

When required, a design engineer, construction supervisor, or building manager may prepare a preliminary job estimate to compare with the contractor's bid. In order to bid on larger jobs, the contractor must usually rely on the services of a special estimator. Independent of the job size, however, a *material take off* must be prepared. This consists of a prepared list of materials that includes all pipe sizes (measured from the engineering plans), fittings and valves (some practical experience is required), and accessories such as insulation, as well as the equipment to be installed. Based on such a list, the contractor may determine the final bid price. The price includes materials, labor, overhead, taxes (as applicable), a small percentage for the unknown, and profit.

PREPARING AN ESTIMATE

When preparing an estimate, keep the following points in mind (excerpted with permission from the R.S. Means Company, Inc.):

1. Be consistent in listing dimensions; for example, length x width x height. This helps in re-checking to ensure that the total length of partitions is appropriate for the building area.
2. Use printed (rather than measured) dimensions where given.
3. Add up multiple printed dimensions for a single entry where possible.
4. Measure all other dimensions carefully.
5. Use each set of dimensions to calculate multiple related quantities.
6. Convert foot and inch measurements to decimal feet when listing pipe lengths. Memorize decimal equivalents to .01 parts of a foot (1/8" equals approximately .01').
7. Do not "round off" quantities until the final summary.
8. Mark drawings with different colors as items are taken off.
9. Keep similar items together, different items separate.
10. Identify location drawing numbers to aid in future checking for completeness.
11. Measure or list everything on the drawings or mentioned in the specifications.
12. It may be necessary to list items not called for to make the job complete.
13. Be alert for: Notes on plans such as N.T.S. (not to scale); changes in scale throughout the drawings; scale of details (usually 1/4" = 1 ft); reduced size drawings; and discrepancies between the specifications and the drawings.

14. Develop a consistent pattern of performing an estimate. For example:
    a. Start the quantity take-off at the lower floor and move to the next elevation.
    b. Proceed from the main section of the building to the wings.
    c. Proceed from south to north or vice versa, clockwise or counterclockwise.
    d. Take off floor plan quantities first, elevations next, then detail drawings.
15. List all gross dimensions that can either be used again for different quantities, or used as a rough check of other quantities for verification (exterior perimeter, gross floor area, individual floor areas, window dimensions, door locations, etc.).
16. Utilize design symmetry or repetition (repetitive floors, repetitive wings, symmetrical design around a center line, similar room layouts, etc.).
17. Do not convert units until the final total is obtained.
18. When figuring alternatives, it is best to total all items involved in the basic system, then total all items involved in the alternatives. This way you work with positive numbers all the time. When additions or subtractions are used, it is often confusing whether to add or subtract a portion of an item; especially on a complicated or involved alternative.

Another point that may be added to this list is that when shorter connections or saving possibilities are included, be sure to make a note for future reference.

SHOP DRAWINGS

After the job contract is secured based on the drawings prepared by the engineer and designer, the contractor prepares the shop drawings. The specification usually contains the following request:

> Shop documents (submissions) shall include drawings, schedules, performance charts, instructions, brochures, diagrams, and other information to illustrate the system requirements and operation of the system. Shop drawings shall be provided for the complete hydronic system including piping layout and location of connections; schematic diagrams and wiring diagrams; or connection and interconnection diagrams.

Shop drawings submitted by the contractor will inform the design engineer whether the concept was completely and correctly followed. The shop drawings serve a dual purpose:

- They allow the engineer to analyze the contractor's intentions in regard to installation details so he/she can check for any discrepancies from the contract drawings. The drawings will also be checked for clarity as well as details and compliance with the codes.
- They allow the contractor to give the architect or engineer the opportunity to develop a more practical approach for pipe routing for

heating and/or cooling equipment connections. These modifications may be based on the code or standard but perhaps with a slightly different interpretation.

The contractor usually has the latitude to choose between some alternative materials or brands of equipment as listed in the specification. The design engineer's approval is sometimes required for material substitution. Certain alternatives may produce a savings. Alternatives are usually acceptable as long as the system presented is in complete compliance with the governing standards or codes as well as the requirements of the owner representative and the local building inspector (if necessary). However, alternatives should not add money to the cost of the job.

REQUIREMENT FOR PERMIT

When installing a hydronic system in a building or structure, a construction permit is not normally required as part of the building permit. However, in order for system connections to be made, potable water and/or gas permits are required. This requires knowledge of local conditions, applicable standards, and local regulations. In some cases, more than one organization may have jurisdiction. These organizations could include the following:

- Public Health and/or Safety Department
- Public Water Authority
- Building Code Enforcing Authority (Building Inspector)
- Gas Company
- Fire Department

These agencies have legal regulations which must be abided by all.

Chapter 10

Energy Conservation

Energy conservation in hydronic systems has three main goals:

- Reduce heat losses
- Reduce fuel consumption
- Maximize equipment efficiency

People became aware of the need for energy conservation (reduction in fuel consumption) mainly during the 1973 oil crisis. The sudden and sharp rise in fuel prices suddenly became a significant part of doing business, and it became necessary to reduce the amount of energy consumed. People started to realize that the world-wide economy could no longer afford to waste natural resources.

Groups such as ASHRAE have issued standards that set guidelines and limitations in the design of heating and cooling systems. There are also mandatory laws, rules, and regulations to reduce pollution. However, the increasing need for air conditioning, fire protection, etc., has negatively influenced the preservation of clean air, water, and soil.

INSULATION

The aim of a hydronic system is to heat or cool water for distribution to people and/or processes. In the process, water temperature must be raised or lowered a number of degrees. The goal is to keep this gain or drop in temperature from changing significantly before it reaches the user. Pipes that carry hot water through an area with a lower temperature will lose a few degrees along the way, which is why these pipes should be insulated. This is an important consideration for pipes conveying hot or cold water in a hydronic system. The sample specification at the end of this book lists pipe sizes, type of insulation based on the temperature of the fluid, and the thickness recommended. This provision is important in the design phase as well as during installation.

Pipe insulation represents an added expenditure, and a specialized subcontractor usually performs the installation per specification requirements. Allowances must be made for the pipe plus the insulation, and adequate supports must be fashioned. Insulation cost, appearance, and surface temperature are other points to consider. A good insulation must:

- have low thermal conductivity;
- be non-combustible;
- not be subject to deterioration with time;
- be of adequate strength.

Insulation efficiency depends on the type of insulating material and its heat transmission (R) factor. The R factor is the thermal resistance or radiant heat flow, which depends on the material and the thickness applied. A better insulation has a higher R factor and lower heat loss. The insulation must have a limit in flame spread and smoke development.

Other equipment (boiler pumps, heat exchanger, domestic water heaters, valves, etc.) must also be insulated to conserve energy. The cost of insulation is offset by the energy savings experienced over time. Applying insulation also protects personnel from burns, controls condensation, and reduces noise.

Example 10-1. Compare the heat loss between bare pipes and insulated pipes for a system with the following data:

Fluid to be conveyed: hot water
Pipe material: Steel Schedule 40
Pipe diameter: 1"
Pipe length: 210 linear feet
Hot water temperature: 200°F
Insulation thickness: 2"
Insulation K factor: 0.50
Temperature of room where pipe is located: 70°F

Solution 10-1. According to information located in Appendix E, the heat loss from bare (uninsulated) pipes is 112 Btuh, while the heat loss for insulated pipes with a K factor of 0.50 is 24 Btuh. For the pipe selected, the heat loss is as follows:

Bare pipe: (210 linear ft) (112 Btuh) = 23,520 Btuh
Insulated pipe: (210 linear ft) (24 Btuh) = 5040 Btuh

The heat loss difference between the bare pipe and insulated pipe is 18,480 Btuh. The insulated pipe as specified has only 21.4% heat loss vs the bare pipe. Considering one gallon of oil equals 140,000 Btuh, .132 gallons of oil is conserved every hour.

REDUCING FUEL CONSUMPTION

Good insulation may promote a reduction in fuel consumption; however, the equipment (boiler, furnace, chiller, pumps, etc.) is where most fuel consumption occurs.

All systems have losses (heat, friction, etc.) that contribute to a lower efficiency. Thermal efficiency is expressed as follows:

$$\text{Efficiency} = \frac{\text{Net work done by a system}}{\text{Heat supplied to a system}} \text{ or } \frac{\text{Brake horsepower}}{\text{Nameplate horsepower}}$$

Boiler efficiency is based on combustion efficiency, as well as overall efficiency. Boilers may be electric or fuel burning. Electric boilers have efficiencies of 92% to 96%, while combustion efficiency for fuel-burning boilers is between 79% and 87%. The overall efficiency for fuel-burning boilers is lower, because other losses are involved such as radiation loss. This efficiency is indicated by the boiler manufacturer based on tests.

Boilers are usually rated by one of the following organizations:

- Hydronics Institute (I=B=R)
- Steel Boiler Institute (SBI)
- American Gas Association (AGA)
- American Boiler Manufacturers Association (ABMA)

The Air Conditioning and Refrigeration Institute (ARI) issues the standards in the refrigeration (cooling) field.

The American Society of Heating, Refrigerating and Air-Conditioning Engineers (ASHRAE) suggests the following possibilities to help conserve energy:

- Run equipment only when needed. Schedule hvac unit(s) and domestic water heater(s) operation for occupied periods only. Run heat at night only to maintain a reduced internal temperature (between 50° and 55°F to prevent freezing). Start morning warm-up as late as possible to achieve design internal temperature by occupancy time (optimum start control), considering residual temperature in space and outdoor temperature (optimum stop control). Calculate shutdown time so that space temperature as well as hot water temperature does not drift out of the selected comfort zone before the end of occupancy.

- Sequence heating and cooling. Do not supply heating and cooling simultaneously. Central fan systems should use cool outdoor air between heating and cooling season. The zoning and system selection should eliminate, or at least minimize, simultaneous heating and cooling. Humidification and dehumidification should not occur concurrently.
- Provide only the heating or cooling actually needed. Generally, reset the supply temperature of hot and cold air (or water) according to actual need. This is especially important on systems or zones that allow simultaneous heating and cooling.
- Supply heating and cooling from the most efficient source. Use free or low-cost energy sources first like solar, then use higher-cost sources, as necessary.
- Apply outdoor air control. Do not use outdoor air for ventilation until the building is occupied, and then use proper outdoor air quantities. When on minimum outdoor air, use no more than that recommended by ASHRAE Standard 62, *Standards for Natural and Mechanical Ventilation*. In the cooling mode (in cost-effective areas), use enthalpy rather than dry bulb to determine whether outdoor or return air is the most energy efficient air source.
- Adjust the air/fuel ratio so that excess air is the minimum recommended for the equipment.
- Clean all heat transfer surfaces regularly (boiler tubes, heat exchangers).
- Do not use oversized boilers or furnaces. Efficiency is less at partial loads.

Some practical energy conservation measures are as follows:

- Reduce the temperature of all occupied space in winter to a reasonable temperature (68°F).
- For spaces such as offices, laboratories, etc., reduce the temperature when not occupied to 50°F.
- Provide a time clock to turn off water heater during night time. Leave it on for the selected periods when working people need hot water.

Tables 10-1 and 10-2 illustrate the importance of setting back a thermostat and applying insulation to a building.

Amount of insulation	Annual heating requirements	Percent savings with night setback 5°F	10°F
None in outside walls (R = 3.3) Two inches in ceiling (R = 8.4)	85.2 x 10^4 Btu	7.6	12.4
Two inches in outside walls (R = 9.3) Four inches in ceiling (R = 13.8)	52.1 x 10^4 Btu	8.7	12.5
3-5/8 inches in outside walls (R = 13.3) Six inches in ceiling (R = 18.4)	37.7 x 10^4 Btu	9.8	12.5

Table 10-1. Percent savings available with varying amounts of insulation and setback temperature (Reprinted by permission of the American Society of Heating, Refrigerating and Air-Conditioning Engineers, Atlanta, Georgia, from the 1972 ASHRAE *Handbook -- Fundamentals*)

City	5°F Setback	10°F Setback
Atlanta	11	15
Boston	7	11
Buffalo	6	10
Chicago	7	11
Cincinnati	8	12
Cleveland	8	12
Dallas	11	15
Denver	7	11
Des Moines	7	11
Detroit	7	11
Kansas City	8	12
Los Angeles	12	16
Louisville	9	13
Milwaukee	6	10
Minneapolis	5	9
New York City	8	12
Omaha	7	11
Philadelphia	8	12
Pittsburgh	7	11
Portland	9	13
St. Louis	8	12
Salt Lake City	7	11
San Francisco	10	14
Seattle	8	12
Washington, D.C.	9	13

Table 10-2. Approximate percent fuel savings with night setback for 25 cities (Reprinted by permission of the American Society of Heating, Refrigerating and Air-Conditioning Engineers, Atlanta, Georgia, from the 1972 ASHRAE *Handbook -- Fundamentals*)

Energy Conservation Categories

Energy conservation in design is divided into four categories: building construction, design criteria, system design, and controls.

The rest of this chapter is reprinted with permission from the book *Air Conditioning Principles and Systems, Second Edition* by E.G. Pita, published by J. Wiley & Sons, 1981.

Building Construction

Methods of achieving conservation are usually considered in the planning stages of a new building. Some of the methods may be applicable to existing buildings through *retrofitting* (changes to the existing system). Whether or not a specific method is practical in existing buildings depends on the nature of each case. In some cases, the decision is obvious (an existing building could not be turned around to reduce solar heat gain).

Reduced hvac energy consumption in building construction usually is a direct result of minimizing heat gains and losses. Some suggestions are:

- Use exterior wall and roof materials with high thermal resistance. An R-20 (resistance coefficient) value is not unreasonable for roofs, considering present energy costs.
- Avoid excessive use of exterior glass (which has a low R-value and high solar heat gain). An exception to this is a residence where there is great solar intensity during most of the winter and a moderate outside temperature. Some homes are designed with a large, south-facing glass expanse to utilize the resulting winter solar radiation heat.
- Plan the site for reduced gains or losses. Trees or other objects may reduce solar heat gains and infiltration.
- Orient the building to minimize solar heat gain in summer (and maximize it in winter).
- Use internal shading devices and external shading overhangs or side reveals.
- Use double or even triple window glazing for areas with severe winter climates.
- Avoid excessive lighting requirements.
- Use window sash that has a tight seal at the frame. Consider possible use of nonoperable windows; however, this prevents use of natural ventilation in case the mechanical system fails.

Example 10-2. Based on the following data, how much gas will be saved in the winter if a roof with an insulation value of R-20 is substituted for a roof with an R-5 value?

Roof area: 20,000 ft^2
Winter indoor design temperature: 70°F
Winter outdoor design temperature: 0°F
Degree days for the location: 6500
Gas for heating purposes: 1000 Btu/ft^3

Solution 10-2. The design heat losses for the two roof types are calculated, and the difference found:

For an R-5 roof, the heat loss is as follows:

$$U = \frac{1}{R} = \frac{1}{5} = 0.20 \text{ Btuh/ft}^2\text{-°F}$$

$$Q = (0.20 \text{ Btuh/ft}^2\text{-°F})\ (20{,}000 \text{ ft}^2)\ (70\text{°F}) = 280{,}000 \text{ Btuh}$$ (Q = heat loss in Btuh)

For an R-20 roof, the heat loss is as follows:

$$U = \frac{1}{R} = \frac{1}{20} = 0.05 \text{ Btuh/ft}^2\text{-°F}$$

$$Q = (0.05 \text{ Btuh/ft}^2\text{-°F})\ (20{,}000 \text{ ft}^2)\ (70\text{°F}) = 70{,}000 \text{ Btuh}$$

The difference between the two is 210,000 Btuh (280,000 Btuh - 70,000 Btuh). To find the annual savings, the degree days are used:

$$\text{Btu saved} = \left(\frac{210{,}000 \text{ Btuh}}{70\text{°F}}\right) (6500 \text{ degree days})(24 \text{ hr/day}) = 468{,}000{,}000 \text{ Btu/year}$$

$$\text{Gas saved} = \frac{468{,}000{,}000 \text{ Btu/year}}{1000 \text{ Btu/ft}^3} = 468{,}000 \text{ ft}^3\text{/year}$$

Design Criteria

Hvac system design values used in the past have often resulted in systems that consume excessive amounts of energy. Some recommended energy conserving design factors are as follows (many are based on ASHRAE Standards):

- Use a 68° to 72°F maximum winter indoor design temperature except for special occupancy, such as lockers, nurseries, toilets, showers, etc.
- Use the 97.5% winter outdoor design temperature found in the latest ASHRAE tables. As an example of the savings, use 15°F winter design temperature for New York City instead of 0°F, which is the recommended value found in old tables. This does not reduce the heat loss directly, because it is proportional to the actual temperature difference; however, it will result in less oversized equipment and therefore lower energy consumption.
- Use a 78°F summer indoor design temperature.
- Use the 2.5% occurrence for summer outdoor design temperature. Many old tables list a temperature that is higher than 1% occurrence. See latest ASHRAE tables.
- Use the summer outdoor wet bulb temperature coinciding with the dry bulb, not the maximum, which usually occurs at a different time.
- Use a design window infiltration rate of no more than 0.5 cfm per foot of crack, for well-fitted windows. Designers often use much larger values based on existing windows.

- Use minimum recommended design ventilation rates consistent with codes or standards. For systems using recirculated air, use ASHRAE Standards.
- Use cooling load calculation procedures and data that account for the building's thermal storage.
- Use correct procedures and data for calculating ducting and piping friction losses, accounting for duct pressure regain if significant.

System Design

The type of hvac system and equipment affects the consumption of energy. Some suggestions that should be considered are as follows:

- Avoid terminal reheat systems or any other types that mix hot and cold air (or water). These are inherently energy wasteful, although steps can be taken to minimize energy losses.
- Choose the highest chilled water (evaporating) temperature and the lowest condensing temperature that result in satisfactory space conditions. This results in minimum refrigeration machine power consumption.
- Choose the largest satisfactory chilled water and hot water temperature range, thereby reducing water flow rates and pumping power.
- Choose the lowest reasonable friction loss in ducts and piping, reducing fan and pumping power.
- Select pumps and fans near the most efficient operating point.
- For equipment that is to be operated at partial load, speed reduction is usually the most efficient means of control.
- Consider reducing chilled water and hot water pump flow rates when load reduces (variable speed pumps).
- Try to avoid using preheat coils for outside air. This often results in heating and recooling, a ridiculous waste of energy. Instead, design the intake plenum so that the outside and return air mix thoroughly.
- Use multiple equipment so that at partial loads some equipment can be shut down. The equipment operating will then be closer to full load, which is often most efficient.
- Use refrigeration condenser water for heating if the temperature is right.
- Compare the energy consumption of a heat pump with separate systems for heating and cooling.
- Examine the feasibility of a solar energy collector system for heating service water or even for space heating or cooling.
- Use a variable air volume (VAV) type air conditioning system, if suitable. This is generally the most energy efficient system, because the reduction in air quantity directly lowers power use.
- Recover waste heat in exhaust air by use of heat recovery devices to heat or cool ventilation air.
- Examine the possible use of a total energy system compared with a conventional system.
- Recover waste heat from hot combustion gases with a heat exchanger.

Controls

Considerable energy conservation can be achieved through selection of proper automatic controls. Some suggestions are as follows:

- Use night and weekend automatic temperature setback for unoccupied spaces.
- Investigate the possible use of a central computerized control system designed to provide the most efficient operation at all times.
- Use automatic time switching to start and stop equipment according to need.
- Use enthalpy control for the supply air. Enthalpy control devices sense and measure the total enthalpy of outside and return air and adjust the dampers to provide the air mixture proportion that will provide the most economical natural cooling, thus reducing or eliminating need for refrigeration at times.

INSTALLATION

The system must be installed so that it does not waste energy. Some suggestions are as follows:

- Install the system as designed, unless some apparent energy wasteful decision has been made; if so, verify before a change is made.
- Provide air-tight duct joints, test for leakage, and seal if necessary.
- Avoid obstructions in ducts.
- Make gradual transitions in ducts.
- Use wide radius turns in ducts and use turning vanes if necessary.
- Provide ample insulation on ducts, piping, and equipment.
- Use dampers that have tight closing features.
- Install air vents at proper locations in water and steam distribution systems.
- Locate thermostats so that they do not sense an abnormal condition. For example, a cooling thermostat exposed to solar radiation might cause the cooling system to operate unnecessarily.
- Use low pressure drop valves in water lines, consistent with function. For example, butterfly valves have a lower pressure loss than globe valves.
- Install ducts and piping with shortest lengths of run and fewest changes of direction.
- Test and balance the complete hvac system in accordance with design requirements.

OPERATION AND MAINTENANCE

No energy conservation design procedure will be successful without follow-up of system operation to achieve and maintain minimum energy consumption. Some suggestions are as follows:

- Retest and rebalance the water and air system completely at regular scheduled intervals during its lifetime. Often, the initial balancing procedures are all a system ever receives, and it gradually loses efficiency.

- Shut down any equipment when not needed. The use of the space must be examined before this decision is made. For example, even though a space is unoccupied, possible freeze-ups must be prevented.
- Set back temperatures when spaces are not occupied. In winter, the temperature can often be set back to 55°F or even 50°F.
- Start equipment shut down or temperature setback shortly before occupants leave.
- Start equipment as late as possible before occupants arrive, consistent with achieving comfort.
- Utilize natural precooling at night from outdoor air (enthalpy control may do this automatically).
- Close outdoor air dampers near the beginning and end of operation each day when equipment is providing heating or cooling.
- Turn off unneeded lighting.
- Check that solar shading devices are used properly.
- Check and replace or clean air filters when resistance reaches design conditions.
- Clean all heat transfer surfaces regularly (coils, tubes).
- Set chilled water temperatures high and condensing water temperatures low, consistent with comfort and design values. Condensing water temperatures must not be so low that the refrigeration equipment malfunctions.
- Adjust burners and draft on furnaces and boilers periodically to ensure complete combustion and minimum excess air.
- Arrange for reliable water treatment services from a specialist for boiler and condensing water systems. Dirt in the system will reduce heat transfer as well as harm equipment.
- Check operation of air vents regularly.
- Limit the maximum electric power demand of the system (utility companies usually charge extra for high demands). This is accomplished by starting each piece of equipment in sequence, with ample time in between. In addition, some equipment (e.g., electrically driven refrigeration compressors) can be furnished with devices that limit their power demand.
- Check ductwork regularly for leaks that may develop.
- Check outside windows and doors regularly for abnormal cracks.
- Perform routine maintenance regularly (lubrication, belt tension, inspection for damage or breakage).
- Check that objects (books, clothing, furniture) are not obstructing air flow through room terminal units.
- Check that air distribution outlets have not been obstructed or tampered with.
- Check that room thermostats or humidistats have not been readjusted by unauthorized personnel. Provide locks if necessary.

Chapter 11

Hydronic System Inspection and Maintenance

After an installation is completed by a contractor but before it is insulated and covered, the owner's representative should visit the site and perform an installation inspection. Based on his/her observations, the inspector must prepare a *punch list*.

A punch list consists of a summary of deficiencies found during the inspection. The inspector must check for the following:

- Properly installed system
- Satisfactory quality of workmanship
- Satisfactory quality of piping, fittings, and equipment
- Leaks
- Properly functioning equipment
- Hot water (or chilled water) temperature set per specification
- Correctly operating valves, pumps, etc.
- Properly installed pipe joints, pipe sizes, fittings, and required accessories
- Equipment accessibility and maintenance space required

Basically, the inspector needs to make sure that everything is installed according to project documents, which consist of the drawings, specifications, and manufacturer's installation manuals. The inspector's punch list contains deficiencies that must be corrected by the contractor-installer. It is difficult to imagine a system that is perfectly installed from the beginning with no deficiencies whatsoever. The first inspection is usually scheduled when the system is completed but uncovered; a second inspection takes place when the system is completely finished.

A completed hydronic system is not visible in its entirety, because pipes, certain valves, fittings, connections, etc., may be hidden in pipe chases, above closets, etc. If necessary, the architect must provide pipe chases as part of building walls or partitions. These chases must have access doors at valve locations, gauges, thermometers, etc., for inspection, adjustment, replacement, or maintenance. It is definitely preferable that all the required accessories be installed close to the equipment and readily accessible. The first punch list must be done before the system is concealed to allow for accurate observation of leaks and to ensure the pipes are correctly installed and that the system is in general agreement with applicable codes.

After the hydronic system is installed, checked again, and final corrections are made, the system is taken over by the owner and used for its intended purpose. The responsibilities of the owner or user do not cease, because they must begin the system maintenance program.

MAINTENANCE

Equipment reliability depends on an adequate and sustained maintenance program. Without proper care, the equipment is bound to fail sooner or later. In order to perform the required maintenance, the staff mechanic or maintenance contractor must be familiar with the equipment installed. This in turn requires manufacturer installation, maintenance, and operation instructions to be available at the site as soon as the materials and equipment are delivered. When floor space is allocated during the design phase, it is important to provide enough space to service and/or dismantle the equipment as the need arises. Chillers, boilers, and pumps must be provided with instrumentation, safety devices, and controls to indicate, monitor, and ensure safety for equipment and system operation.

The National Fire Protection Association (NFPA) defines scheduled maintenance as "a thorough system check, intended to give maximum assurance that the system will operate effectively and safely. It includes any necessary repair or replacement of components."[1]

Three factors contribute to the performance of the hydronic system:

- Cleaning the system when installation is completed and proper filling
- Minimum addition of new water to the system during its life
- Regular inspection and maintenance of the mechanical components to prevent failures

Despite the fact that the hydronic system is not as "complicated" as other building services, a building would be rendered inoperable if the heating system were interrupted. A hydronic system will possibly last as long as the building itself if it is installed and maintained properly. Maintenance for the system components, such as boilers, burners, pumps, terminal heating equipment, controls, and accessories should be done as outlined in the operating and maintenance equipment manuals.

Maintenance activities can be divided into five categories:

- Inspecting
- Testing
- Filling and cleaning
- Preventive maintenance
- Repair and replacement

Inspecting

Inspection schedules are usually generated based on manufacturer recommendations for the particular equipment and must be conducted to identify warning signs of possible failure.

Maintenance instructions for various pieces of equipment must be integrated into the general maintenance plan. The maintenance inspection must include any exposed parts, piping, hangers and supports, valves, gauges, thermometers, compression tank(s), pumps, etc., which must be observed for leaks, abnormal noise, rust, or incorrect operating position. The equipment operation should be checked within the manufacturer's range.

Testing

All equipment testing must be scheduled to include checking for performance and safety. Testing must be accomplished periodically as determined by the responsible person in charge, based on practical experience and the manufacturer's recommendations in order to ensure the equipment meets specification requirements.

Filling and Cleaning

The initial filling and cleaning of a hydronic (water) system should include the following:

1. Thoroughly clean all equipment and piping of scraps and other foreign matter. Give particular attention to:
   a. pump packing glands or mechanical seals;
   b. valve seats and glands;
   c. flange and union faces or seats;
   d. strainers, orifices, gauges, etc.
2. Fill the system completely with clean (domestic preferable) water and circulate it without the heat on. Ensure pumps are properly aligned and securely anchored before start-up to prevent damage.
3. After water circulation, drain the system completely to flush out foreign matter.
4. If excessive dirt is found, repeat flushing procedure.

5. Refill the system with clean water, venting all high points and equipment of air.
6. Apply heat to the system slowly, with pumps operating, to produce 180°F (82.2°C) water temperature. Recheck all vent points during the heating process and remove all air.
7. Check all strainer baskets ahead of pumps, control valves, etc. If a heavy accumulation of dirt is noted, reflush the system and repeat Steps 5 and 6.

When the installation is completed, it is necessary to make final adjustments to the system. These adjustments should include: (1) balancing of all piping circuits and terminal heating/cooling units to prescribed flow rates; (2) adjusting of all pumps to operate at design flows and heads; (3) checking of all controls for proper setting and operation; (4) checking of boilers' combustion efficiency; and (5) balancing of all zones in a multizone system.[2]

A regular overall cleaning program is required and should include all equipment and appurtenances installed throughout the premises of the respective residential, commercial, industrial, or institutional facility.

As hydronic systems have evolved, improvements have been made in safety, economy, and reduced sizes. Unclean equipment collects dust, which may hide or cover cracks, rust, or other problems; therefore, equipment now has better finishes for enclosures for easier cleaning. It should be noted that the cleaning is easier when it is done on a regular schedule with the right materials and tools.

Cleaning and preventive maintenance for a domestic water heater includes regular purging due to scale accumulation at the bottom or on the heating element. Scale accumulation reduces efficiency, which means more fuel must be used than normally required. Scale formation may ultimately lead to equipment failure and must be prevented. If neglected, scale becomes harder to purge. Scale deposits result from precipitation of otherwise dissolved calcium and magnesium salts. A water softener additive may prevent scaling.

The higher temperature of the water results in a higher amount of precipitation of mineral and salts contained in the water. Hotter surfaces induce the higher precipitation. Once deposited, the salts act like an insulating material on the hot surface, such as the electric heating elements. The existence of these deposits may be detected by turning the heater off one hour to allow water to seep into the deposit's natural cracks. Then turn the heater on. Part of the trapped water will become steam, resulting in a cracking noise. This means mechanical cleaning is required.

Preventive Maintenance

All equipment such as controls, pumps, motors, valves, etc., must be scheduled for regular inspection and preventive maintenance. There are various levels of maintenance, and the owner must establish the particular level of maintenance to prevent undesired breakdowns. On the flip side of preventive maintenance is *breakdown maintenance*, which is nothing more than running the equipment until it breaks. This is not a good policy or good management.

Replacing valves, thermometers, gauges, thermostats, and safety devices is part of an educated and intelligent way to avoid costly interruptions. An important part of safety maintenance also includes checking pressure and temperature relief valves on boilers and water heaters. As soon as a valve shows signs of sticking or leaking, it requires immediate replacement.

An example of preventive maintenance for pumps is as follows:

- ***Every month*** — Check bearing temperature with thermometer (not by hand). If anti-friction bearings are running hot, they probably have too much lubricant. Eliminate the excess. If sleeve bearings are running hot, they probably need lubricant. If adding lubricant doesn't correct condition immediately, find the reason by disassembling and inspecting the bearings. Another reason may be the misalignment of the pump and motor. Check and correct.
- ***Every 3 months*** — Drain lubricant, wash out oil wells and bearings with special cleaning fluid (see manufacturer recommendations). Check sleeve bearings to ensure oil rings are free to turn with the shaft. Replace if defective. Refill with lubricant as recommended by pump manufacturer.
- ***Every 6 months*** — Replace packing, using only the grade recommended by pump manufacturer. Be sure seal cages are centered in stuffing boxes at the entrance of water seal piping.
- ***Every year*** — Remove rotating element and inspect thoroughly for wear. Order replacement parts where necessary. Check wearing ring clearances. Remove any deposits or scaling. Clean out water seal piping. Inspect foot valves and check valves (especially check valves, which must safeguard against water hammer when pump stops). Annual maintenance on the boiler and burner must include:
  - brushing and vacuuming boiler flue surfaces;
  - cleaning the flue;
  - replacing the oil filter;
  - servicing the oil burner;
  - replacing the oil burner nozzles;
  - cleaning the oil pump strainer;
  - cleaning and adjusting the electrodes;
  - firing off the burner;
  - making any needed adjustments;
  - checking the safety controls;
  - taking complete efficiency tests in low and high fire;
  - reporting the efficiency test results to the owner.

Other preventive maintenance procedures are as follows:

- Every month operate safety valves on the water heaters and boilers to ensure they are working properly. Replace defective safety valves.
- Every month check all gauges and thermometers in the system and make adjustments as necessary.

- Every two months open and close each equipment isolation valve. Repair or replace leaky valves.
- Every six months open and purge the water heater.

The maintenance schedule may vary in different parts of the country due to varying water quality conditions. Hard water, salty water, water containing sand, etc., may require a more aggressive preventive maintenance program.

Repair and Replacement

As the installation ages (ASHRAE indicates the life of a boiler between 24 years for tube boilers to 35 years for cast iron), there will be a need at some point for repair or replacement of equipment. It is important to store and maintain spare parts for these interim repairs.

Over time, impurities in fuel and water deposit on various parts, pipes, tubes, etc., reducing equipment efficiency. Preventive maintenance helps restore the original efficiency. In certain cases, old and inefficient equipment might be replaced with more economical models. For example, a single large boiler may be replaced with two boilers with 65% (of the total) capacity each. Unseen and undetected corrosion in piping and fittings might also occur. When any sign of corrosion becomes apparent, immediate remedial action is required.

When occupants complain about noise or interruptions in operation, the equipment must be promptly inspected and serviced. The maintenance personnel team is usually composed of qualified, well trained people who can accomplish this kind of work quickly and efficiently. The maintenance supervisor may hire individuals to assemble a team, or a contract may be given to a facility management company.

Each type of pipe in a hydronic installation is different, and the corresponding repair and replacement will also be different. For example, if a break occurs in a long stretch of pipe, it may be possible to save money by replacing only the damaged section. Cut out the damaged piece, then connect a new piece of pipe with a coupling and a union fitting. Before operating a hot water installation, always check the relief valve setting of the water heater; it should be maintained at or below 100 psi at a maximum 180°F. Connect the pipe to the heater with a heat dissipating fitting.

NOTES

[1] National Fire Protection Association Standard No. 10, Batterymarch Park, Quincy, MA 02269.

[2] American Society of Heating, Refrigerating and Air-Conditioning Engineers, 1980 ASHRAE *Handbook.*

Chapter 12

CONTROL SYSTEMS

Modern automatic heating or cooling systems must have corresponding automatic control systems. Controls do not measure the variables but transmit signals that change the variables. A control for a hydronic system includes three distinct elements:

- Sensing device, which measures the value
- Control device, which compares the measured value with the set point
- Adjusting element, which is usually a controlling (motor operated) valve

There is a difference between a control device and an instrument. A thermostat is a control, while a gauge or thermometer is an instrument. The thermostat is a temperature sensitive electrically operated switch. One type of thermostat contains a bimetallic strip, which changes its shape with a temperature change. Attached to the bimetal element is a set of electrical contacts. When room temperature increases or drops, the strip either closes or breaks the contacts, activating a valve, burner, fan, etc.

The ability to control the heating or cooling system means controlling water temperature, flow, or both. The heating or cooling control system must:

- maintain design condition temperature and humidity;
- perform mechanical operations directly connected to heating and cooling, such as starting or stopping a fan, opening/closing a damper, etc.;

- minimize energy consumption by allowing equipment to operate only when required;
- act as safety controls by maintaining the operation of equipment within a safe range or turning the system off or letting off excess pressure.

The capacity of a heating or cooling system may be controlled either through bypassing or mixing, or by throttling the amount of hot water, steam, or chilled water to a terminal unit or user (radiator, coil, etc.).

Before detailing the control system, it is first necessary to provide some definitions:

- ***Set point*** — Value on the controller scale at which the control indicator is set.
- ***Throttling range*** — Total amount of change in the controlled variable required for the controller to move the controlled device through its complete stroke, from one extreme to the other.
- ***Differential*** — Applied to two position control, difference between the high and low settings at which the controller operates.
- ***Primary element*** — Controller that first uses energy from the controlled medium to produce a condition that represents the value of the controlled variable (e.g., thermostat).
- ***Final control element*** — Mechanism which directly acts to change conditions of the controlled variable.

VALVES

Controls may be manual (mechanical or electrical) or automatic (pneumatic or electric). The usual controlling element is the control valve, which may be single seated (normally open), double seated (normally closed), or of the three-way mixing type.

A *single-seated valve* consists of one seat and one disc, Figure 12-1a. It works well when tight shutoff is required and is not recommended where high differential exists.

A *double-seated valve* consists of two seats and two discs, Figure 12-1b. It requires less power than a single-seated valve and is more expensive. It is beneficial when closing against high pressures; however, it is not recommended when a tight shutoff is required.

A three-way mixing valve is generally suited for mixing service only, Figure 12-1c.

Control of Water Flow[1]

Valve control, either single-seated or three-way, is the most common means of varying water flow to terminals. Single-seated valves vary the flow by throttling, while three-way valves control the flow by diverting water around the terminal. Both produce the same result within the terminal: reduction in

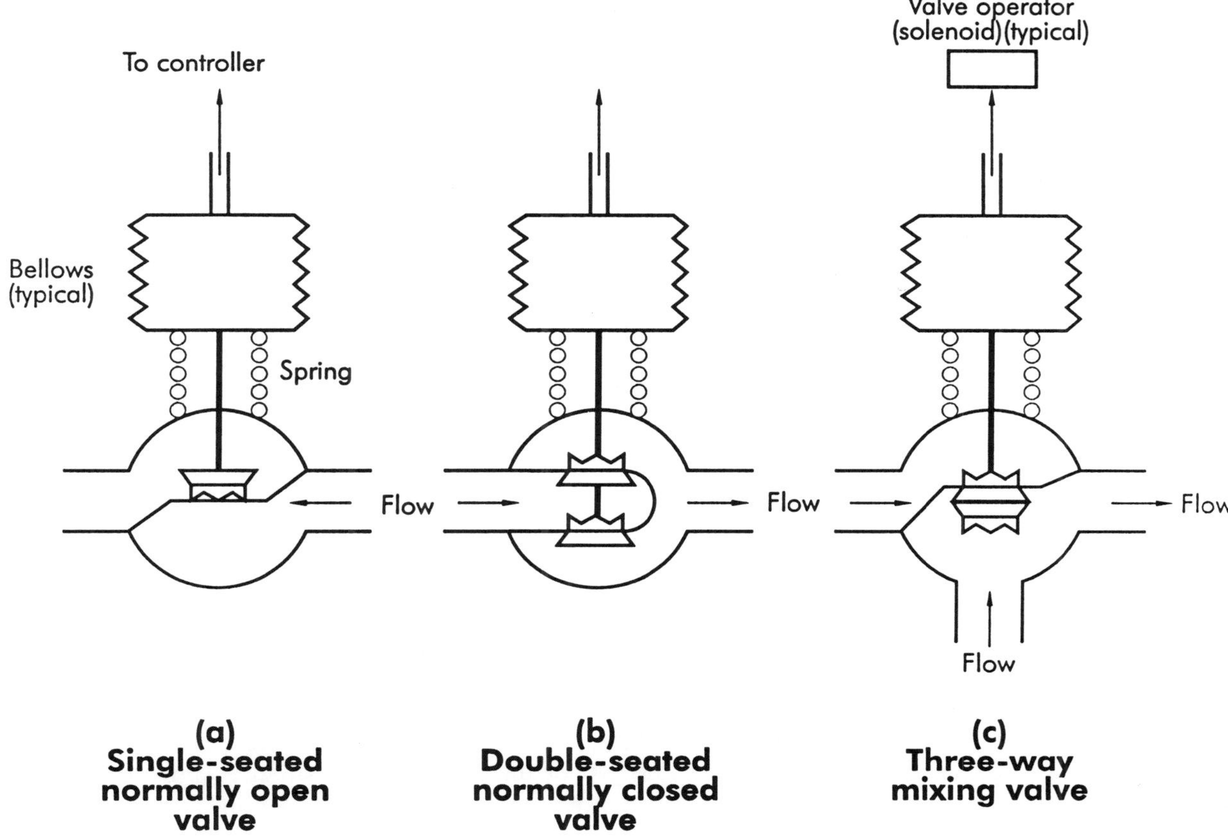

Figure 12-1. Control valves

capacity by reducing flow within the terminal. They differ, however, in their effect on system flow rates and pressures.

Control Valve Selection[2]

The proper selection and application of control valves in hydronic systems is essential to good performance. Such control valves must, at minimum, meet two requirements: 1) they must be specifically characterized for water service; and 2) they must be sized for a pressure drop at design flow rate, which is a significant portion of the maximum available head. Proportioning valves (diverting) must have equal percentage characteristics, and the valve pressure drop at design flow should be no less (and preferably more) than 5 psi (34.5 kPa), or 25% of the maximum pump head, whichever is greater. The higher the design water temperature, the more critical such requirements for valve selection become.

Effect of Flow Variations

Control valve performance may be adversely affected by wide variations in differential pressure. Sluggish circulation may reduce capacity due to air binding or deposit of solids in various parts of the system. Consequently, it is important to maintain flows in the system at partial loads that are comparable to design flow conditions. Since the capacity of hot water heat exchangers is nonlinear compared to the flow, little capacity reduction occurs until temperature drops in excess of 50% of the original temperature difference between the supply water and the heated medium are reached. The control valve, for optimum stability, must have the opposite characteristic (equal percentage).

Proper control performance requires that the valve provide an initial pressure drop high enough to ensure that control is not destroyed by the increased pressure differential due to flow reduction. Also, the valve must have a relatively high turn-down ration (ratio of maximum usable flow to minimum controllable flow) and a minimum leakage in its closed position. The valve must be designed to meet the temperature and pressure requirements of the system in which it is used.

Single-Seated vs Three-Way Valves

The choice between single-seated (two-way valve) and three-way valves must always be made in terms of the specific system in question. Single-seated valves have the advantage of lower cost and better flow-lift characteristics, and they can be obtained with equal-percentage-characteristic plugs. Single-seated valves are available in smaller port sizes and are, therefore, more readily selected for the proper pressure drops at small design flow rates. All these factors tend to improve the control performance of single-seated valves as far as the control of the terminal heat transfer element is concerned.

Three-way valves have the advantage of maintaining a more or less constant flow in the system. They minimize variations in developed pump pressure and system resistance with variations in load, while maintaining the flow through the pump. Single-seated valves with other provisions for maintaining flow, such as differential pressure controllers, are usually preferred, especially in MTW or HTW systems or in process applications where control is critical.

A two-way throttling valve reduces the flow rate and increases the pressure differentials across the pump and load circuits. Throttling requires more attention to valve and pump selection. Because of the more exacting design requirements of two-way valve control at terminal units, the three-way valve has been widely used until recently. Three-way valve control requires more energy because the flow rate is constant; thus, pump horsepower is constant, regardless of cooling load. With constant flow, there are many operating hours when the temperature difference across a chiller is 2° to 3°F relative to design chilled water ranges of 6° to 10°F. When it comes to selecting an actual valve, the designer or installer should discuss the details and operation characteristics with a specialty manufacturer's representative.

The three-way valves are designated as either mixing valves or diverting valves. The internal construction of the valve determines the designation. The mixing valve has two inlets and one outlet while the diverting valve has one inlet and two outlets. Either one, however, may perform one or the other function depending on its location. Figure 12-2 shows the three-way valve performance function depending on its location.

Controlling the water flow in a system can be done by either varying the flow or by varying the (media) water temperature. Keep in mind that the variations are not directly proportional. A usual method to obtain zone control for hot water is to vary the water temperature inversely to the outdoor temperature.

When space heating is the principal load, resetting the supply temperature with the outdoor temperature change is almost always desirable. Outdoor reset may provide all the necessary control in small systems or may be used in combination with zone or individual room control in larger systems. With hot water boilers, outdoor reset can be accomplished by varying the boiler water temperature or by varying the flow from the boiler to the system. Variable water flow may be accomplished with a three-way valve, Figure 12-3, or with a primary-secondary pumping arrangement, Figure 12-4. This may be

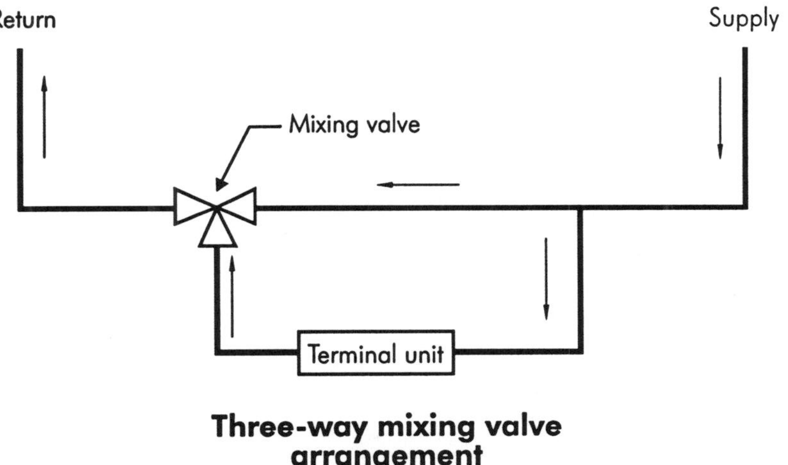

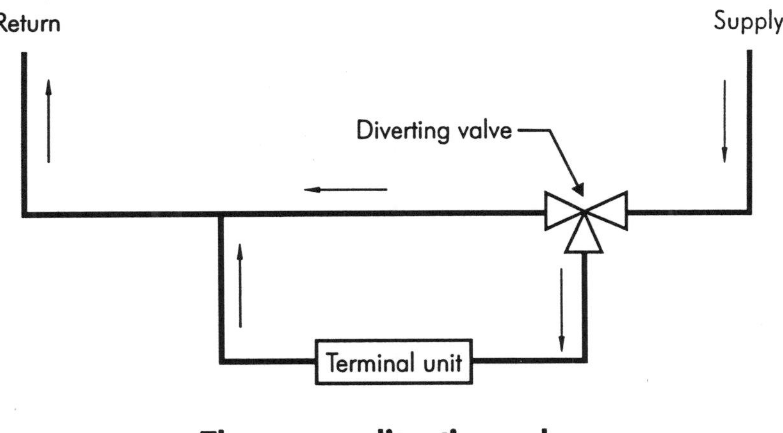

Figure 12-2. Three-way valve arrangements

accomplished with a submaster thermostat in the water supply line working in conjunction with an outdoor master thermostat.

For chiller operation, the schematic in Figure 12-5 shows the solution to the variable water flow to terminal units and constant flow through the chiller while operating the chiller plant efficiently. Although only two chillers are shown in Figure 12-5, any number of chillers can be used. At full load, both chillers are on-line and full flow goes to the terminal units. As terminal unit valves modulate from decreased load, flow decreases and the pressure drop from supply to return mains increases. The pressure differential controller senses this change and partially opens the bypass valve to compensate. The bypass valve is sized to match the flow through one chiller, enabling one chiller and pump to shut down when the bypass is fully open. As load increases to a point where the bypass is closed completely, the second chiller and pump can restart.

Example 12-1. Determine the flow in gpm required for a one-pipe series loop system. The heat loss calculation was already prepared, and the result was 36,000 Btuh. The water supply is 200°F, and the water return is 180°F.

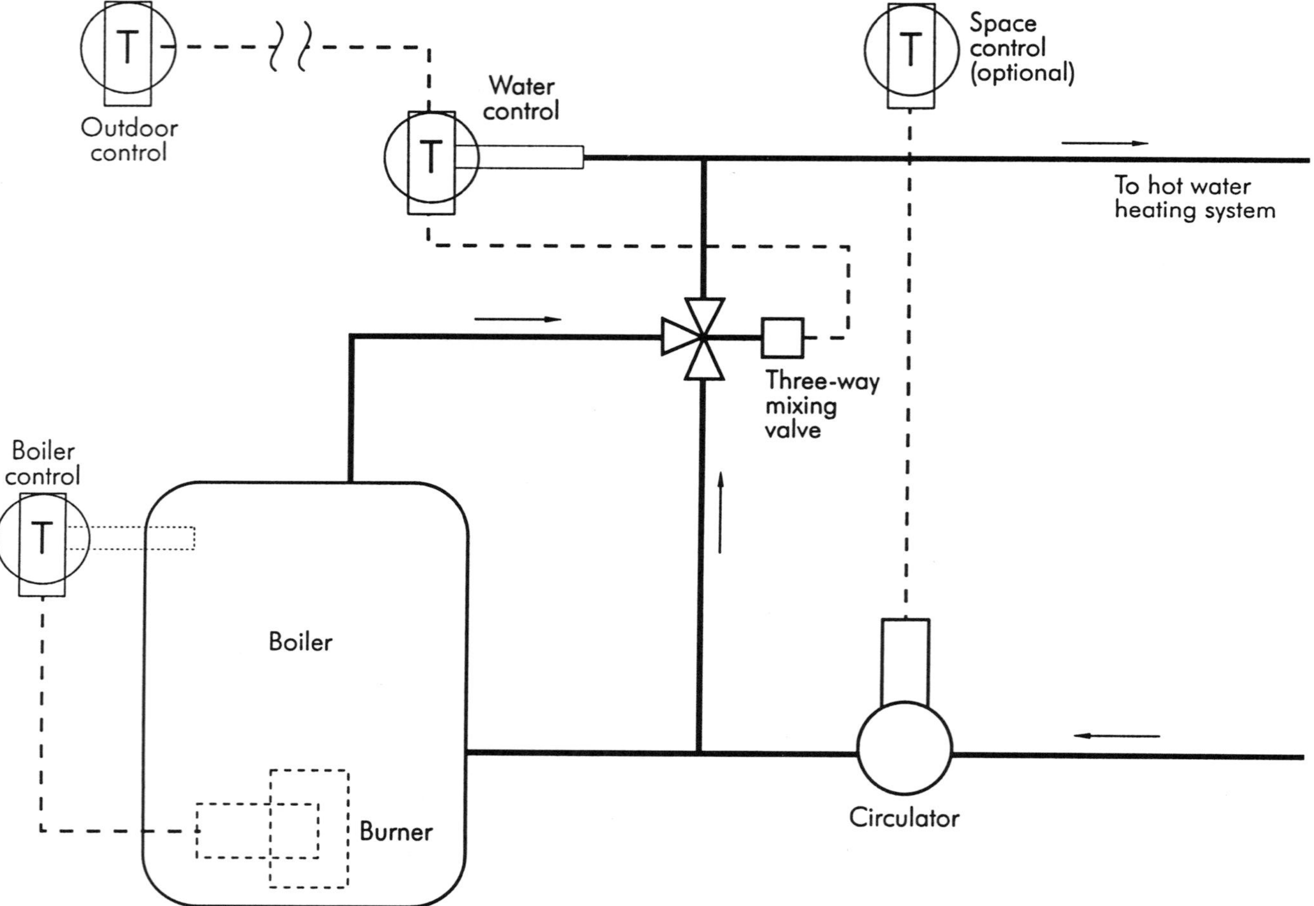

Figure 12-3. Control of hot water heating (Reprinted by permission of the American Society of Heating, Refrigerating and Air-Conditioning Engineers, Atlanta, Georgia, from the 1980 ASHRAE *Handbook--Systems*)

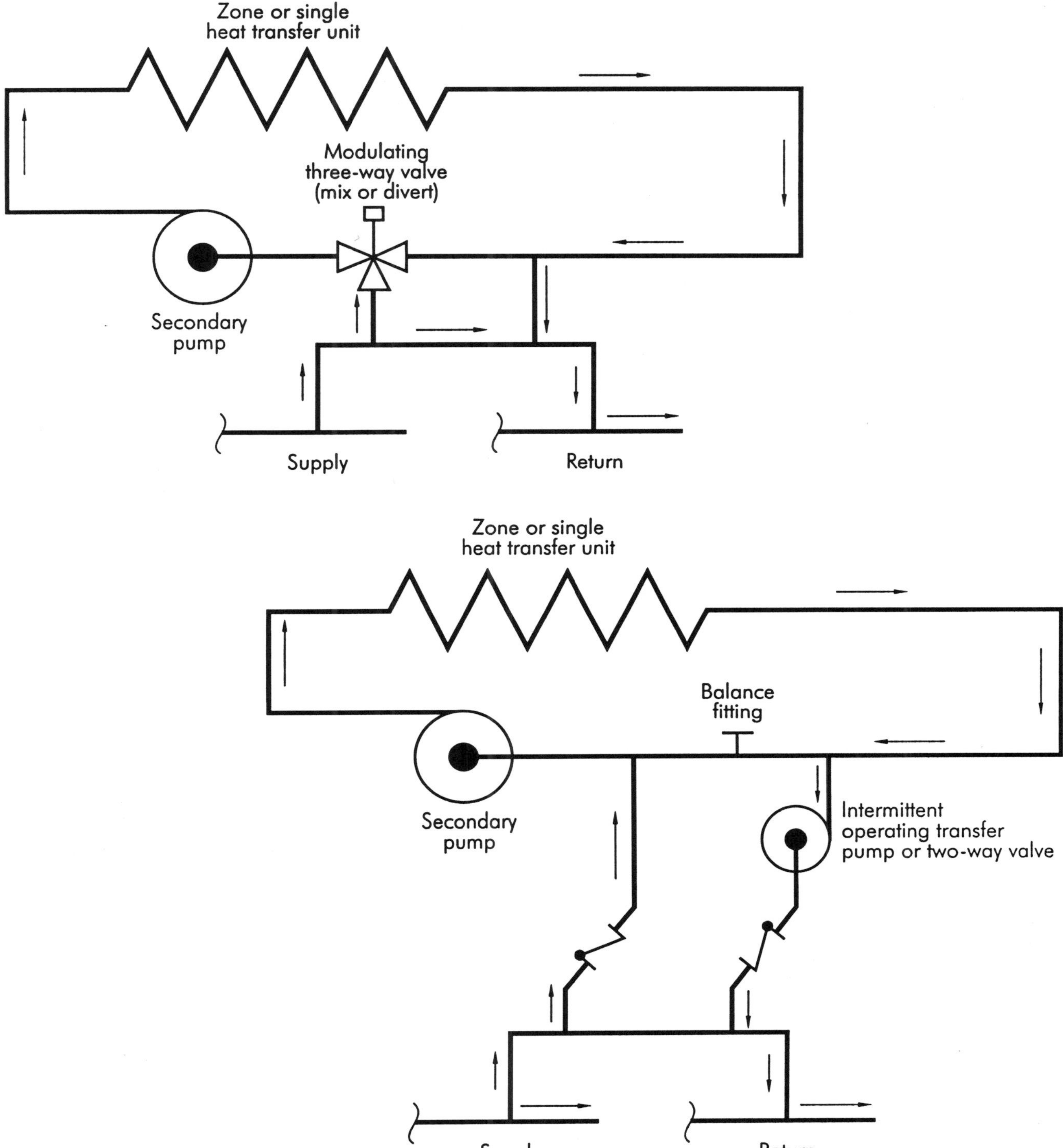

Figure 12-4. Control of water temperature (Reprinted by permission of the American Society of Heating, Refrigerating and Air-Conditioning Engineers, Atlanta, Georgia, from the 1980 ASHRAE *Handbook--Systems*)

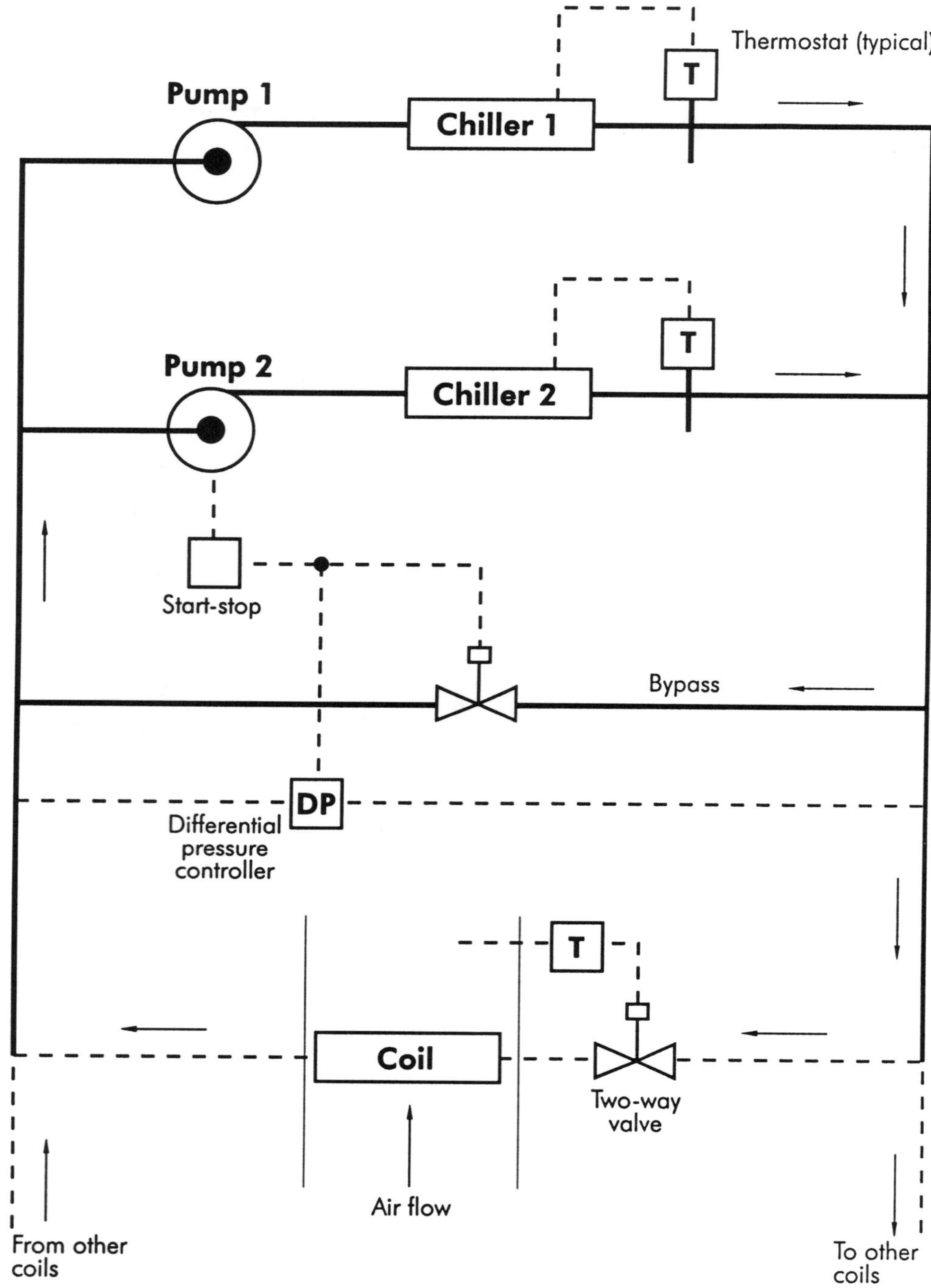

Figure 12-5. Two-way valve with pump bypass (Reprinted by permission of the American Society of Heating, Refrigerating and Air-Conditioning Engineers, Atlanta, Georgia, from the 1991 ASHRAE *Handbook--HVAC Applications*)

Solution 12-1. The gpm may be calculated as follows:

$$\text{gpm} = \frac{\text{Btu}}{(8.33 \text{ lb/gal} \times 60 \text{ min/hr})(\text{TD})} = \frac{36{,}000 \text{ Btu}}{(500)(20°\text{F})} = 3.6 \text{ gpm}$$

This flow is constant so the actual temperature entering at the next user can be calculated.

Zone Control

A larger building with various occupancies and long lengths of pipes from boiler to last user requires zoning and zone controls. Zone control helps to deliver the amount of heat or cooling required when there are considerable amounts of glass (single or double glazing windows) or other specific requirements.

A zone is an area that requires certain environmental conditions that might be different from those of an adjacent zone. This could include a room, a group of rooms, or even an entire floor. The hydronic system can be calculated, sized, and installed to deliver the temperature of water required, the amount required, and at the time it is needed for each particular area. Either control valves or small pumps may be used to control zones for heating and cooling. Figures 12-6 and 12-7 show zone control with valves and pumps.

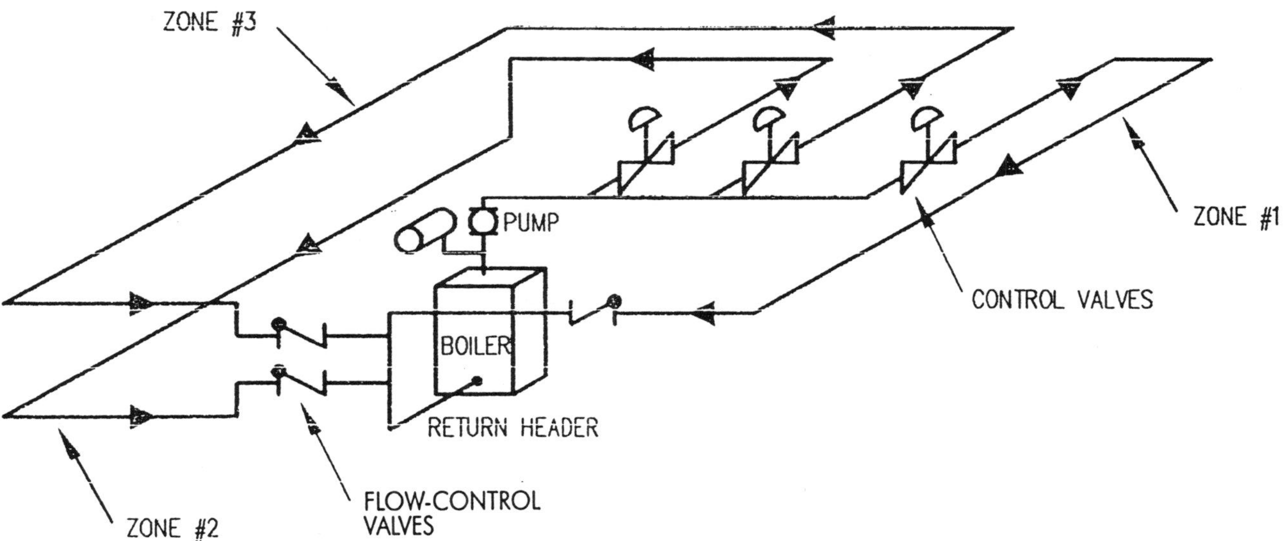

Figure 12-6. Zoning with two-way valves (Courtesy, ITT Fluid Handling)

The system using valves may use two-way or three-way valves. Valves must be of good quality in order to close tightly. A system using pumps for zone control presents more flexibility, because different zones may have different conditions, such as friction loss, capacity, etc. Each zone or area that requires independent control has its own secondary pump or valve, as well as its own thermostat conveniently located within the area.

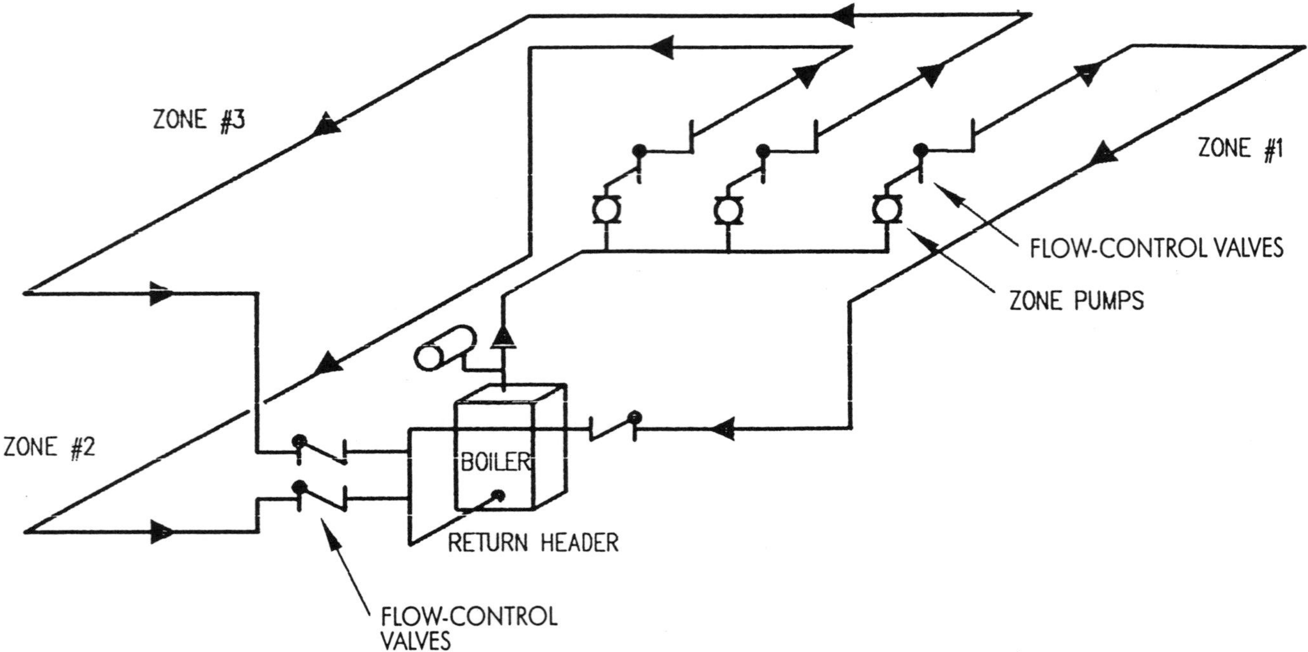

Figure 12-7. Zoning with pumps (Courtesy, ITT Fluid Handling)

If there is one primary pump and only a control (on-off) valve, the thermostat controls the valve. The valve position is either normally open (NO) or normally closed (NC). "Normally" in this case is the valve position when no signal from a thermostat is sent to the valve. Figures 12-8, 12-9, and 12-10 illustrate different valve configurations.

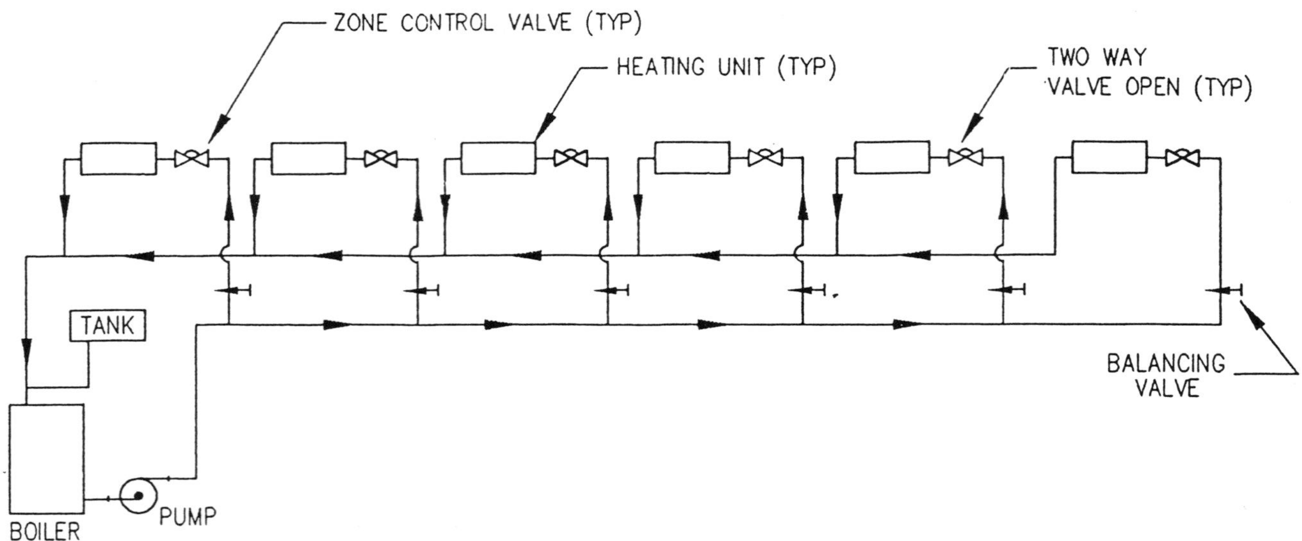

Figure 12-8. System with two-way valves (Courtesy, ITT Fluid Handling)

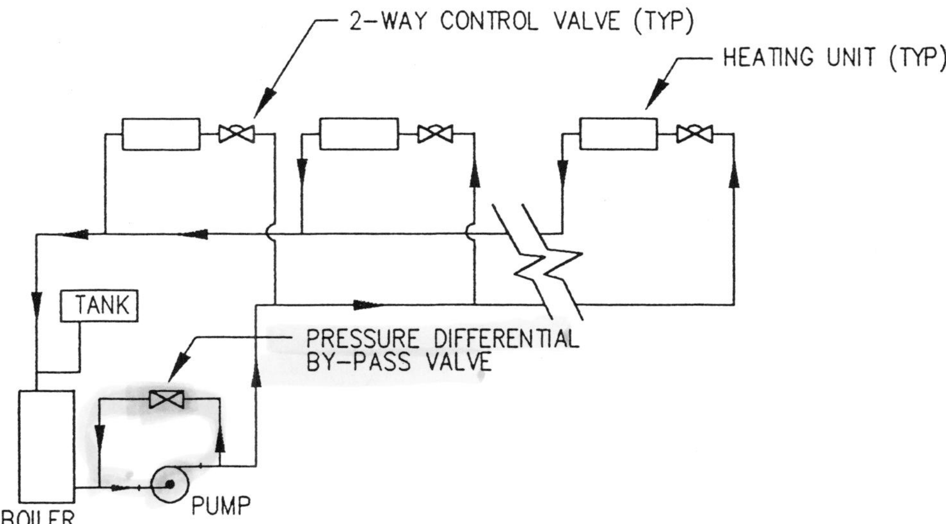

Figure 12-9. System using two-way valves, bypass, and pump (Courtesy, ITT Fluid Handling)

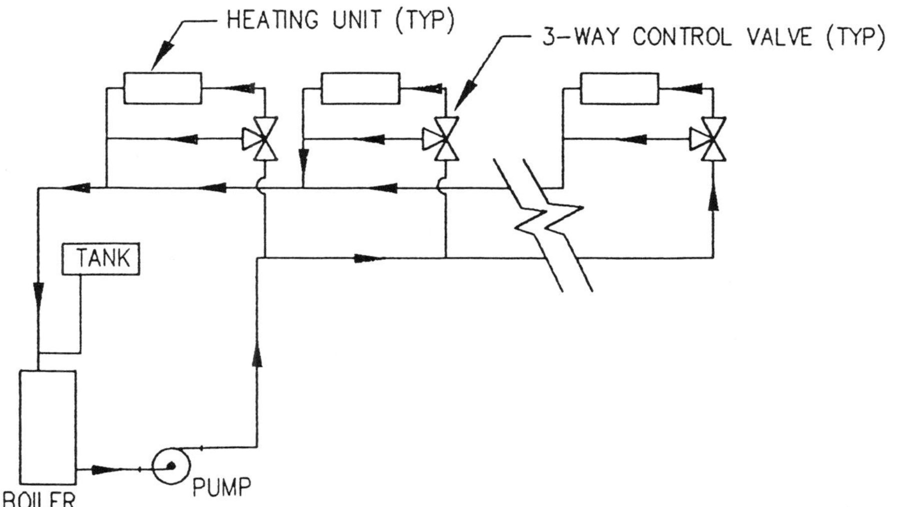

Figure 12-10. Constant flow system using three-way bypass valves (Courtesy, ITT Fluid Handling)

Figures 12-8 and 12-9 are slightly less expensive systems but are no longer recommended due to their associated noise, large pressure drop, etc. In Figure 12-10, the water has a constant flow and eliminates some of the disadvantages mentioned in the other two figures. Figure 12-11 shows a simple system with a primary and secondary pump, which is applicable to a single-pipe system as well as a two-pipe system. Figure 12-12 shows a two-pipe direct return circuit. Figures 12-13 and 12-14 show a common pipe system used for summer cooling and winter heating, respectively.

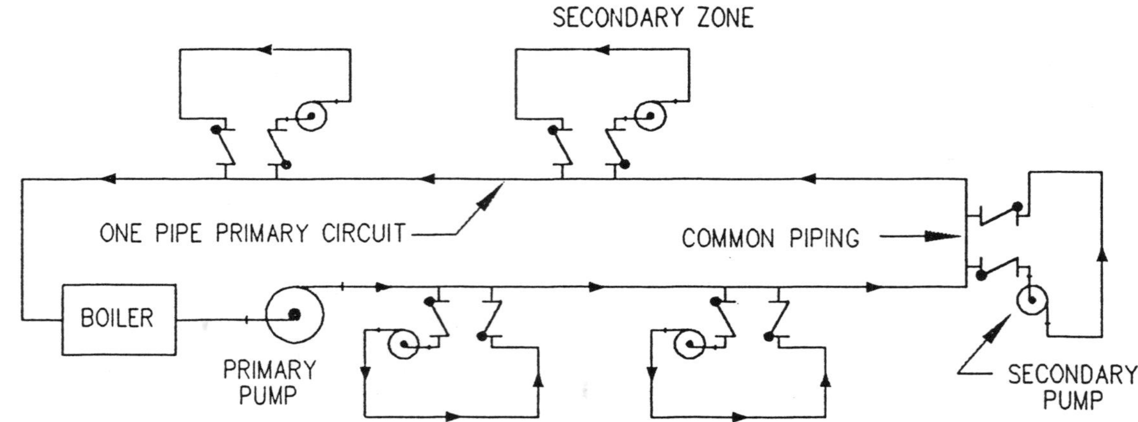

Figure 12-11. One-pipe primary circuit (Courtesy, ITT Fluid Handling)

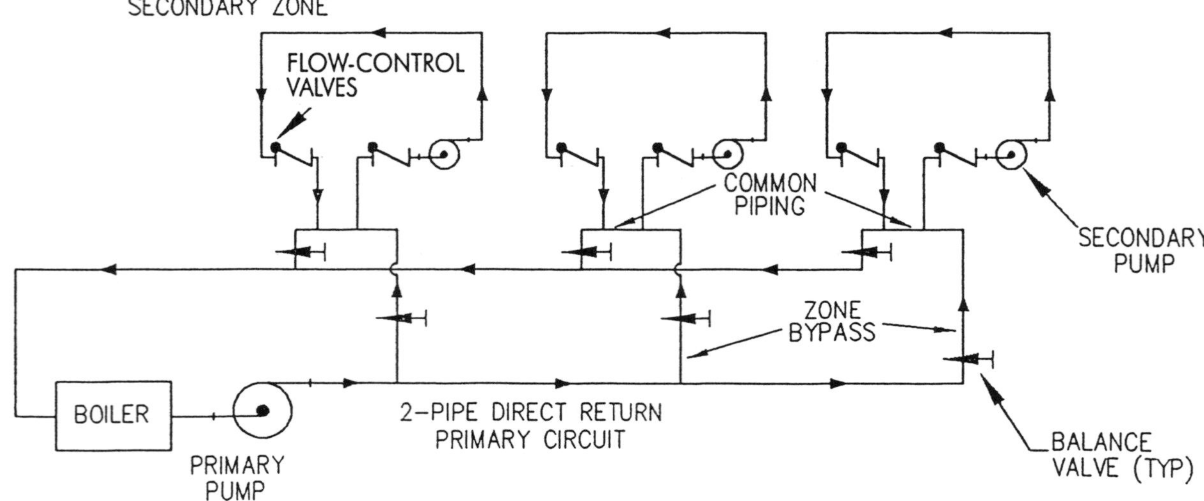

Figure 12-12. Two-pipe direct return circuitry (Courtesy, ITT Fluid Handling)

Zone temperature control is frequently used in space heating or cooling applications with zoning requirements established by exposure, occupancy, or other factors. Individual zone control may be achieved by intermittent circulator operation or through single-seated or three-way valves, often combined with outdoor reset control. In large buildings, continuous water circulation is usually needed to minimize the hazard of freezing. In such systems, control should be accomplished through secondary pumping, not by intermittent operation. *Zone control is mostly applicable to convection heating terminals.*

Specialty equipment and system manufacturer representatives may help select the best system for a certain installation. The following are general recommendations for hydronic control systems:

- For large systems, use commercial or industrial grade control equipment when possible.

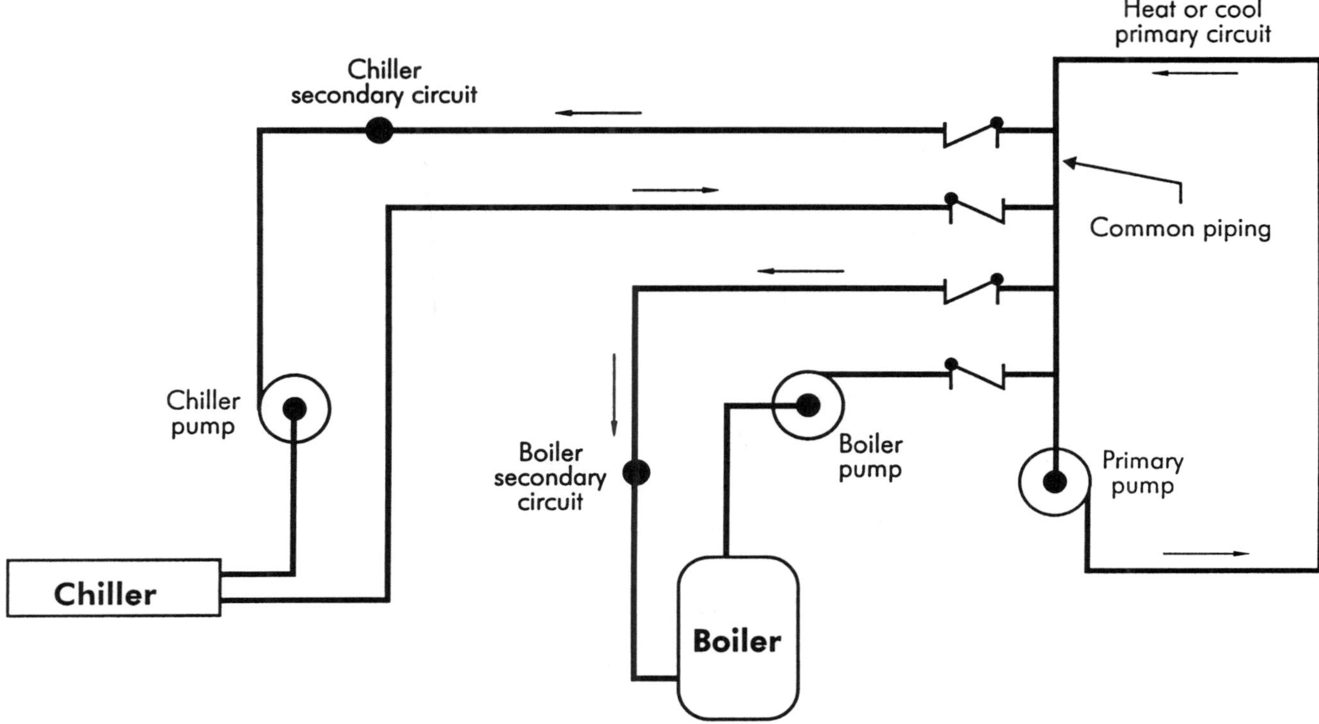

Figure 12-13. Common pipe system used for summer cooling

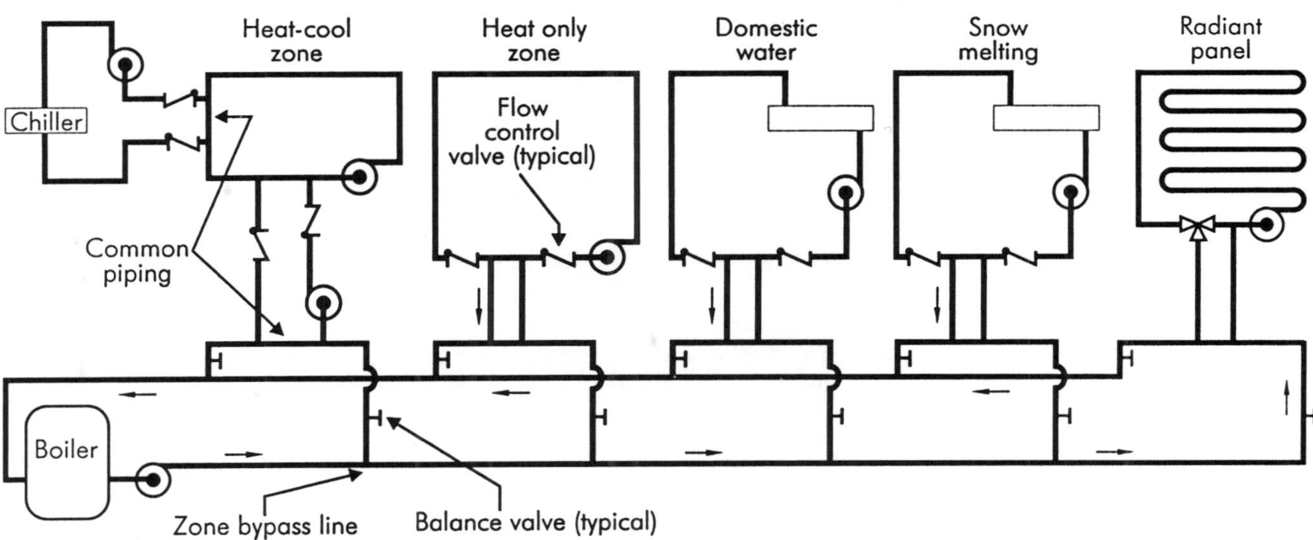

Figure 12-14. Common pipe system used for winter heating

- Select control valves having a range to adapt to the minimum expected flow.
- Do not oversize control valves. Applying 10% of system resistance is satisfactory.
- Do not use on-off controls except for very small pipe sizes, as water hammer may occur.

- When tight control is important, do not use three-way valves either for output control in constant-volume-flow systems or for mixing control in any variable-volume-flow systems, since their characteristics are of linear type. Provide equal-percentage valves instead.
- Do not use three-way valves for throttling when pressure under one port is higher than under the other.
- Do not use three-way mixing valves for flow diversion or diverting valves for mixing since the valves tend to slam at reduced flows.
- When specifying control valves, include operating characteristics, such as maximum and minimum flow rates, internal pressure, pressure difference at design load, and pressure difference at minimum load equal to pump head or controlled differential pressure in variable-volume-flow systems. For valve size selection consult with the valve manufacturer.
- Provide a simple control system. To add extra controls in order to correct conceptual errors is not good practice. The result will be a cumbersome system. Control systems are broken down into simple, well-defined circuits and subcircuits. Analyze the basic function of each and their interrelation.
- Install a fine mesh strainer for protection ahead of each control valve.

SYSTEM PRESSURE CONTROL[3]

The pressure in a hydronic system must be controlled within certain maximum and minimum limits. This is a subject that often is not understood correctly, leading to operating difficulties and seldom to equipment damage.

The maximum allowable pressures are usually based on the permissible equipment pressures. In a low temperature hydronic heating system, for example, the boiler relief valve is usually set at 30 psig, which would therefore be the maximum allowable pressure at the boiler.

The minimum pressure requirement is based on two factors:

1. The pressure at any location must not be lower than the saturation pressure of the water. If this happens, the water will boil and the steam will cause operating problems. This may occur particularly at the pump suction.
2. The pressure at any location should not be lower than atmospheric pressure. If this happens, air may enter the system.

Control of maximum and minimum pressures to ensure that none of these problems occur is achieved by proper sizing and location of the compression tank and by correctly pressurizing the system when filling. To know how to accomplish this we must understand how the compression tank functions.

The compression tank acts similarly to a spring or an air cushion. The water in the tank will be at the same pressure as the gas in the tank. The value of this pressure will depend on how much the gas in the tank is compressed. Once the system is filled with water and the water is heated to operating temperature, the total volume of water in the system remains constant. The volume of gas in the tank therefore also remains constant, and its pressure does not change. This holds true regardless of where the tank is located and whether or not the pump is operating. *The point at which the compression tank is connected to the system is the point of no pressure change.*

By assuming two different tank locations and utilizing this principle, we can see what effect the tank's location has on controlling system pressure. Consider first the compression tank located at the discharge side of the pump for the system shown in Figure 12-15. All of the piping is on the same level. Assume that the pressure throughout the system initially is 10 psig (25 psia) without the pump running (Figure 12-15a). Now consider what happens to the pressures when the pump runs (Figure 12-15b). Assume the pump has a head of 20 psi. The pressure at the tank must be the point of no pressure change, so when the pump runs the pressure at this point is still 25 psia. The pressure at the pump suction must therefore be 20 psi less than this value, or 5 psia, because the pump adds 20 psi. But 5 psia is equal to -10 psig, which is far below atmospheric pressure. Air would undoubtedly leak into the system at the pump suction. Cavitation in the pump might also occur in a heating system, because the boiling point of water at 5 psia is only 160°F.

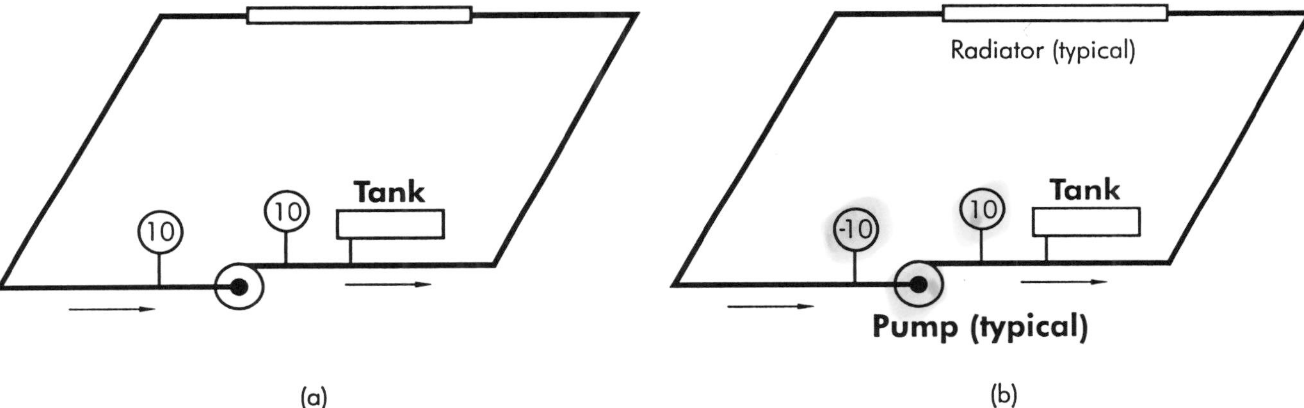

Figure 12-15. Compression tank location

Let us see what happens if the tank is located at the suction side of the pump. The initial pressure is at 25 psia throughout the system, as before. When the pump runs, the pressure at the tank location remains at 25 psia. As the pump adds 20 psi, the pressure at the pump discharge must be 45 psia (30 psig). The pressure throughout the system is well above atmospheric.

This example shows that the compression tank should be connected to the system at the pump suction, not the pump discharge. If the pump head is very low, the pressure at the pump suction might not fall below atmospheric, even if the tank is located at the pump discharge (this arrangement is not advisable).

To keep the pressure exerted on the boiler as low as possible, the pump should be located to discharge away from the boiler, so that the boiler is not subject to the pump discharge pressure. In a high-rise building, the static head at the boiler (if it is in the basement) might be above the maximum pressure. In this case the boiler could be located in a penthouse, or a steam boiler and hot water heat exchanger might be used.

NOTES

[1] This section reprinted with permission by the American Society of Heating, Refrigerating and Air-Conditioning Engineers, Atlanta, Georgia, from the 1991 ASHRAE *Handbook --HVAC Applications.*

[2] Ibid

[3] Reprinted with permission from *Air Conditioning Principles and Systems, Second Edition*, E.G. Pita, Wiley & Sons, 1981.

Chapter 13

Common Hydronic Problems and Solutions

Heat Loss

In order to install a system that will use the right amount of hot water heat in any room, a careful calculation of the heat loss of each room must be performed. A quick estimate of heat loss should only be used as a general guide and for a ballpark figure. To ensure all the rooms in any one zone are heated evenly, make sure the temperature drop from the beginning of the heating system to the end (return to boiler) is not excessive. A rule of thumb for residential applications is a maximum baseboard length per zone of 50 feet.

Baseboard heaters are normally installed on outside walls and under any windows (to counter-balance cold air infiltration). A baseboard heater may not be used in an area that does not have the required amount of wall space, such as certain kitchens or baths. In such cases, floor vectors (in-floor, sunk-type, fin-tube heaters) may be used in front of cabinets, sliding glass doors, or large windows in general areas that do not permit the installation of baseboard heaters. An alternative is a radiant panel installed in the ceiling. Any installation of this type where the space above a heated space is not heated may cause the coil within the panel to freeze.

The hot water boiler in an installation must compensate for the heat losses, infiltration, ventilation, etc. The boiler must also be installed to allow for maintenance, combustion air, etc. The boiler must be the right size, which means it must either be slightly larger or have some safety capacity added (safety factor). The ideally sized boiler would run continuously on the coldest day of the year. Multiple boiler installations should be investigated, as this arrangement can save fuel.

Manufacturers furnish specific instructions for the installation of boilers. The following points must also be considered:

- Allow ample size openings and passages into the boiler room for the boiler. The architect must be informed of these dimensions. If an existing building requires a new boiler but does not have adequate openings for a tubular boiler, a sectional cast iron boiler may be the solution.
- Provide ample space on all sides of the boiler for maintenance, and allow adequate distance in front of the boiler for the tube cleaning and removal.
- Locate the fuel close to the boiler. Install the breaching to the flue without offsets.
- Provide sufficient openings to the outdoors for both combustion air and ventilation air. Fixed grilles in walls or doors are one method. This is extremely important, because if the openings are not adequate, the boiler may be starved of sufficient air for combustion, resulting in the production of toxic carbon monoxide.

LEAKING SAFETY VALVE

The pressure safety valve on a residential boiler is normally set at 30 psi (the rule is 10 to 15 psi above working pressure). If this valve should start leaking, it may be due to one of the following reasons:

- ***Faulty safety valve*** — The valve is worn and starting to relieve at a lower pressure. Replace the valve.
- ***Expansion tank full*** — Every hot water system must allow for the expansion of heated water without causing excessive pressure in the system components. The expansion tank performs this function and must be correctly sized according to system capacity. If the expansion tank becomes full and there is no more room for water expansion, the pressure in the system will rise and cause the safety valve to relieve. Stop the operation and empty 50% of the water in the tank. Restart the system.
- ***Faulty pressure reducing valve*** — Each hot water heating system has a control valve called the pressure reducing valve (PRV), which is on the water supply to the system. This is sometimes referred to as the automatic feed for the system. The typical PRV on a residential heating system allows water into the system only at a pressure equal to about 12 to 15 psi. If the PRV should fail and allow greater pressures into the system, the safety relief valve will start relieving this pressure at its set pressure (30 psi). Replace the faulty valve.
- ***Tankless-type heater*** — Some hot water heating systems heat the domestic water by means of a tankless-type heater. The tankless heater is a copper pipe coil with its ends, supply and return, entering and leaving the boiler. The incoming cold water that travels through this pipe is heated within the boiler and exits as hot water for domestic use. The domestic water does not come into contact with the boiler water; it

remains inside its system. If this coiled pipe develops a leak, the domestic water will begin to fill the heating system, which in turn, will develop a higher pressure, causing the safety valve to leak. Repair or replace the domestic water coil within the boiler.

No Heat or Excessive Heat

If the boiler is working, a no heat call for a particular zone in a heating system can be narrowed down to one of the following reasons:

- Faulty thermostat
- Air pockets in the heating pipes
- Faulty zone valve or zone pump (not working)
- Other, such as, valve shut, frozen pipe, electrical problem

An excessive heat service call seems unlikely, but it is actually fairly common in certain apartment buildings. The cause is usually one of the following:

- Open flow control valve. If the flow control valve is opened manually or struck open for any reason, the zone may heat by gravity circulation.
- Faulty thermostat
- Faulty zone valve or zone pump

Other Problems

The problems that follow are taken from letters received by columnist Dan Holohan published in *Plumbing and Mechanical* magazine.[1]

Problem: I installed a heating system in a nursing home about a year ago that has stumped me as to why so many problems have arisen. It's a forced hot water system, which is usually easier to diagnose than a steam system.

Originally, there was an old oil-fired steel boiler with an input of 1500 MBtu located in the basement. On the ground floor where the patients sleep are convectors throughout the building, 36 to be exact. This building is one-story high and fairly insulated. This old system ran on a 2" steel supply header with 1/2" branches to and from each convector, and an AB&S monoflow tee at the end of each convector and the return side.

Here I come along and rip out the entire system (except convectors) and install two oil-fired Weil-McLain #488 hot water boilers. In addition, I have one boiler feeding the west side of the building and the other is feeding the east. These two boilers combined exceed the original boiler input by 120 MBtu. I have always liked the idea of oversizing my boilers slightly. So I then ran a 2" copper header from each boiler to their respective zones, hitting each convector with 1/2" copper branches (very short runs) and the return side having a venturi tee for each unit. Both boilers are a pretty typical setup with flow controls, B & G circulator pumps, expansion tanks, Airtrol™ fittings, backflow preventer, and dual control valve.

Now here is the problem - they constantly complain the heat is not enough. The system operates on 200°F high limit and 180°F low limit at 10 to 12 lb. There are separate thermostats for west and east boilers.

Also, air is always getting into the system, and I lose the water in the expansion tanks, and sometimes the air, too. When I go to recharge the tanks, it's under a vacuum!

I realize that with elderly people, a room at 65°F might feel like it's -10°F, while you and I feel fine. But how can this system accumulate so much air and of course not give enough heat? Even when fully purged there seems to be an absence of the hot element. Figure 13-1 shows the diagram of the system

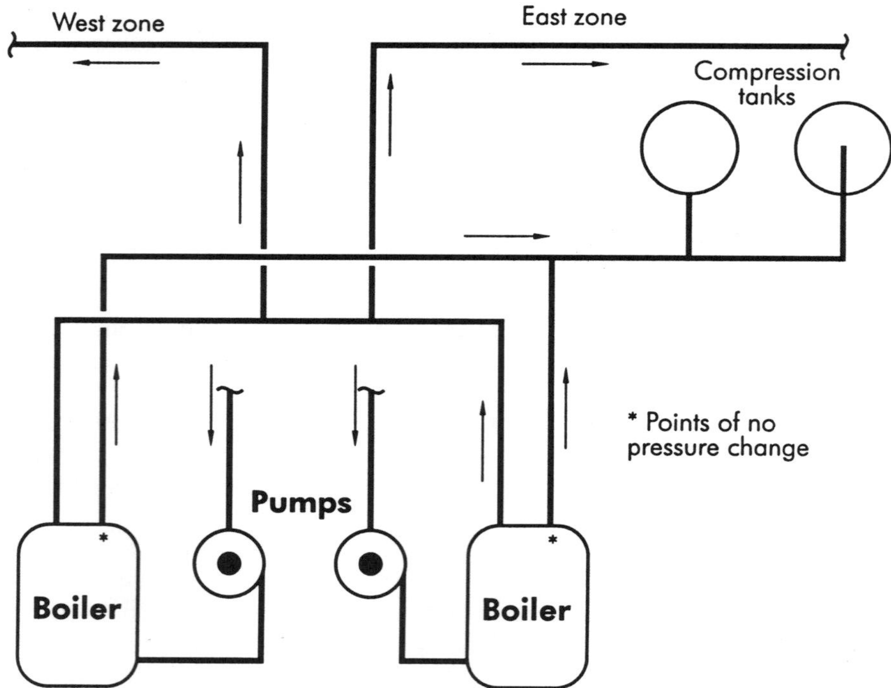

Figure 13-1. Existing system

Dan's Solution: I think you definitely have an air problem here. It's tough enough getting the air out of a venturi tee system, but things get much worse when the circulator is on the return side of the system. That's because the point where the compression is connected is the "point of no pressure change." This is the one place in the system that can't be affected by the pressure differential created by the circulator. When you pump away from the tank, the pump's pressure is added to the system fill pressure. But when you pump toward the tank, the pump's pressure is subtracted from the fill pressure. That drop in pressure causes the trapped air bubbles in the convectors. When you move the pumps, just the opposite happens - the bubbles get smaller and are more easily moved.

Adding to the problem on this job is the way you have the two steel tanks connected into the system. You can use as many tanks as you'd like, but they can only be connected to one point. That connection becomes the "point of no pressure change." The pumps use that point to "know" what to do with their pressure differential. Should they add it to the fill pressure or subtract it from the fill pressure? It depends on where that "point" is.

As things are now, your tanks are connected to two places, giving you two "points of no pressure change." When that happens, the pumps will set up a circulation through the tanks. Before long, all the air gets moved out of the tanks and winds up in the system. The new built-in air separators the boiler manufacturers give us are nice, but if you use them this way, you'll have a problem every time.

My suggestion is that you re-pipe the job. Put an in-line air separator in the main trunk line coming from the two boilers. Connect the two compression tanks into the air separator through a manifold. Put the pumps on the supply side, pumping away from the separator, Figure 13-2.

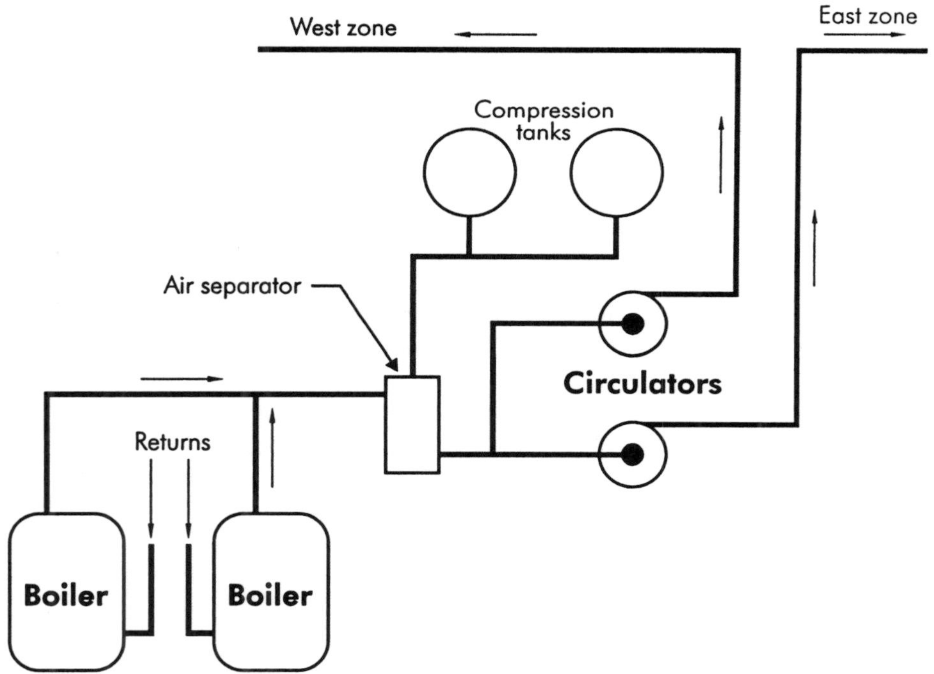

Figure 13-2. Dan's solution

Problem: I want to tell you about a problem I solved on a Monoflow system. This house had one convector that would not heat, and the rest of the loop didn't heat very well either. After taking the circulator apart to see if the impeller was okay and the balancing valves apart, I decided the convector must be clogged.

When I took the convector apart, I found someone had put a metal disc in the union with a 1/8" hole in it. This very effectively stopped the flow, which of course affected the rest of the loop. Also, an old-timer who worked for me many years ago said "you don't need Monoflow tees, just reduce the pipe one size between the two tees for each convector." He put in a job for me that way. It worked fine; however, all the convectors were above the main.

Dan's Solution: Your story is a good one. I've heard of guys using orifice plates in gravity hot water jobs to slow water to the top floor (that's the path of least resistance). In fact, these things drove me crazy for a time before I knew about them. You see, when you add a circulator to a gravity hot water job, the path of least resistance shifts from the upper floors to the lower floors - and it's helped along beautifully by the orifices in the top-floor radiators. It's enough to make you cry; looks just like an air problem.

I've never seen orifice plates used on a Monoflow system, though. Sounds like perhaps the installer wanted to use them in the run of the tee instead of the radiator. That would have increased the resistance to flow along the main (probably by *way* too much) and sent more water through the radiator.

You're right about reducing the pipe size between the tees. This stuff is all about resistance to flow and water taking the path of least resistance. Monoflow tees just make it a bit more predictable. Reminds me of an old-timer I met once. He didn't believe in Monoflow tees or reducing the pipe size. He just grabbed hold of that pipe between the tees with a channel lock and gave it a squeeze. "How much do you squeeze?" I asked. "It depends, kid, it depends." Nothing like experience, eh?

Problem: I have read many articles relating to the placement of the feed water connection, compression tank, and circulator on a hot water heating system. We have been called to troubleshoot several large systems (10 and 12 stories) that became air bound when the system operated. The fix usually involved relocating the compression tank and the feed water connection on the heating system.

So now the question: If we all agree that the proper pipe arrangement would be to pump out of the boiler and away from the compression tank, with the feed water connected to the famous "point of no pressure change," why are boiler manufacturers piping the circulating pump on the return to the boiler?

Dan's Solution: You're asking a very valid and timely question. As far as I can tell, here's why:

Years ago, circulator manufacturers had to be concerned about the temperatures hitting their mechanical seals. They didn't have the ceramics we have today so their concern was justified. They got into the habit of putting the circulator on the return side of the system because that's where the water was coolest. They did this on both large and small installations.

Many of those larger jobs that used high-head pumps were prone to air problems. No one was quite sure what was causing these problems until Gil Carlson, an engineer working for Bell & Gossett, realized that there was this invisible "point of no pressure change" in every closed hydronic system. That's the place where the compression tank is located. He wrote a paper that changed the way heating engineers designed large jobs. Gil Carlson proved that a high-head circulator, when piped on the return side of the system and pumping toward the compression tank, can actually suck air in through the air vents and valve stem packing.

Unfortunately, the idea didn't immediately catch on with the guys doing the smaller jobs. Probably for two reasons: 1) we traditionally used low-head pumps for small jobs (we don't anymore!); and 2) non-engineers couldn't understand Gil Carlson's technical paper on the "point of no pressure change." So back to the boiler manufacturers. They kept doing what they were doing - putting the circulators on the return, mostly because of habit. I frequently get boiler manufacturers and their reps showing up at my hydronics seminars. I always ask them why they do it this way. Here are their reasons:

- Packaging
- They stack packaged boilers in their warehouses.
- If they moved the circulator to the top of the boiler, they'd have to build a stronger crate.
- If the crate was higher (because of the circulator), they couldn't stack as many boilers on the trailer.
- Everyone's happy with the way things are now.
- There's no financial incentive to changing the manufacturing process. Why? Because when the circulator pumps are away from the compression tank, the system will start up faster, you won't have to bleed the radiators and your customers won't hear the water moving through the radiators. It's to *your* advantage to change, not the boiler manufacturers'.

Today, there's no real reason for having the circulator on the return. The only reason circulators are down there is because of habit and blessed tradition, which drive so much of what we do. But don't look to the boiler manufacturers to change things; there's nothing in it for them. The guy who gains by moving the circulator to the supply side, pumping away from the compression tank, is the contractor. It's like one guy said to me recently, "Dan, I haven't bled a radiator in three years! That's when I moved the circulator to the supply side."

Problem: An orifice plate installed on the top floor radiators drives water to the lower floors. What happens when you add a circulator to the system? Remember, forced circulation is the mirror image of gravity circulation. When you add a circulator, the path of least resistance naturally shifts to the first floor because that's the shortest run back to the boiler.

The water doesn't want to go to the top floor any more. And you know what's going to aggravate this bad situation? Those orifice plates in the top-floor radiators. Now they're going to *make sure* the resistance through the top-floor radiators is always greater than it is through the lower floors. In fact, you'll probably have no flow at all through the top-floor radiators. That's a no-heat call, for sure! When you get there, it's going to look *exactly* like an air problem. It is not; it's a flow problem.

Dan's Solution: What's the answer? Yank the orifice plates out of the top-floor radiators and stick them into the first-floor radiators. The system will go into balance and that phantom "air" problem will be just a bad memory.

In the same vein, if you have a sudden no-heat call on the top floor of a gravity hot water system, check to see if the radiators or the walls behind the radiators have recently been painted. Painters often close the hand valve and disconnect the radiators to make their job easier. When they do, the orifice plate usually falls out of the hand-valve union. Since the average painter doesn't know about heating (gravity or otherwise), he doesn't know what to do with the orifice plate. He'll throw it in the garbage and you know what? It will look *exactly* like an air problem!

Check for fresh paint and new wallpaper. Nothing happens by itself. There's *always* a reason. So, when you bleed a radiator and don't get air, guess what? It isn't an air problem! Stop bleeding.

As for piping, the old boiler probably has two outlets and two inlets because the idea in the old days was to get the greatest possible gravity induced flow of water across the boiler. But when you add the new circulator, you won't need to use such big pipes coming and going out of the boiler. In fact, it's a good idea to reduce the size of the near-boiler piping to give the circulator something to "push" against. Otherwise, the circulator will move so much water, it will probably kick itself off on its internal overload protector.

Here's a good rule of thumb for you on the near-boiler piping size. Take the largest pipe, divide its size in half and go down one size. That becomes the size of your new near-boiler piping. For instance, let's say your largest pipe size is 2". Near your new boiler it should be 3/4". It will look odd, and it might make you feel uncomfortable, but it will work.

NOTE

[1] Published by *Plumbing and Mechanical* magazine, 3150 River Rd., #101, Des Plaines, IL 60018, December 1991, 1992.

Appendix A

SYMBOLS AND LEGEND

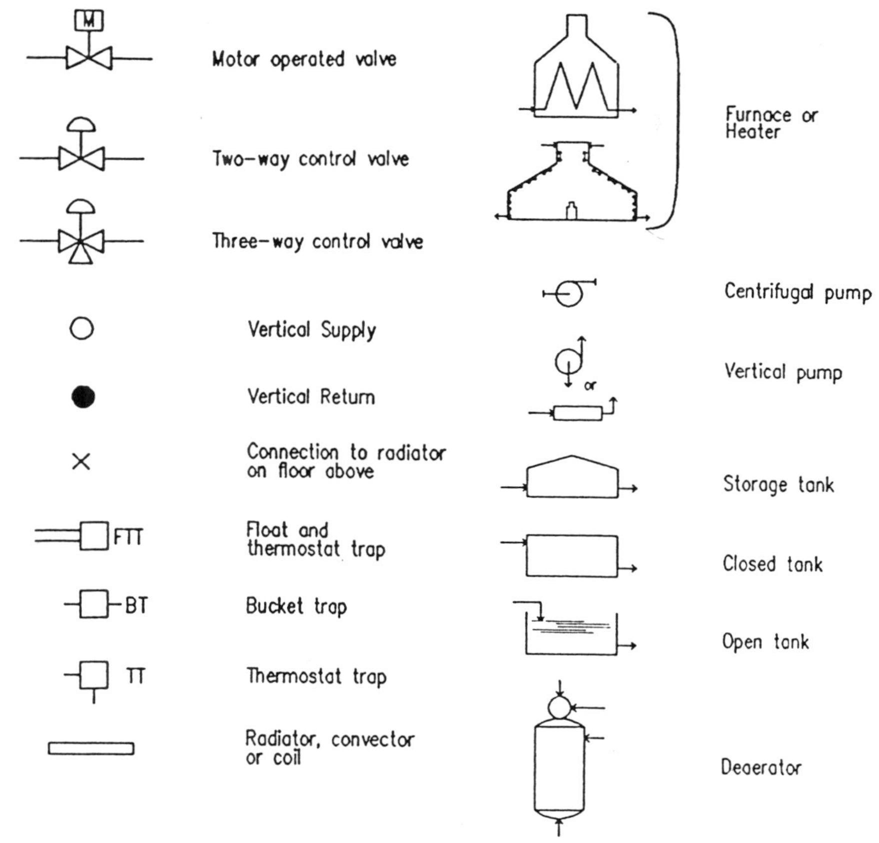

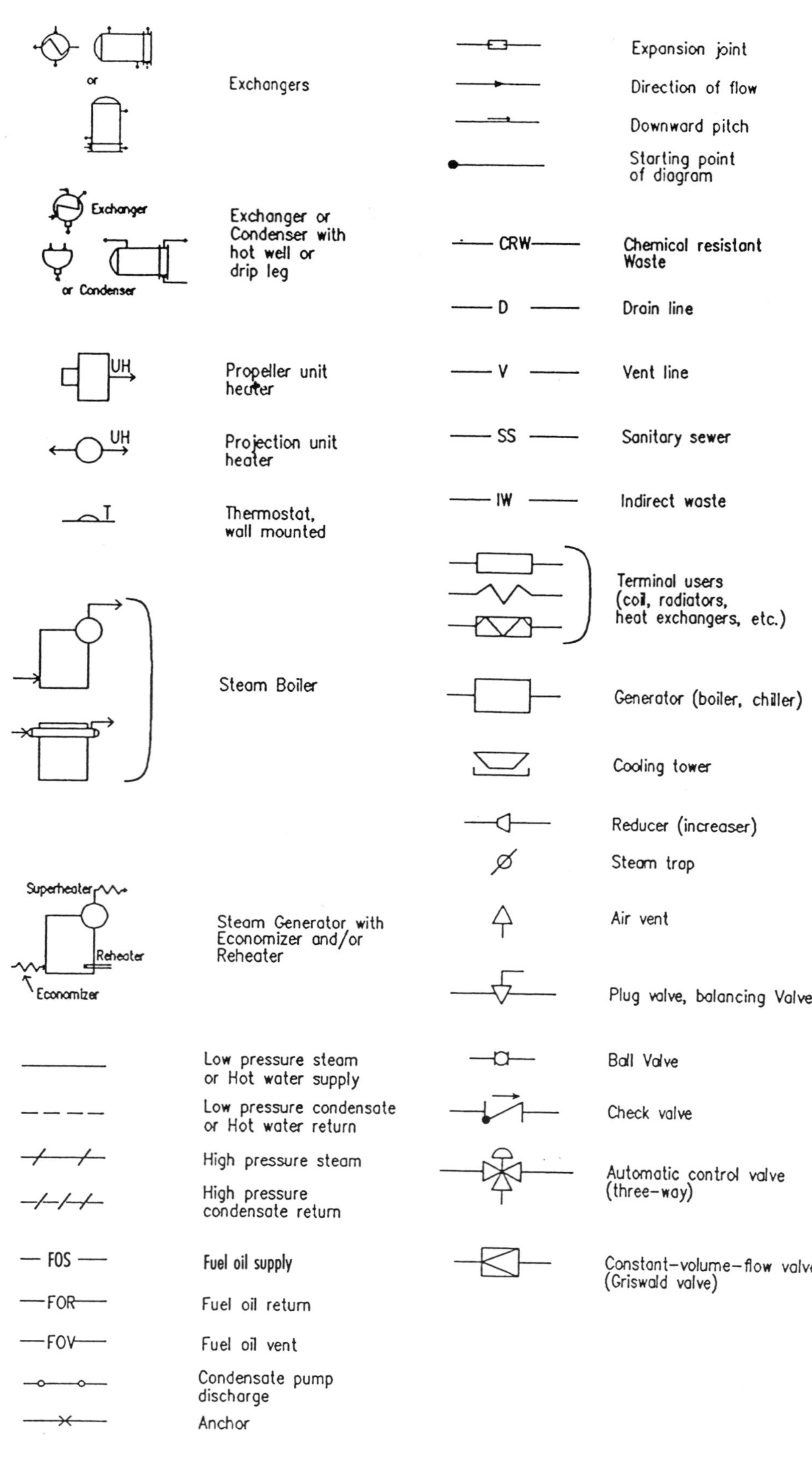
or
Exchangers
Exchanger
or Condenser
Exchanger or Condenser with hot well or drip leg
UH
Propeller unit heater
UH
Projection unit heater
T
Thermostat, wall mounted
Steam Boiler
Superheater
Reheater
Economizer
Steam Generator with Economizer and/or Reheater
Low pressure steam or Hot water supply
Low pressure condensate or Hot water return
High pressure steam
High pressure condensate return
FOS
Fuel oil supply
FOR
Fuel oil return
FOV
Fuel oil vent
Condensate pump discharge
Anchor
Expansion joint
Direction of flow
Downward pitch
Starting point of diagram
CRW
Chemical resistant Waste
D
Drain line
V
Vent line
SS
Sanitary sewer
IW
Indirect waste
Terminal users (coil, radiators, heat exchangers, etc.)
Generator (boiler, chiller)
Cooling tower
Reducer (increaser)
Steam trap
Air vent
Plug valve, balancing Valve
Ball Valve
Check valve
Automatic control valve (three-way)
Constant-volume-flow valve (Griswald valve)

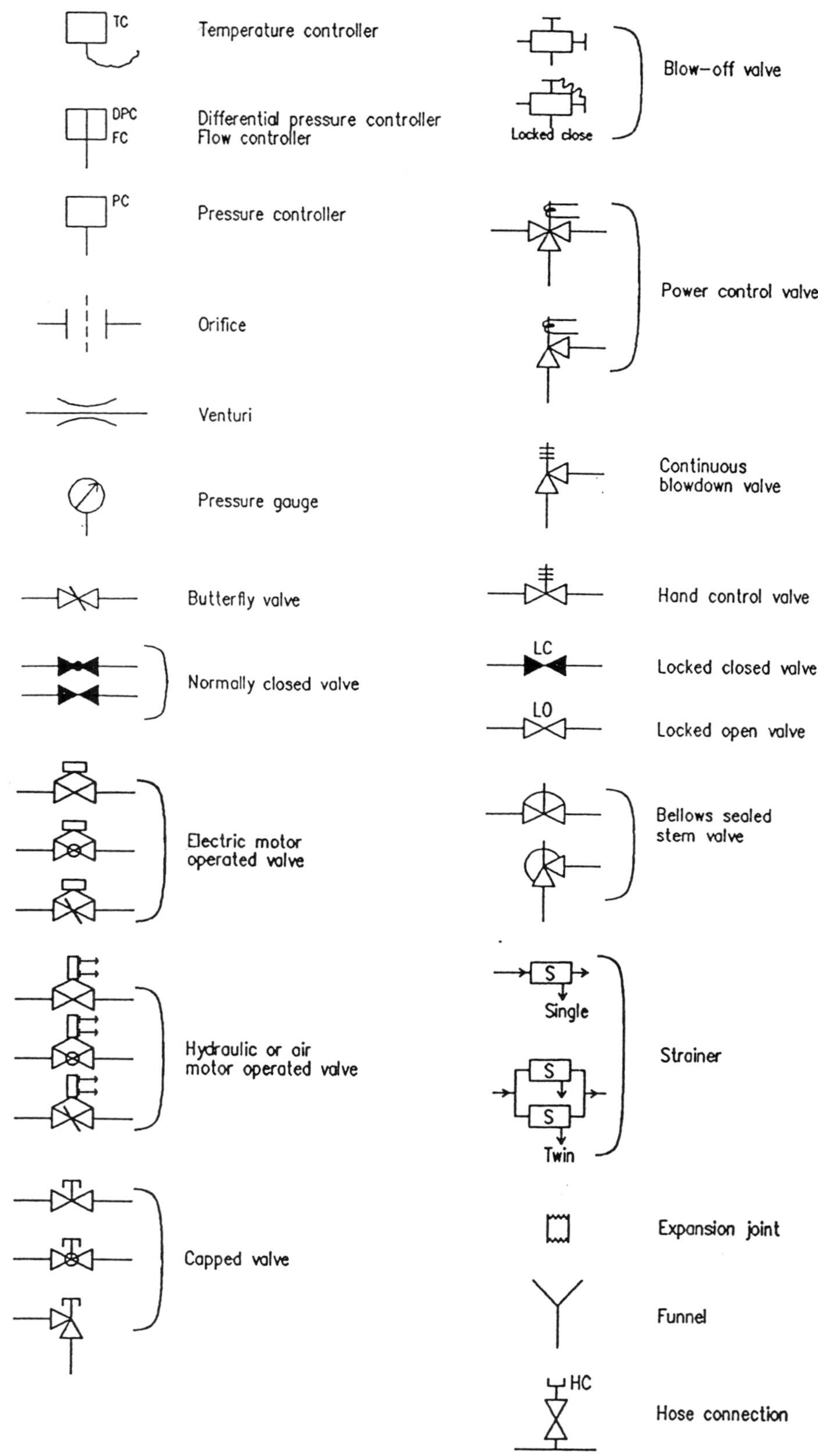

TC
Temperature controller
DPC
FC
Differential pressure controller
Flow controller
PC
Pressure controller
Orifice
Venturi
Pressure gauge
Butterfly valve
Normally closed valve
Electric motor operated valve
Hydraulic or air motor operated valve
Capped valve
Locked close
Blow-off valve
Power control valve
Continuous blowdown valve
Hand control valve
LC
Locked closed valve
LO
Locked open valve
Bellows sealed stem valve
S
Single
S
S
Twin
Strainer
Expansion joint
Funnel
HC
Hose connection

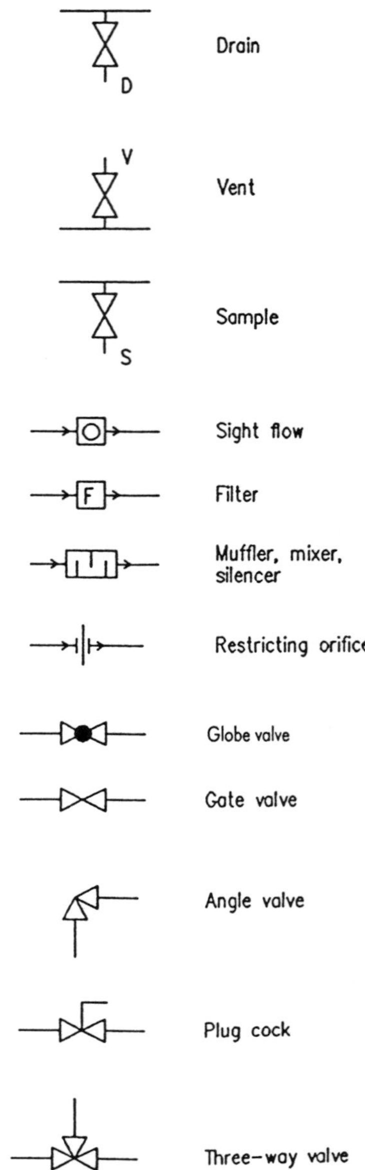
D
Drain
V
Vent
S
Sample
Sight flow
F
Filter
Muffler, mixer, silencer
Restricting orifice
Globe valve
Gate valve
Angle valve
Plug cock
Three-way valve

Appendix B

VARIOUS USEFUL TABLES

This appendix contains the following information:

Part One: Friction of Water

- Copper and Brass Tubing
- New Steel Pipe
- Asphalt-dipped Cast Iron Pipe

Part Two: Friction Losses in Pipe Fittings

Part Three: Net Positive Suction Head (NPSH) for Pumps

Part Four: Pressure Conversion Table

PART ONE: FRICTION OF WATER

Friction of Water

(Based on Darcy's Formula)

Copper Tubing—*S.P.S. Copper and Brass Pipe

3/8 Inch

Flow — US gal per min	Type K tubing .402" inside dia .049" wall thk		Type L tubing .430" inside dia .035" wall thk		Type M tubing .450" inside dia .025" wall thk		*Pipe .494" inside dia .0905" wall thk		Flow — US gal per min
	Velocity ft/sec	Head loss ft/100 ft	Velocity ft/sec	Head loss ft/100 ft	Velocity ft/sec	Head loss ft/100 ft	Velocity ft/sec	Head loss ft/100 ft	
0.2	0.51	**0.66**	0.44	**0.48**	0.40	**0.39**	0.34	**0.26**	0.2
0.4	1.01	**2.15**	0.88	**1.57**	0.81	**1.27**	0.67	**0.82**	0.4
0.6	1.52	**4.29**	1.33	**3.12**	1.21	**2.52**	1.00	**1.63**	0.6
0.8	2.02	**7.02**	1.77	**5.11**	1.61	**4.12**	1.34	**2.66**	0.8
1	2.52	**10.32**	2.20	**7.50**	2.01	**6.05**	1.68	**3.89**	1
1½	3.78	**20.86**	3.30	**15.15**	3.02	**12.21**	2.51	**7.84**	1½
2	5.04	**34.48**	4.40	**20.03**	4.02	**20.16**	3.35	**12.94**	2
2½	6.30	**51.03**	5.50	**37.01**	5.03	**29.80**	4.19	**19.11**	2½
3	7.55	**70.38**	6.60	**51.02**	6.04	**41.07**	5.02	**26.32**	3
3½	8.82	**92.44**	7.70	**66.98**	7.04	**53.90**	5.86	**34.52**	3½
4	10.1	**117.1**	8.80	**84.85**	8.05	**68.26**	6.70	**43.70**	4
4½	11.4	**144.4**	9.90	**104.6**	9.05	**84.11**	7.53	**53.82**	4½
5	12.6	**174.3**	11.0	**126.1**	10.05	**101.4**	8.36	**64.87**	5

Note: No allowance has been made for age, difference in diameter, or any abnormal condition of interior surface. Any factor of safety must be estimated from the local conditions and the requirements of each particular installation. It is recommended that for most commercial design purposes a safety factor of 15 to 20% be added to the values in the tables

Reprinted with permission from *Cameron Hydraulic Data* book.

Friction of Water

(Based on Darcy's Formula)

Copper Tubing—*S.P.S. Copper and Brass Pipe

½ Inch

Flow — U S gal per min	Type K tubing .527" inside dia .049" wall thk Velocity ft/sec	Head loss ft/100 ft	Type L tubing .545" inside dia .040" wall thk Velocity ft/sec	Head loss ft/100 ft	Type M tubing .569" inside dia .028" wall thk Velocity ft/sec	Head loss ft/100 ft	*Pipe .625" inside dia .1075" wall thk Velocity ft/sec	Head loss ft/100 ft	Flow — U S gal per min
½	0.74	**0.88**	0.69	**0.75**	0.63	**0.62**	0.52	**0.40**	½
1	1.47	**2.87**	1.38	**2.45**	1.26	**2.00**	1.04	**1.28**	1
1½	2.20	**5.77**	2.06	**4.93**	1.90	**4.02**	1.57	**2.58**	1½
2	2.94	**9.52**	2.75	**8.11**	2.53	**6.61**	2.09	**4.24**	2
2½	3.67	**14.05**	3.44	**11.98**	3.16	**9.76**	2.61	**6.25**	2½
3	4.40	**19.34**	4.12	**16.48**	3.79	**13.42**	3.13	**8.59**	3
3½	5.14	**25.36**	4.81	**21.61**	4.42	**17.59**	3.66	**11.25**	3½
4	5.87	**32.09**	5.50	**27.33**	5.05	**22.25**	4.18	**14.22**	4
4½	6.61	**39.51**	6.19	**33.65**	5.68	**27.39**	4.70	**17.50**	4½
5	7.35	**47.61**	6.87	**40.52**	6.31	**32.99**	5.22	**21.07**	5
6	8.81	**65.79**	8.25	**56.02**	7.59	**45.57**	6.26	**29.09**	6
7	10.3	**86.57**	9.62	**73.69**	8.84	**59.93**	7.31	**38.23**	7
8	11.8	**109.9**	11.0	**93.50**	10.1	**76.03**	8.35	**48.47**	8
9	13.2	**135.6**	12.4	**115.4**	11.4	**93.82**	9.40	**59.79**	9
10	14.7	**163.8**	13.8	**139.4**	12.6	**113.3**	10.4	**72.16**	10

⅝ Inch

Flow — U S gal per min	Type K tubing .652" inside dia .049" wall thk Velocity ft/sec	Head loss ft/100 ft	Type L tubing .666" inside dia .042" wall thk Velocity ft/sec	Head loss ft/100 ft	Type M tubing .690" inside dia .030" wall thk Velocity ft/sec	Head loss ft/100 ft	*Pipe		Flow — U S gal per min
½	0.48	**0.31**	0.46	**0.29**	0.43	**0.24**			½
1	0.96	**1.05**	0.92	**0.95**	0.86	**0.76**			1
1½	1.44	**2.11**	1.38	**1.91**	1.29	**1.53**			1½
2	1.92	**3.47**	1.84	**3.14**	1.72	**2.51**			2
2½	2.40	**5.11**	2.30	**4.62**	2.14	**3.68**			2½
3	2.88	**7.02**	2.75	**6.35**	2.57	**5.07**			3
3½	3.36	**9.20**	3.21	**8.32**	3.00	**6.64**			3½
4	3.84	**11.63**	3.67	**10.51**	3.43	**8.40**			4
4½	4.32	**14.30**	4.13	**12.93**	3.86	**10.35**			4½
5	4.80	**17.22**	4.59	**15.56**	4.29	**12.49**			5
6	5.75	**23.76**	5.51	**21.47**	5.15	**17.21**			6
7	6.71	**31.22**	6.42	**28.21**	6.00	**22.58**			7
8	7.67	**39.58**	7.35	**35.75**	6.85	**28.54**			8
9	8.64	**48.81**	8.25	**44.09**	7.71	**35.35**			9
10	9.60	**58.90**	9.18	**53.19**	8.57	**42.48**			10
11	10.6	**69.83**	10.1	**63.06**	9.43	**50.47**			11
12	11.5	**81.59**	11.0	**73.67**	10.3	**59.1**			12
13	12.5	**94.18**	11.9	**85.03**	11.2	**68.8**			13

Note: No allowance has been made for age, difference in diameter, or any abnormal condition of interior surface. Any factor of safety must be estimated from the local conditions and the requirements of each particular installation. It is recommended that for most commercial design purposes a safety factor of 15 to 20% be added to the values in the tables

Friction of Water

(Based on Darcy's Formula)

Copper Tubing—*S.P.S. Copper and Brass Pipe

¾ Inch

Flow — U S gal per min	Type K tubing		Type L tubing		Type M tubing		*Pipe		Flow — U S gal per min
	.745" inside dia .065" wall thk		.785 inside dia .045" wall thk		.811" inside dia .032" wall thk		.822" inside dia .114" wall thk		
	Velocity ft/sec	**Head loss ft/100 ft**	Velocity ft sec	**Head loss ft/100 ft**	Velocity ft/sec	**Head loss ft/100 ft**	Velocity ft/sec	**Head loss ft/100 ft**	
1	0.74	**0.56**	0.66	**0.44**	0.62	**0.38**	0.60	**0.35**	1
2	1.47	**1.84**	1.33	**1.44**	1.24	**1.23**	1.21	**1.16**	2
3	2.21	**3.73**	1.99	**2.91**	1.86	**2.49**	1.81	**2.34**	3
4	2.94	**6.16**	2.65	**4.81**	2.48	**4.12**	2.42	**3.86**	4
5	3.67	**9.12**	3.31	**7.11**	3.10	**6.09**	3.02	**5.71**	5
6	4.41	**12.57**	3.98	**9.80**	3.72	**8.39**	3.62	**7.86**	6
7	5.14	**16.51**	4.64	**12.86**	4.34	**11.01**	4.23	**10.32**	7
8	5.88	**20.91**	5.30	**16.28**	4.96	**13.94**	4.83	**13.07**	8
9	6.61	**25.77**	5.96	**20.06**	5.59	**17.17**	5.44	**16.10**	9
10	7.35	**31.08**	6.62	**24.19**	6.20	**20.70**	6.04	**19.41**	10
11	8.09	**36.83**	7.29	**28.66**	6.82	**24.52**	6.64	**22.99**	11
12	8.83	**43.01**	7.95	**33.47**	7.44	**28.63**	7.25	**26.84**	12
13	9.56	**49.62**	8.61	**38.61**	8.06	**33.02**	7.85	**30.96**	13
14	10.3	**56.66**	9.27	**44.07**	8.68	**37.69**	8.45	**35.33**	14
15	11.0	**64.11**	9.94	**49.86**	9.30	**42.64**	9.05	**39.97**	15
16	11.8	**71.97**	10.6	**55.97**	9.92	**47.86**	9.65	**44.86**	16
17	12.5	**80.24**	11.25	**62.39**	10.55	**53.35**	10.25	**50.00**	17
18	13.2	**88.92**	11.92	**69.13**	11.17	**59.10**	10.85	**55.40**	18

1 Inch

Flow — U S gal per min	Type K tubing		Type L tubing		Type M tubing		*Pipe		Flow — U S gal per min
	.995" inside dia .065" wall thk		1.025" inside dia .050" wall thk		1.055" inside dia .035" wall thk		1.062" inside dia .1265" wall thk		
	Velocity ft/sec	**Head loss ft/100 ft**	Velocity ft sec	**Head loss ft 100 ft**	Velocity ft/sec	**Head loss ft/100 ft**	Velocity ft/sec	**Head loss ft/100 ft**	
2	0.82	**0.47**	0.78	**0.41**	0.73	**0.36**	0.72	**0.35**	2
3	1.24	**0.95**	1.17	**0.82**	1.10	**0.72**	1.08	**0.70**	3
4	1.65	**1.56**	1.56	**1.35**	1.47	**1.18**	1.45	**1.14**	4
5	2.06	**2.30**	1.95	**2.00**	1.83	**1.74**	1.81	**1.69**	5
6	2.48	**3.17**	2.34	**2.75**	2.20	**2.40**	2.17	**2.32**	6
7	2.89	**4.15**	2.72	**3.60**	2.56	**3.14**	2.53	**3.04**	7
8	3.30	**5.25**	3.11	**4.56**	2.93	**3.97**	2.89	**3.85**	8
9	3.71	**6.47**	3.50	**5.61**	3.30	**4.89**	3.25	**4.74**	9
10	4.12	**7.79**	3.89	**6.76**	3.66	**5.89**	3.61	**5.71**	10
12	4.95	**10.76**	4.67	**9.33**	4.40	**8.13**	4.34	**7.88**	12
14	5.77	**14.15**	5.45	**12.27**	5.13	**10.69**	5.05	**10.36**	14
16	6.60	**17.94**	6.22	**15.56**	5.86	**13.55**	5.78	**13.13**	16
18	7.42	**22.14**	7.00	**19.20**	6.60	**16.72**	6.50	**16.20**	18
20	8.24	**26.73**	7.78	**23.18**	7.33	**20.18**	7.22	**19.55**	20
25	10.30	**39.87**	9.74	**34.56**	9.16	**30.09**	9.03	**29.15**	25
30	12.37	**55.33**	11.68	**47.96**	11.00	**41.74**	10.84	**40.43**	30
35	14.42	**73.06**	13.61	**63.31**	12.82	**55.09**	12.65	**53.37**	35
40	16.50	**93.00**	15.55	**80.58**	14.66	**70.11**	14.45	**67.90**	40
45	18.55	**115.1**	17.50	**99.72**	16.50	**86.75**	16.25	**84.02**	45
50	20.60	**139.4**	19.45	**120.7**	18.32	**105.0**	18.05	**101.7**	50

Note: No allowance has been made for age, difference in diameter, or any abnormal condition of interior surface. Any factor of safety must be estimated from the local conditions and the requirements of each particular installation. It is recommended that for most commercial design purposes a safety factor of 15 to 20% be added to the values in the tables

Reprinted with permission from *Cameron Hydraulic Data* book.

Friction of Water

(Based on Darcy's Formula)

Copper Tubing—*S.P.S. Copper and Brass Pipe

1¼ Inch

Flow — US gal per min	Type K tubing 1.245" inside dia .065" wall thk		Type L tubing 1.265" inside dia .055" wall thk		Type M tubing 1.291" inside dia .042" wall thk		*Pipe 1.368" inside dia .146" wall thk		Flow — US gal per min
	Velocity ft/sec	**Head loss ft/100 ft**	Velocity ft/sec	**Head loss ft/100 ft**	Velocity ft/sec	**Head loss ft/100 ft**	Velocity ft/sec	**Head loss ft/100 ft**	
5	1.31	**0.79**	1.28	**0.74**	1.22	**0.67**	1.09	**0.51**	5
6	1.58	**1.09**	1.53	**1.01**	1.47	**0.92**	1.31	**0.70**	6
7	1.84	**1.43**	1.79	**1.32**	1.71	**1.20**	1.53	**0.91**	7
8	2.11	**1.81**	2.04	**1.67**	1.96	**1.52**	1.75	**1.15**	8
9	2.37	**2.22**	2.30	**2.06**	2.20	**1.87**	1.96	**1.42**	9
10	2.63	**2.67**	2.55	**2.48**	2.45	**2.25**	2.18	**1.71**	10
12	3.16	**3.69**	3.06	**3.42**	2.93	**3.10**	2.62	**2.35**	12
15	3.95	**5.47**	3.83	**5.07**	3.66	**4.60**	3.27	**3.49**	15
20	5.26	**9.13**	5.10	**8.46**	4.89	**7.67**	4.36	**5.81**	20
25	6.58	**13.59**	6.38	**12.59**	6.11	**11.42**	5.46	**8.65**	25
30	7.90	**18.83**	7.65	**17.44**	7.33	**15.82**	6.55	**11.98**	30
35	9.21	**24.83**	8.94	**23.00**	8.55	**20.86**	7.65	**15.79**	35
40	10.5	**31.57**	10.2	**29.24**	9.77	**26.51**	8.74	**20.06**	40
45	11.8	**38.03**	11.5	**36.15**	11.0	**32.77**	9.83	**24.80**	45
50	13.2	**47.20**	12.8	**43.71**	12.2	**39.63**	10.9	**29.98**	50
60	15.8	**65.65**	15.3	**60.78**	14.7	**55.10**	13.1	**41.66**	60
70	18.4	**86.82**	17.9	**80.38**	17.1	**72.86**	15.3	**55.07**	70
80	21.1	**110.7**	20.4	**102.5**	19.6	**92.85**	17.5	**70.16**	80
90	23.7	**137.2**	23.0	**127.0**	22.0	**115.1**	19.6	**86.91**	90
100	26.3	**166.3**	25.5	**153.9**	24.4	**139.4**	21.8	**105.3**	100

1½ Inch

Flow — US gal per min	Type K tubing 1.481" inside dia .072" wall thk		Type L tubing 1.505" inside dia .060" wall thk		Type M tubing 1.527" inside dia .049" wall thk		*Pipe 1.600" inside dia .150" wall thk		Flow — US gal per min
	Velocity ft/sec	**Head loss ft/100 ft**	Velocity ft/sec	**Head loss ft/100 ft**	Velocity ft/sec	**Head loss ft/100 ft**	Velocity ft/sec	**Head loss ft/100 ft**	
8	1.49	**0.79**	1.44	**0.73**	1.40	**0.68**	1.27	**0.55**	8
9	1.67	**0.97**	1.62	**0.90**	1.57	**0.84**	1.43	**0.67**	9
10	1.86	**1.17**	1.80	**1.08**	1.75	**1.01**	1.59	**0.81**	10
12	2.23	**1.61**	2.16	**1.49**	2.10	**1.39**	1.91	**1.12**	12
15	2.79	**2.39**	2.70	**2.21**	2.63	**2.07**	2.39	**1.65**	15
20	3.72	**3.98**	3.60	**3.68**	3.50	**3.44**	3.19	**2.75**	20
25	4.65	**5.91**	4.51	**5.48**	4.38	**5.11**	3.98	**4.09**	25
30	5.58	**8.19**	5.41	**7.58**	5.25	**7.07**	4.78	**5.65**	30
35	6.51	**10.79**	6.31	**9.99**	6.13	**9.31**	5.58	**7.45**	35
40	7.44	**13.70**	7.21	**12.68**	7.00	**11.83**	6.37	**9.45**	40
45	8.37	**16.93**	8.11	**15.67**	7.88	**14.61**	7.16	**11.68**	45
50	9.30	**20.46**	9.01	**18.94**	8.76	**17.66**	7.96	**14.11**	50
60	11.2	**28.42**	10.8	**26.30**	10.5	**24.53**	9.56	**19.59**	60
70	13.0	**37.55**	12.6	**34.74**	12.3	**32.40**	11.2	**25.87**	70
80	14.9	**47.82**	14.4	**44.24**	14.0	**41.25**	12.8	**32.93**	80
90	16.7	**59.21**	16.2	**54.78**	15.8	**51.07**	14.4	**40.76**	90
100	18.6	**71.70**	18.0	**66.34**	17.5	**61.84**	15.9	**49.34**	100
110	20.5	**85.29**	19.8	**78.90**	19.3	**73.55**	17.5	**58.67**	110
120	22.3	**99.95**	21.6	**92.46**	21.0	**86.18**	19.1	**68.74**	120
130	24.2	**115.7**	23.4	**107.0**	22.8	**99.73**	20.7	**79.53**	130

Note No allowance has been made for age, difference in diameter, or any abnormal condition of interior surface. Any factor of safety must be estimated from the local conditions and the requirements of each particular installation. It is recommended that for most commercial design purposes a safety factor of 15 to 20% be added to the values in the tables.

Reprinted with permission from *Cameron Hydraulic Data* book.

Friction of Water
(Based on Darcy's Formula)

Copper Tubing—*S.P.S. Copper and Brass Pipe

2 Inch

Flow — US gal per min	Type K tubing 1.959" inside dia .083" wall thk		Type L tubing 1.985" inside dia .070" wall thk		Type M tubing 2.009" inside dia .058" wall thk		*Pipe 2.062" inside dia .1565" wall thk		Flow — US gal per min
	Velocity ft/sec	**Head loss ft/100 ft**	Velocity ft/sec	**Head loss ft/100 ft**	Velocity ft/sec	**Head loss ft/100 ft**	Velocity ft/sec	**Head loss ft/100 ft**	
10	1.07	**0.31**	1.04	**0.29**	1.01	**0.27**	.96	**0.24**	10
12	1.28	**0.43**	1.24	**0.40**	1.21	**0.38**	1.15	**0.33**	12
14	1.49	**0.56**	1.45	**0.52**	1.42	**0.50**	1.34	**0.44**	14
16	1.70	**0.71**	1.66	**0.66**	1.62	**0.63**	1.53	**0.55**	16
18	1.92	**0.87**	1.87	**0.82**	1.82	**0.77**	1.72	**0.68**	18
20	2.13	**1.05**	2.07	**0.98**	2.02	**0.93**	1.92	**0.82**	20
25	2.66	**1.55**	2.59	**1.46**	2.53	**1.38**	2.39	**1.22**	25
30	3.19	**2.15**	3.11	**2.01**	3.03	**1.90**	2.87	**1.68**	30
35	3.73	**2.82**	3.62	**2.65**	3.54	**2.50**	3.35	**2.21**	35
40	4.26	**3.58**	4.14	**3.36**	4.05	**3.17**	3.83	**2.80**	40
45	4.79	**4.42**	4.66	**4.15**	4.55	**3.92**	4.30	**3.46**	45
50	5.32	**5.34**	5.17	**5.01**	5.05	**4.73**	4.80	**4.17**	50
60	6.39	**7.40**	6.21	**6.95**	6.06	**6.56**	5.75	**5.79**	60
70	7.45	**9.76**	7.25	**9.16**	7.07	**8.65**	6.70	**7.63**	70
80	8.52	**12.42**	8.28	**11.65**	8.09	**11.00**	7.65	**9.70**	80
90	9.58	**15.36**	9.31	**14.41**	9.10	**13.60**	8.61	**12.00**	90
100	10.65	**18.58**	10.4	**17.43**	10.1	**16.45**	9.57	**14.51**	100
110	11.71	**22.07**	11.4	**20.71**	11.1	**19.55**	10.5	**17.24**	110
120	12.78	**25.84**	12.4	**24.25**	12.1	**22.88**	11.5	**20.18**	120
130	13.85	**29.88**	13.4	**28.04**	13.1	**26.45**	12.5	**23.33**	130
140	14.9	**34.18**	14.5	**32.07**	14.2	**30.26**	13.4	**26.69**	140
150	16.0	**38.75**	15.5	**36.36**	15.2	**34.30**	14.4	**30.25**	150
160	17.0	**43.58**	16.5	**40.89**	16.2	**38.58**	15.3	**34.01**	160
170	18.1	**48.67**	17.6	**45.66**	17.2	**43.08**	16.3	**37.98**	170
180	19.2	**54.01**	18.6	**50.67**	18.2	**47.81**	17.2	**42.15**	180
190	20.2	**59.61**	19.6	**55.92**	19.2	**52.76**	18.2	**46.51**	190
200	21.3	**65.46**	20.7	**61.41**	20.2	**57.94**	19.2	**51.07**	200
210	22.4	**71.57**	21.7	**67.14**	21.2	**63.34**	20.1	**55.83**	210
220	23.4	**77.93**	22.8	**73.10**	22.2	**68.96**	21.0	**60.78**	220
230	24.5	**84.53**	23.8	**79.29**	23.2	**74.80**	22.0	**65.93**	230
240	25.6	**91.38**	24.8	**85.72**	24.3	**80.86**	23.0	**71.26**	240
250	26.6	**98.43**	25.9	**92.37**	25.3	**87.14**	23.9	**76.79**	250
260	27.7	**105.8**	26.9	**99.26**	26.3	**93.63**	24.9	**82.51**	260
270	28.8	**113.4**	27.9	**106.4**	27.3	**100.3**	25.8	**88.42**	270
280	29.8	**121.3**	29.0	**113.7**	28.3	**107.3**	26.8	**94.52**	280
290	30.9	**129.3**	30.0	**121.3**	29.4	**114.4**	27.8	**100.8**	290
300	32.0	**137.6**	31.1	**129.1**	30.4	**121.8**	28.7	**107.3**	300

Note: No allowance has been made for age, difference in diameter, or any abnormal condition of interior surface. Any factor of safety must be estimated from the local conditions and the requirements of each particular installation. It is recommended that for most commercial design purposes a safety factor of 15 to 20% be added to the values in the tables

Reprinted with permission from *Cameron Hydraulic Data* book.

Friction of Water

(Based on Darcy's Formula)

Copper Tubing—*S.P.S. Copper and Brass Pipe

2½ Inch

Flow — U S gal per min	Type K tubing 2.435" inside dia .095" wall thk		Type L tubing 2.465" inside dia .080" wall thk		Type M tubing 2.495" inside dia .065" wall thk		*Pipe 2.500" inside dia .1875" wall thk		Flow — U S gal per min
	Velocity ft/sec	**Head loss ft/100 ft**	Velocity ft/sec	**Head loss ft/100 ft**	Velocity ft/sec	**Head loss ft/100 ft**	Velocity ft/sec	**Head loss ft/100 ft**	
20	1.38	**0.37**	1.34	**0.35**	1.31	**0.33**	1.31	**0.33**	20
25	1.72	**0.55**	1.68	**0.52**	1.64	**0.49**	1.63	**0.49**	25
30	2.07	**0.76**	2.02	**0.72**	1.97	**0.68**	1.96	**0.67**	30
35	2.41	**1.00**	2.35	**0.94**	2.30	**0.89**	2.29	**0.88**	35
40	2.76	**1.26**	2.69	**1.19**	2.62	**1.13**	2.61	**1.12**	40
45	3.10	**1.56**	3.02	**1.47**	2.95	**1.39**	2.94	**1.38**	45
50	3.45	**1.88**	3.36	**1.77**	3.28	**1.68**	3.26	**1.66**	50
60	4.14	**2.61**	4.03	**2.46**	3.93	**2.32**	3.92	**2.30**	60
70	4.82	**3.43**	4.70	**3.24**	4.59	**3.06**	4.57	**3.03**	70
80	5.51	**4.36**	5.37	**4.12**	5.25	**3.88**	5.22	**3.85**	80
90	6.20	**5.39**	6.04	**5.08**	5.90	**4.80**	5.88	**4.75**	90
100	6.89	**6.52**	6.71	**6.15**	6.55	**5.80**	6.53	**5.74**	100
110	7.58	**7.74**	7.38	**7.30**	7.21	**6.89**	7.19	**6.82**	110
120	8.27	**9.06**	8.05	**8.54**	7.86	**8.05**	7.84	**7.98**	120
130	8.96	**10.46**	8.73	**9.87**	8.52	**9.31**	8.49	**9.22**	130
140	9.65	**11.97**	9.40	**11.28**	9.18	**10.64**	9.14	**10.54**	140
150	10.35	**13.56**	10.1	**12.78**	9.83	**12.06**	9.79	**11.94**	150
160	11.0	**15.24**	10.8	**14.36**	10.5	**13.55**	10.45	**13.42**	160
170	11.7	**17.01**	11.4	**16.03**	11.1	**15.12**	11.1	**14.98**	170
180	12.4	**18.87**	12.1	**17.79**	11.8	**16.78**	11.8	**16.61**	180
190	13.1	**20.81**	12.8	**19.62**	12.5	**18.51**	12.4	**18.33**	190
200	13.8	**22.85**	13.4	**21.54**	13.1	**20.31**	13.1	**20.12**	200
220	15.2	**27.18**	14.8	**25.61**	14.4	**24.16**	14.4	**23.93**	220
240	16.5	**31.84**	16.1	**30.01**	15.7	**28.31**	15.7	**28.03**	240
260	17.9	**36.85**	17.5	**34.73**	17.1	**32.75**	17.0	**32.44**	260
280	19.3	**42.19**	18.8	**39.76**	18.4	**37.50**	18.3	**37.13**	280
300	20.7	**47.86**	20.1	**45.10**	19.7	**42.53**	19.6	**42.12**	300
320	22.1	**53.86**	21.5	**50.75**	21.0	**47.86**	20.9	**47.40**	320
340	23.4	**60.18**	22.8	**56.71**	22.3	**53.48**	22.2	**52.96**	340
360	24.8	**66.83**	24.2	**62.97**	23.6	**59.38**	23.5	**58.81**	300
380	26.2	**73.80**	25.5	**69.54**	24.9	**65.57**	24.8	**64.94**	380
400	27.6	**81.09**	26.9	**76.41**	26.2	**72.04**	26.1	**71.35**	400
420	29.0	**88.70**	28.2	**83.57**	27.5	**78.80**	27.4	**78.04**	420
440	30.3	**96.62**	29.5	**91.04**	28.8	**85.83**	28.7	**85.00**	440
460	31.7	**104.9**	30.9	**98.80**	30.2	**93.15**	30.0	**92.24**	460
480	33.1	**113.4**	32.2	**106.8**	31.5	**100.7**	31.4	**99.76**	480
500	34.5	**122.3**	33.6	**115.2**	32.8	**108.6**	32.6	**107.5**	500

Note: No allowance has been made for age, difference in diameter, or any abnormal condition of interior surface. Any factor of safety must be estimated from the local conditions and the requirements of each particular installation. It is recommended that for most commercial design purposes a safety factor of 15 to 20% be added to the values in the tables

Reprinted with permission from *Cameron Hydraulic Data* book.

Friction of Water
(Based on Darcy's Formula)
Copper Tubing—*S.P.S. Copper and Brass Pipe
3 Inch

Flow — US gal per min	Type K tubing 2.907" inside dia .109" wall thk		Type L tubing 2.945" inside dia .090" wall thk		Type M tubing 2.981" inside dia 072" wall thk		*Pipe 3.062" inside dia .219" wall thk		Flow — US gal per min
	Velocity ft/sec	**Head loss ft/100 ft**	Velocity ft sec	**Head loss ft/100 ft**	Velocity ft/sec	**Head loss ft/100 ft**	Velocity ft/sec	**Head loss ft/100 ft**	
20	0.96	**0.16**	0.94	**0.15**	0.92	**0.14**	0.87	**0.13**	20
30	1.45	**0.33**	1.41	**0.31**	1.37	**0.29**	1.30	**0.25**	30
40	1.93	**0.54**	1.88	**0.51**	1.83	**0.48**	1.74	**0.42**	40
50	2.41	**0.81**	2.35	**0.76**	2.29	**0.72**	2.17	**0.63**	50
60	2.89	**1.12**	2.82	**1.05**	2.75	**0.99**	2.61	**0.87**	60
70	3.38	**1.47**	3.29	**1.38**	3.20	**1.30**	3.04	**1.15**	70
80	3.86	**1.87**	3.76	**1.75**	3.66	**1.65**	3.48	**1.45**	80
90	4.34	**2.30**	4.23	**2.16**	4.12	**2.04**	3.91	**1.80**	90
100	4.82	**2.78**	4.70	**2.61**	4.59	**2.47**	4.35	**2.17**	100
110	5.30	**3.30**	5.17	**3.10**	5.05	**2.93**	4.79	**2.57**	110
120	5.79	**3.86**	5.64	**3.63**	5.50	**3.42**	5.21	**3.01**	120
130	6.27	**4.46**	6.11	**4.19**	5.95	**3.95**	5.65	**3.47**	130
140	6.75	**5.10**	6.58	**4.79**	6.41	**4.52**	6.09	**3.97**	140
150	7.24	**5.77**	7.05	**5.42**	6.87	**5.12**	6.52	**4.50**	150
160	7.72	**6.49**	7.52	**6.09**	7.34	**5.75**	6.95	**5.05**	160
170	8.20	**7.24**	7.99	**6.80**	7.79	**6.41**	7.39	**5.64**	170
180	8.69	**8.03**	8.46	**7.54**	8.25	**7.11**	7.82	**6.25**	180
190	9.16	**8.85**	8.93	**8.32**	8.70	**7.84**	8.25	**6.89**	190
200	9.64	**9.71**	9.40	**9.13**	9.16	**8.61**	8.70	**7.56**	200
220	10.6	**11.55**	10.3	**10.85**	10.1	**10.23**	9.56	**8.99**	220
240	11.6	**13.52**	11.3	**12.70**	11.0	**11.98**	10.4	**10.52**	240
260	12.6	**15.64**	12.2	**14.69**	11.9	**13.85**	11.3	**12.17**	260
280	13.5	**17.90**	13.2	**16.81**	12.8	**15.85**	12.2	**13.93**	280
300	14.5	**20.30**	14.1	**19.06**	13.7	**17.97**	13.0	**15.79**	300
320	15.4	**22.83**	15.0	**21.44**	14.7	**20.22**	13.9	**17.76**	320
340	16.4	**25.50**	16.0	**23.95**	15.6	**22.58**	14.8	**19.83**	340
360	17.4	**28.30**	16.9	**26.58**	16.5	**25.06**	15.7	**22.01**	360
380	18.3	**31.24**	17.9	**29.34**	17.4	**27.66**	16.5	**24.29**	380
400	19.3	**34.32**	18.8	**32.22**	18.3	**30.38**	17.4	**26.68**	400
450	21.7	**42.58**	21.2	**39.98**	20.6	**37.69**	19.6	**33.09**	450
500	24.1	**51.65**	23.5	**48.50**	22.9	**45.72**	21.7	**40.14**	500
550	26.6	**61.54**	25.8	**57.77**	25.2	**54.46**	23.9	**47.81**	550
600	29.0	**72.22**	28.2	**67.80**	27.5	**63.91**	26.1	**56.10**	600
650	31.4	**83.69**	30.6	**78.56**	29.8	**74.05**	28.2	**65.00**	650
700	33.8	**95.95**	32.9	**90.06**	32.1	**84.89**	30.4	**74.50**	700
750	36.2	**109.0**	35.2	**102.3**	34.4	**96.41**	32.6	**84.61**	750
800	38.6	**122.8**	37.6	**115.3**	36.6	**108.6**	34.8	**95.31**	800

Note: No allowance has been made for age, difference in diameter, or any abnormal condition of interior surface. Any factor of safety must be estimated from the local conditions and the requirements of each particular installation. It is recommended that for most commercial design purposes a safety factor of 15 to 20% be added to the values in the tables

Friction of Water New Steel Pipe

(Based on Darcy's Formula)

¼ Inch

Flow US gal per min	Standard wt steel—sch 40			Extra strong steel—sch 80		
	0.364″ inside dia			0.302″ inside dia		
	Velocity ft per sec	Velocity head-ft	**Head loss ft per 100 ft**	Velocity ft per sec	Velocity head-ft	**Head loss ft per 100 ft**
0.4	1.23	0.024	**3.7**	1.79	0.05	**9.18**
0.6	1.85	0.053	**7.6**	2.69	0.11	**19.0**
0.8	2.47	0.095	**12.7**	3.59	0.20	**32.3**
1.0	3.08	0.148	**19.1**	4.48	0.31	**48.8**
1.2	3.70	0.213	**26.7**	5.38	0.45	**68.6**
1.4	4.32	0.290	**35.6**	6.27	0.61	**91.7**
1.6	4.93	0.378	**45.6**	7.17	0.80	**118.1**
1.8	5.55	0.479	**56.9**	8.07	1.01	**147.7**
2.0	6.17	0.591	**69.4**	8.96	1.25	**180.7**
2.4	7.40	0.850	**98.1**	10.75	1.79	**256**
2.8	8.63	1.157	**132**	12.54	2.44	**345**

⅜ Inch

Flow US gal per min	Standard wt steel—sch 40			Extra strong steel—sch 80		
	0.493″ inside dia			0.423″ inside dia		
	Velocity ft per sec	Velocity head-ft	**Head loss ft per 100 ft**	Velocity ft per sec	Velocity head-ft	**Head loss ft per 100 ft**
0.5	0.84	0.011	**1.26**	1.14	0.02	**2.63**
1.0	1.68	0.044	**4.26**	2.28	0.08	**9.05**
1.5	2.52	0.099	**8.85**	3.43	0.18	**19.0**
2.0	3.36	0.176	**15.0**	4.57	0.32	**32.4**
2.5	4.20	0.274	**22.7**	5.71	0.51	**49.3**
3.0	5.04	0.395	**32.0**	6.85	0.73	**69.6**
3.5	5.88	0.538	**42.7**	8.00	0.99	**93.3**
4.0	6.72	0.702	**55.0**	9.14	1.30	**120**
5.0	8.40	1.097	**84.2**	11.4	2.0	**185**
6.0	10.08	1.58	**119**	13.7	2.9	**263**

Note: No allowance has been made for age, difference in diameter, or any abnormal condition of interior surface. Any factor of safety must be estimated from the local conditions and the requirements of each particular installation. It is recommended that for most commercial design purposes a safety factor of 15 to 20% be added to the values in the tables

Reprinted with permission from *Cameron Hydraulic Data* book.

Friction of Water New Steel Pipe

(Based on Darcy's Formula)

½ Inch

Flow US gal per min	Standard wt steel—sch 40			Extra strong steel—sch 80			Schedule 160		
	.622" inside dia			.546" inside dia			.464" inside dia		
	Velocity ft per sec	Velocity head ft	**Head loss ft per 100 ft**	Velocity ft per sec	Velocity head ft	**Head loss ft per 100 ft**	Velocity ft per sec	Velocity head ft	**Head loss ft per 100 ft**
0.7	0.739	.008	**0.74**	.96	.01	**1.39**			
1.0	1.056	.017	**1.86**	1.37	.03	**2.58**	1.90	.056	**1.68**
1.5	1.58	.039	**2.82**	2.06	.07	**5.34**	2.85	.126	**5.73**
2.0	2.11	.069	**4.73**	2.74	.12	**9.02**	3.80	.224	**12.0**
2.5	2.64	.108	**7.10**	3.43	.18	**13.6**	4.74	.349	**20.3**
3.0	3.17	.156	**9.94**	4.11	.26	**19.1**	5.69	.503	**30.8**
3.5	3.70	.212	**13.2**	4.80	.36	**25.5**	6.64	.684	**43.5**
4.0	4.22	.277	**17.0**	5.48	.47	**32.7**	7.59	.894	**58.2**
4.5	4.75	.351	**21.1**	6.17	.59	**40.9**	8.54	1.13	**75.0**
5.0	5.28	.433	**25.8**	6.86	.73	**50.0**	9.49	1.40	**94.0**
5.5	5.81	.524	**30.9**	7.54	.88	**59.9**	10.44	1.69	**115**
6.0	6.34	.624	**36.4**	8.23	1.05	**70.7**	11.38	2.01	**138**
6.5	6.86	.732	**42.4**	8.91	1.23	**82.4**	12.33	2.36	**163**
7.0	7.39	.849	**48.8**	9.60	1.43	**95.0**	13.28	2.74	**190**
7.5	7.92	.975	**55.6**	10.3	1.6	**109**	14.23	3.14	**220**
8.0	8.45	1.109	**63.0**	11.0	1.9	**123**			
8.5	8.98	1.25	**70.7**	11.6	2.1	**138**			
9.0	9.50	1.40	**78.9**	12.3	2.4	**154**			
9.5	10.03	1.56	**87.6**	13.0	2.6	**171**			
10	10.56	1.73	**96.6**	13.7	2.9	**189**			

¾ Inch

Flow US gal per min	Standard wt steel—sch 40			Extra strong steel—sch 80			Steel—schedule 160		
	.824" inside dia			.742" inside dia			.612" inside dia		
	Velocity ft per sec	Velocity head ft	**Head loss ft per 100 ft**	Velocity ft per sec	Velocity head ft	**Head loss ft per 100 ft**	Velocity ft per sec	Velocity head ft	**Head loss ft per 100 ft**
1.5	0.90	.013	**0.72**	1.11	.02	**1.19**	1.64	.042	**3.05**
2.0	1.20	.023	**1.19**	1.48	.03	**1.99**	2.18	.074	**5.12**
2.5	1.50	.035	**1.78**	1.86	.05	**2.97**	2.73	.115	**7.70**
3.0	1.81	.051	**2.47**	2.23	.08	**4.14**	3.27	.166	**10.8**
3.5	2.11	.069	**3.26**	2.60	.11	**5.48**	3.82	.226	**14.3**
4.0	2.41	.090	**4.16**	2.97	.14	**7.01**	4.36	.295	**18.4**
4.5	2.71	.114	**5.17**	3.34	.17	**8.72**	4.91	.374	**22.9**
5.0	3.01	.141	**6.28**	3.71	.21	**10.6**	5.45	.462	**28.0**
6	3.61	.203	**8.80**	4.45	.31	**14.9**	6.54	.665	**39.5**
7	4.21	.276	**11.7**	5.20	.42	**19.9**	7.64	.905	**53.0**
8	4.81	.360	**15.1**	5.94	.55	**25.6**	8.73	1.18	**68.4**
9	5.42	.456	**18.8**	6.68	.69	**32.1**	9.82	1.50	**85.8**
10	6.02	.563	**23.0**	7.42	.86	**39.2**	10.91	1.85	**105**
11	6.62	.681	**27.6**	8.17	1.04	**47.0**	12.00	2.23	**126**
12	7.22	.722	**32.5**	8.91	1.23	**55.5**	13.09	2.66	**149**
13	7.82	.951	**37.9**	9.63	1.44	**64.8**	14.18	3.13	**175**
14	8.42	1.103	**43.7**	10.4	1.7	**74.7**	15.27	3.62	**202**
16	9.63	1.44	**56.4**	11.9	2.2	**96.7**	17.45	4.73	**261**
18	10.8	1.82	**70.8**	13.4	2.8	**121**			
20	12.0	2.25	**86.8**	14.8	3.4	**149**			

Note: No allowance has been made for age, difference in diameter, or any abnormal condition of interior surface. Any factor of safety must be estimated from the local conditions and the requirements of each particular installation. It is recommended that for most commercial design purposes a safety factor of 15 to 20% be added to the values in the tables

Reprinted with permission from *Cameron Hydraulic Data* book.

Friction of Water New Steel Pipe

(Based on Darcy's Formula)

1 Inch

Flow US gal per min	Standard wt steel—sch 40			Extra strong steel—sch 80			Schedule 160 steel		
	1.049" inside dia			.957" inside dia			.815" inside dia		
	Velocity ft per sec	Velocity head ft	**Head loss ft per 100 ft**	Velocity ft per sec	Velocity head ft	**Head loss ft per 100 ft**	Velocity ft per sec	Velocity head ft	**Head loss ft per 100 ft**
2	0.74	.009	**.385**	.89	.01	**.599**	1.23	.023	**1.26**
3	1.11	.019	**.787**	1.34	.03	**1.19**	1.85	.053	**2.60**
4	1.48	.034	**1.270**	1.79	.05	**1.99**	2.46	.094	**4.40**
5	1.86	.054	**1.90**	2.23	.08	**2.99**	3.08	.147	**6.63**
6	2.23	.077	**2.65**	2.68	.11	**4.17**	3.69	.211	**9.30**
8	2.97	.137	**4.50**	3.57	.20	**7.11**	4.92	.376	**15.9**
10	3.71	.214	**6.81**	4.46	.31	**10.8**	6.15	.587	**24.3**
12	4.45	.308	**9.58**	5.36	.45	**15.2**	7.38	.845	**34.4**
14	5.20	.420	**12.8**	6.25	.61	**20.4**	8.61	1.15	**46.2**
16	5.94	.548	**16.5**	7.14	.79	**26.3**	9.84	1.50	**59.7**
18	6.68	.694	**20.6**	8.03	1.00	**32.9**	11.07	1.90	**74.9**
20	7.42	.857	**25.2**	8.92	1.24	**40.3**	12.30	2.35	**91.8**
22	8.17	1.036	**30.3**	9.82	1.50	**48.4**	13.53	2.84	**110**
24	8.91	1.23	**35.8**	10.7	1.8	**57.2**	14.76	3.38	**131**
26	9.65	1.45	**41.7**	11.6	2.1	**66.8**	15.99	3.97	**153**
28	10.39	1.68	**48.1**	12.5	2.4	**77.1**			
30	11.1	1.93	**55.0**	13.4	2.8	**88.2**			
35	13.0	2.62	**74.1**	15.6	3.8	**119**			
40	14.8	3.43	**96.1**	17.9	5.0	**154**			
45	16.7	4.33	**121**	20.1	6.3	**194**			

1¼ Inch

Flow US gal per min	Standard wt steel—sch 40			Extra strong steel—sch 80			Schedule 160—steel		
	1.380" inside dia			1.278" inside dia			1.160" inside dia		
	Velocity ft per sec	Velocity head ft	**Head loss ft per 100 ft**	Velocity ft per sec	Velocity head ft	**Head loss ft per 100 ft**	Velocity ft per sec	Velocity head ft	**Head loss ft per 100 ft**
4	.858	.011	**.35**	1.00	.015	**.51**	1.21	.023	**.806**
5	1.073	.018	**.52**	1.25	.024	**.75**	1.52	.036	**1.20**
6	1.29	.026	**.72**	1.50	.034	**1.04**	1.82	.051	**1.61**
7	1.50	.035	**.95**	1.75	.048	**1.33**	2.13	.070	**2.14**
8	1.72	.046	**1.20**	2.00	.062	**1.69**	2.43	.092	**2.73**
10	2.15	.072	**1.74**	2.50	.097	**2.55**	3.04	.143	**4.12**
12	2.57	.103	**2.45**	3.00	.140	**3.57**	3.64	.206	**5.78**
14	3.00	.140	**3.24**	3.50	.190	**4.75**	4.25	.280	**7.72**
16	3.43	.183	**4.15**	4.00	.249	**6.10**	4.86	.366	**9.92**
18	3.86	.232	**5.17**	4.50	.315	**7.61**	5.46	.463	**12.4**
20	4.29	.286	**6.31**	5.00	.388	**9.28**	6.07	.572	**15.1**
25	5.36	.431	**9.61**	6.25	.607	**14.2**	7.59	.894	**23.2**
30	6.44	.644	**13.6**	7.50	.874	**20.1**	9.11	1.29	**32.9**
35	7.51	.876	**18.2**	8.75	1.19	**27.0**	10.63	1.75	**44.2**
40	8.58	1.14	**23.5**	10.0	1.55	**34.9**	12.14	2.29	**57.3**
50	10.7	1.79	**36.2**	12.5	2.43	**53.7**	15.18	3.58	**88.3**
60	12.9	2.57	**51.5**	15.0	3.50	**76.5**	18.22	5.15	**126**
70	15.0	3.50	**69.5**	17.5	4.76	**103**	21.25	7.01	**170**
80	17.2	4.53	**90.2**	20.0	6.21	**134**	24.29	9.16	**221**
90	19.3	5.79	**114**	22.5	7.86	**168**	27.32	11.59	**279**

Note: No allowance has been made for age, difference in diameter, or any abnormal condition of interior surface. Any factor of safety must be estimated from the local conditions and the requirements of each particular installation. It is recommended that for most commercial design purposes a safety factor of 15 to 20% be added to the values in the tables

Reprinted with permission from *Cameron Hydraulic Data* book.

Friction of Water New Steel Pipe
(Based on Darcy's Formula)
1½ Inch

Flow U S gal per min	Standard wt steel—sch 40			Extra strong steel—sch 80			Schedule 160—steel		
	1.610" inside dia			1.500" inside dia			1.338" inside dia		
	Velocity ft per sec	Velocity head ft	**Head loss ft per 100 ft**	Velocity ft per sec	Velocity head ft	**Head loss ft per 100 ft**	Velocity ft per sec	Velocity head ft	**Head loss ft per 100 ft**
4	.63	.006	**.166**	.73	.01	**.233**	.913	.013	**.404**
5	.79	.010	**.246**	.91	.01	**.346**	1.14	.020	**.601**
6	.95	.014	**.340**	1.09	.02	**.478**	1.37	.029	**.832**
7	1.10	.019	**.447**	1.27	.03	**.630**	1.60	.040	**1.10**
8	1.26	.025	**.567**	1.45	.03	**.800**	1.83	.052	**1.35**
9	1.42	.031	**.701**	1.63	.04	**.990**	2.05	.065	**1.67**
10	1.58	.039	**.848**	1.82	.05	**1.20**	2.28	.081	**2.03**
12	1.89	.056	**1.18**	2.18	.07	**1.61**	2.74	.116	**2.84**
14	2.21	.076	**1.51**	2.54	.10	**2.14**	3.20	.158	**3.78**
16	2.52	.099	**1.93**	2.90	.13	**2.74**	3.65	.207	**4.85**
18	2.84	.125	**2.40**	3.27	.17	**3.41**	4.11	.262	**6.04**
20	3.15	.154	**2.92**	3.63	.20	**4.15**	4.56	.323	**7.36**
22	3.47	.187	**3.48**	3.99	.25	**4.96**	5.02	.391	**8.81**
24	3.78	.222	**4.10**	4.36	.30	**5.84**	5.48	.465	**10.4**
26	4.10	.261	**4.76**	4.72	.35	**6.80**	5.93	.546	**12.1**
28	4.41	.303	**5.47**	5.08	.40	**7.82**	6.39	.634	**13.9**
30	4.73	.347	**6.23**	5.45	.46	**8.91**	6.85	.727	**15.9**
32	5.04	.395	**7.04**	5.81	.52	**10.1**	7.30	.828	**18.0**
34	5.36	.446	**7.90**	6.17	.59	**11.3**	7.76	.934	**20.2**
36	5.67	.500	**8.80**	6.54	.66	**12.6**	8.22	1.05	**22.5**
38	5.99	.577	**9.76**	6.90	.74	**14.0**	8.67	1.17	**25.0**
40	6.30	.618	**10.8**	7.26	.82	**15.4**	9.13	1.29	**27.6**
42	6.62	.681	**11.8**	7.63	.90	**16.9**	9.58	1.43	**30.3**
44	6.93	.747	**12.9**	7.99	.99	**18.5**	10.04	1.57	**33.1**
46	7.25	.817	**14.0**	8.35	1.08	**20.1**	10.50	1.71	**36.1**
48	7.56	.889	**15.2**	8.72	1.18	**21.8**	10.95	1.86	**39.2**
50	7.88	.965	**16.5**	9.08	1.28	**23.6**	11.41	2.02	**42.4**
55	8.67	1.17	**19.8**	9.99	1.55	**28.4**	12.55	2.45	**51.0**
60	9.46	1.39	**23.4**	10.9	1.8	**33.6**	13.69	2.91	**60.4**
65	10.24	1.63	**27.3**	11.8	2.2	**39.2**	14.83	3.41	**70.6**
70	11.03	1.89	**31.5**	12.7	2.5	**45.3**	15.97	3.96	**81.5**
75	11.8	2.17	**36.0**	13.6	2.9	**51.8**	17.11	4.55	**93.2**
80	12.6	2.47	**40.8**	14.5	3.3	**58.7**	18.25	5.17	**106**
85	13.4	2.79	**45.9**	15.4	3.7	**66.0**	19.40	5.84	**119**
90	14.2	3.13	**51.3**	16.3	4.1	**73.8**	20.54	6.55	**133**
95	15.0	3.48	**57.0**	17.2	4.6	**82.0**	21.68	7.29	**148**
100	15.8	3.86	**63.0**	18.2	5.1	**90.7**	22.82	8.08	**164**
110	17.3	4.67	**75.8**	20.0	6.2	**109.3**	25.10	9.78	**197**
120	18.9	5.56	**89.9**	21.8	7.4	**129.6**	27.38	11.6	**234**
130	20.5	6.52	**105**	23.6	8.7	**151.6**	29.66	13.7	**274**
140	22.1	7.56	**122**	25.4	10.0	**175**			
150	23.6	8.68	**139**	27.2	11.5	**201**			
160	25.2	9.88	**158**	29.0	13.1	**228**			
170	26.8	11.15	**178**	30.9	14.8	**257**			
180	28.4	12.50	**199**	32.7	16.6	**288**			

Note: No allowance has been made for age, difference in diameter, or any abnormal condition of interior surface. Any factor of safety must be estimated from the local conditions and the requirements of each particular installation. It is recommended that for most commercial design purposes a safety factor of 15 to 20% be added to the values in the tables

Reprinted with permission from *Cameron Hydraulic Data* book.

Friction of Water New Steel Pipe

(Based on Darcy's Formula)

2 Inch

Flow U S gal per min	Standard wt steel—sch 40			Extra strong steel—sch 80			Schedule 160—steel		
	2.067" inside dia			1.939" inside dia			1.687" inside dia		
	Velocity ft per sec	Velocity head ft	**Head loss ft per 100 ft**	Velocity ft per sec	Velocity head ft	**Head loss ft per 100 ft**	Velocity ft per sec	Velocity head ft	**Head loss ft per 100 ft**
5	.478	.004	**.074**	.54	.00	**.101**	.718	.008	**.197**
6	.574	.005	**.102**	.65	.01	**.139**	.861	.012	**.271**
7	.669	.007	**.134**	.76	.01	**.182**	1.01	.016	**.357**
8	.765	.009	**.170**	.87	.01	**.231**	1.15	.020	**.452**
9	.860	.012	**.209**	.98	.01	**.285**	1.29	.026	**.559**
10	.956	.014	**.252**	1.09	.02	**.343**	1.44	.032	**.675**
12	1.15	.021	**.349**	1.30	.03	**.476**	1.72	.046	**.938**
14	1.34	.028	**.461**	1.52	.04	**.629**	2.01	.063	**1.20**
16	1.53	.036	**.586**	1.74	.05	**.800**	2.30	.082	**1.53**
18	1.72	.046	**.725**	1.96	.06	**.991**	2.58	.104	**1.90**
20	1.91	.057	**.878**	2.17	.07	**1.16**	2.87	.128	**2.31**
22	2.10	.069	**1.05**	2.39	.09	**1.38**	3.16	.155	**2.76**
24	2.29	.082	**1.18**	2.61	.11	**1.62**	3.45	.184	**3.25**
26	2.49	.096	**1.37**	2.83	.12	**1.88**	3.73	.216	**3.77**
28	2.68	.111	**1.57**	3.04	.14	**2.16**	4.02	.251	**4.33**
30	2.87	.128	**1.82**	3.26	.17	**2.46**	4.31	.288	**4.93**
35	3.35	.174	**2.38**	3.80	.22	**3.28**	5.02	.392	**6.59**
40	3.82	.227	**3.06**	4.35	.29	**4.21**	5.74	.512	**8.49**
45	4.30	.288	**3.82**	4.89	.37	**5.26**	6.46	.648	**10.6**
50	4.78	.355	**4.66**	5.43	.46	**6.42**	7.18	.799	**13.0**
55	5.26	.430	**5.58**	5.98	.56	**7.70**	7.89	.967	**15.6**
60	5.74	.511	**6.58**	6.52	.66	**9.09**	8.61	1.15	**18.4**
65	6.21	.600	**7.66**	7.06	.77	**10.59**	9.33	1.35	**21.5**
70	6.69	.696	**8.82**	7.61	.90	**12.2**	10.05	1.57	**24.8**
75	7.17	.799	**10.1**	8.15	1.03	**13.9**	10.77	1.80	**28.3**
80	7.65	.909	**11.4**	8.69	1.17	**15.8**	11.48	2.05	**32.1**
85	8.13	1.03	**12.8**	9.03	1.27	**17.7**	12.20	2.31	**36.1**
90	8.60	1.15	**14.3**	9.78	1.49	**19.8**	12.92	2.59	**40.3**
95	9.08	1.28	**15.9**	10.3	1.6	**22.0**	13.64	2.89	**44.8**
100	9.56	1.42	**17.5**	10.9	1.8	**24.3**	14.35	3.20	**49.5**
110	10.52	1.72	**21.0**	12.0	2.2	**29.2**	15.79	3.87	**59.6**
120	11.5	2.05	**24.9**	13.0	2.6	**34.5**	17.22	4.61	**70.6**
130	12.4	2.40	**29.1**	14.1	3.1	**40.3**	18.66	5.40	**82.6**
140	13.4	2.78	**33.6**	15.2	3.6	**46.6**	20.10	6.27	**95.5**
150	14.3	3.20	**38.4**	16.3	4.1	**53.3**	21.53	7.20	**109**
160	15.3	3.64	**43.5**	17.4	4.7	**60.5**	22.97	8.19	**124**
170	16.3	4.11	**49.0**	18.5	5.3	**68.1**	24.40	9.24	**140**
180	17.2	4.60	**54.8**	19.6	6.0	**76.1**	25.84	10.36	**156**
190	18.2	5.13	**60.9**	20.6	6.6	**84.6**	27.27	11.54	**174**
200	19.1	5.68	**67.3**	21.7	7.3	**93.6**	28.71	12.79	**192**
220	21.0	6.88	**81.1**	23.9	8.9	**113**			
240	22.9	8.18	**96.2**	26.9	10.6	**134**			
260	24.9	9.60	**113**	28.3	12.4	**157**			
280	26.8	11.14	**130**	30.4	14.4	**181**			
300	28.7	12.8	**149**	32.6	16.5	**208**			

Note: No allowance has been made for age, difference in diameter, or any abnormal condition of interior surface. Any factor of safety must be estimated from the local conditions and the requirements of each particular installation. It is recommended that for most commercial design purposes a safety factor of 15 to 20% be added to the values in the tables

Reprinted with permission from *Cameron Hydraulic Data* book.

Friction of Water New Steel Pipe

(Based on Darcy's Formula)

2½ Inch

Flow US gal per min	Standard wt steel—sch 40			Extra strong steel—sch 80			Schedule 160—steel		
	2.469" inside dia			2.323" inside dia			2.125" inside dia		
	Velocity ft per sec	Velocity head ft	**Head loss ft per 100 ft**	Velocity ft per sec	Velocity head ft	**Head loss ft per 100 ft**	Velocity ft per sec	Velocity head ft	**Head loss ft per 100 ft**
8	.536	.005	**.072**	.61	.01	**.097**	.724	.008	**.149**
10	.670	.007	**.107**	.76	.01	**.144**	.905	.013	**.221**
12	.804	.010	**.148**	.91	.01	**.199**	1.09	.018	**.305**
14	.938	.014	**.195**	1.06	.02	**.261**	1.27	.025	**.403**
16	1.07	.018	**.247**	1.21	.02	**.332**	1.45	.033	**.512**
18	1.21	.023	**.305**	1.36	.03	**.411**	1.63	.041	**.634**
20	1.34	.028	**.369**	1.51	.04	**.497**	1.81	.051	**.767**
22	1.47	.034	**.438**	1.67	.04	**.590**	1.99	.061	**.912**
24	1.61	.040	**.513**	1.82	.05	**.691**	2.17	.073	**1.03**
26	1.74	.047	**.593**	1.97	.06	**.800**	2.35	.086	**1.20**
28	1.88	.055	**.679**	2.12	.07	**.915**	2.53	.100	**1.37**
30	2.01	.063	**.770**	2.27	.08	**1.00**	2.71	.114	**1.56**
35	2.35	.086	**0.99**	2.65	.11	**1.33**	3.17	.156	**2.08**
40	2.68	.112	**1.26**	3.03	.14	**1.71**	3.62	.203	**2.66**
45	3.02	.141	**1.57**	3.41	.18	**2.13**	4.07	.257	**3.32**
50	3.35	.174	**1.91**	3.79	.22	**2.59**	4.52	.318	**4.05**
55	3.69	.211	**2.28**	4.16	.27	**3.10**	4.98	.384	**4.85**
60	4.02	.251	**2.69**	4.54	.32	**3.65**	5.43	.457	**5.72**
65	4.36	.295	**3.13**	4.92	.38	**4.25**	5.88	.537	**6.66**
70	4.69	.342	**3.60**	5.30	.44	**4.89**	6.33	.622	**7.67**
75	5.03	.393	**4.10**	5.68	.50	**5.58**	6.79	.714	**8.75**
80	5.36	.447	**4.64**	6.05	.57	**6.31**	7.24	.813	**9.90**
85	5.70	.504	**5.20**	6.43	.64	**7.08**	7.69	.918	**11.1**
90	6.03	.565	**5.80**	6.81	.72	**7.89**	8.14	1.03	**12.4**
95	6.37	.630	**6.43**	7.19	.80	**8.76**	8.59	1.15	**13.8**
100	6.70	.698	**7.09**	7.57	.89	**9.66**	9.05	1.27	**15.2**
110	7.37	.844	**8.51**	8.33	1.08	**11.6**	9.95	1.54	**18.3**
120	8.04	1.00	**10.1**	9.08	1.28	**13.7**	10.86	1.83	**21.6**
130	8.71	1.18	**11.7**	9.84	1.50	**16.0**	11.76	2.15	**25.2**
140	9.38	1.37	**13.5**	10.6	1.7	**18.5**	12.67	2.49	**29.1**
150	10.05	1.57	**15.5**	11.3	2.0	**21.1**	13.57	2.86	**33.3**
160	10.7	1.79	**17.5**	12.1	2.3	**23.9**	14.47	3.25	**37.8**
170	11.4	2.02	**19.7**	12.9	2.6	**26.9**	15.38	3.67	**42.5**
180	12.1	2.26	**22.0**	13.6	2.9	**30.1**	16.28	4.12	**47.5**
190	12.7	2.52	**24.4**	14.4	3.2	**33.4**	17.19	4.59	**52.8**
200	13.4	2.79	**27.0**	15.1	3.5	**36.9**	18.09	5.08	**58.4**
220	14.7	3.38	**32.5**	16.7	4.3	**44.4**	19.90	6.15	**70.3**
240	16.1	4.02	**38.5**	18.2	5.1	**52.7**	21.71	7.32	**83.4**
260	17.4	4.72	**45.0**	19.7	6.0	**61.6**	23.52	8.59	**97.6**
280	18.8	5.47	**52.3**	21.2	7.0	**71.2**	25.33	9.96	**113**
300	20.1	6.28	**59.6**	22.7	8.0	**81.6**	27.14	11.43	**129**
350	23.5	8.55	**80.6**	26.5	10.9	**110**	31.66	15.56	**175**
400	26.8	11.2	**105**	30.3	14.3	**144**	36.19	20.32	**228.**
450	30.2	14.1	**132**	34.1	18.1	**181**	40.71	25.72	**288**
500	33.5	17.4	**163**	37.9	22.3	**223**	45.23	31.75	**354**

Note: No allowance has been made for age, difference in diameter, or any abnormal condition of interior surface. Any factor of safety must be estimated from the local conditions and the requirements of each particular installation. It is recommended that for most commercial design purposes a safety factor of 15 to 20% be added to the values in the tables

Reprinted with permission from *Cameron Hydraulic Data* book.

Friction of Water — Asphalt-dipped Cast Iron and New Steel Pipe

(Based on Darcy's Formula)

3 Inch

Flow US gal per min	Asphalt-dipped cast iron			Std wt steel sch 40			Extra strong steel sch 80			Schedule 160—steel		
	3.0" inside dia			3.068" inside dia			2.900" inside dia			2.624" inside dia		
	Velocity ft per sec	Velocity head ft	**Head loss ft per 100 ft**	Velocity ft per sec	Velocity head ft	**Head loss ft per 100 ft**	Velocity ft per sec	Velocity head ft	**Head loss ft per 100 ft**	Velocity ft per sec	Velocity head ft	**Head loss ft per 100 ft**
10	.454	.00	**.042**	.434	.003	**.038**	.49	.00	**.050**	.593	.005	**.080**
15	.681	.01	**.088**	.651	.007	**.077**	.73	.01	**.101**	.890	.012	**.164**
20	.908	.01	**.149**	.868	.012	**.129**	.97	.02	**.169**	1.19	.022	**.275**
25	1.13	.02	**.225**	1.09	.018	**.192**	1.21	.02	**.253**	1.48	.034	**.411**
30	1.36	.03	**.316**	1.30	.026	**.267**	1.45	.03	**.351**	1.78	.049	**.572**
35	1.59	.04	**.421**	1.52	.036	**.353**	1.70	.04	**.464**	2.08	.067	**.757**
40	1.82	.05	**.541**	1.74	.047	**.449**	1.94	.06	**.592**	2.37	.087	**.933**
45	2.04	.06	**.676**	1.95	.059	**.557**	2.18	.07	**.734**	2.67	.111	**1.16**
50	2.27	.08	**.825**	2.17	.073	**.676**	2.43	.09	**.860**	2.97	.137	**1.41**
55	2.50	.10	**.990**	2.39	.089	**.776**	2.67	.11	**1.03**	3.26	.165	**1.69**
60	2.72	.12	**1.17**	2.60	.105	**.912**	2.91	.13	**1.21**	3.56	.197	**1.99**
65	2.95	.14	**1.36**	2.82	.124	**1.06**	3.16	.15	**1.40**	3.86	.231	**2.31**
70	3.18	.16	**1.57**	3.04	.143	**1.22**	3.40	.18	**1.61**	4.15	.268	**2.65**
75	3.40	.18	**1.79**	3.25	.165	**1.38**	3.64	.21	**1.83**	4.45	.307	**3.02**
80	3.63	.21	**2.03**	3.47	.187	**1.56**	3.88	.23	**2.07**	4.75	.350	**3.41**
85	3.86	.23	**2.28**	3.69	.211	**1.75**	4.12	.26	**2.31**	5.04	.395	**3.83**
90	4.08	.26	**2.55**	3.91	.237	**1.95**	4.37	.29	**2.58**	5.34	.443	**4.27**
95	4.31	.29	**2.83**	4.12	.264	**2.16**	4.61	.33	**2.86**	5.63	.493	**4.73**
100	4.54	.32	**3.12**	4.34	.293	**2.37**	4.85	.36	**3.15**	5.93	.546	**5.21**
110	4.99	.39	**3.75**	4.77	.354	**2.84**	5.33	.44	**3.77**	6.53	.661	**6.25**
120	5.45	.46	**4.45**	5.21	.421	**3.35**	5.81	.52	**4.45**	7.12	.787	**7.38**
130	5.90	.54	**5.19**	5.64	.495	**3.90**	6.30	.62	**5.19**	7.71	.923	**8.61**
140	6.35	.63	**6.00**	6.08	.574	**4.50**	6.79	.71	**5.98**	8.31	1.07	**9.92**
150	6.81	.72	**6.87**	6.51	.659	**5.13**	7.28	.82	**6.82**	8.90	1.23	**11.3**
160	7.26	.82	**7.79**	6.94	.749	**5.80**	7.76	.93	**7.72**	9.49	1.40	**12.8**
180	8.17	1.04	**9.81**	7.81	.948	**7.27**	8.72	1.01	**9.68**	10.68	1.77	**16.1**
200	9.08	1.28	**12.1**	8.68	1.17	**8.90**	9.70	1.46	**11.86**	11.87	2.19	**19.8**
220	9.98	1.55	**14.5**	9.55	1.42	**10.7**	10.7	1.78	**14.26**	13.05	2.64	**23.8**
240	10.9	1.84	**17.3**	10.4	1.69	**12.7**	11.6	2.07	**16.88**	14.24	3.15	**28.2**
260	11.8	2.16	**20.2**	11.3	1.98	**14.8**	12.6	2.46	**19.71**	15.43	3.69	**32.9**
280	12.7	2.51	**23.4**	12.2	2.29	**17.1**	13.6	2.88	**22.77**	16.61	4.28	**38.0**
300	13.6	2.88	**26.8**	13.0	2.63	**19.5**	14.5	3.26	**26.04**	17.80	4.92	**43.5**
320	14.5	3.28	**30.4**	13.9	3.00	**22.1**	15.5	3.77	**29.53**	18.99	5.59	**49.4**
340	15.4	3.70	**34.3**	14.8	3.38	**24.9**	16.5	4.22	**33.24**	20.17	6.32	**55.6**
360	16.3	4.15	**38.4**	15.6	3.79	**27.8**	17.5	4.73	**37.16**	21.36	7.08	**62.2**
380	17.2	4.62	**42.7**	16.5	4.23	**30.9**	18.4	5.27	**41.31**	22.55	7.89	**69.2**
400	18.2	5.12	**47.3**	17.4	4.68	**34.2**	19.4	5.81	**45.67**	23.73	8.74	**76.5**
420	19.1	5.65	**52.1**	18.2	5.16	**37.6**	20.4	6.43	**50.25**	24.92	9.64	**84.2**
440	20.0	6.20	**57.1**	19.1	5.67	**41.2**	21.4	7.13	**55.05**	26.11	10.58	**92.2**
460	20.9	6.77	**62.4**	20.0	6.19	**44.9**	22.3	7.75	**60.06**	27.29	11.56	**101**
480	21.8	7.38	**67.9**	20.8	6.74	**48.8**	23.3	8.37	**65.30**	28.48	12.59	**109**
500	22.7	8.00	**73.6**	21.7	7.32	**52.9**	24.2	9.15	**70.75**	29.66	13.66	**119**
550	25.0	9.68	**88.9**	23.9	8.85	**63.8**	26.7	11.1	**85.33**	32.63	16.53	**143**
600	27.2	11.5	**106**	26.0	10.5	**75.7**	29.1	13.1	**101**	35.60	19.67	**170**
650	29.5	13.5	**124**	28.2	12.4	**88.6**	31.6	15.5	**119**	38.56	23.08	**199**

Note: No allowance has been made for age, difference in diameter, or any abnormal condition of interior surface. Any factor of safety must be estimated from the local conditions and the requirements of each particular installation. It is recommended that for most commercial design purposes a safety factor of 15 to 20% be added to the values in the tables

Reprinted with permission from *Cameron Hydraulic Data* book.

PART TWO: FRICTION LOSSES IN PIPE FITTINGS

Friction of Water

Friction Loss in Pipe Fittings

Resistance coefficient K $\left(\text{use in formula } h_f = K \frac{V^2}{2g}\right)$

Note: Fittings are standard with full openings.

Fitting	L/D	Nominal pipe size ½	¾	1	1¼	1½	2	2½–3	4	6	8–10	12–16	18–24
		K value											
Gate Valves	8	0.22	0.20	0.18	0.18	0.15	0.15	0.14	0.14	0.12	0.11	0.10	0.10
Globe Valves	340	9.2	8.5	7.8	7.5	7.1	6.5	6.1	5.8	5.1	4.8	4.4	4.1
Angle Valves	55	1.48	1.38	1.27	1.21	1.16	1.05	0.99	0.94	0.83	0.77	0.72	0.66
Angle Valves	150	4.05	3.75	3.45	3.30	3.15	2.85	2.70	2.55	2.25	2.10	1.95	1.80
Ball Valves	3	0.08	0.08	0.07	0.07	0.06	0.06	0.05	0.05	0.05	0.04	0.04	0.04

Calculated from data in Crane Co. Technical Paper No. 410.

Reprinted with permission from *Cameron Hydraulic Data* book.

Friction of Water

Friction Losses in Pipe Fittings

Resistance coefficient K (use in formula $h_f = K \frac{V^2}{2g}$)

Note: Fittings are standard with full openings.

Fitting		L/D	Nominal pipe size											
			½	¾	1	1¼	1½	2	2½–3	4	6	8–10	12–16	18–24
			K value											
Butterfly Valve								0.86	0.81	0.77	0.68	0.63	0.35	**0.30**
Plug Valve straightway		18	0.49	0.45	0.41	0.40	0.38	0.34	0.32	0.31	0.27	0.25	0.23	**0.22**
Plug Valve 3-way thru-flo		30	0.81	0.75	0.69	0.66	0.63	0.57	0.54	0.51	0.45	0.42	0.39	**0.36**
Plug Valve branch-flo		90	2.43	2.25	2.07	1.98	1.89	1.71	1.62	1.53	1.35	1.26	1.17	**1.08**
Standard elbow	90°	30	0.81	0.75	0.69	0.66	0.63	0.57	0.54	0.51	0.45	0.42	0.39	**0.36**
	45°	16	0.43	0.40	0.37	0.35	0.34	0.30	0.29	0.27	0.24	0.22	0.21	**0.19**
	long radius 90°	16	0.43	0.40	0.37	0.35	0.34	0.30	0.29	0.27	0.24	0.22	0.21	**0.19**

Calculated from data in Crane Co., Technical Paper No. 410.

Reprinted with permission from *Cameron Hydraulic Data* book.

Friction of Water

Friction Losses in Pipe Fittings

Resistance coefficient K $\left(\text{use in formula } h_f = K\frac{V^2}{2g}\right)$

Note: Fittings are standard with full openings.

Fitting	Type of bend	L/D	Nominal pipe size ½	¾	1	1¼	1½	2	2½–3	4	6	8–10	12–16	18–24
			K value											
Close Return Bend		50	1.35	1.25	1.15	1.10	1.05	0.95	0.90	0.85	0.75	0.70	0.65	0.60
Standard Tee	thru flo	20	0.54	0.50	0.46	0.44	0.42	0.38	0.36	0.34	0.30	0.28	0.26	0.24
	thru branch	60	1.62	1.50	1.38	1.32	1.26	1.14	1.08	1.02	0.90	0.84	0.78	0.72
90° Bends. Pipe bends, flanged elbows, butt welded elbows	r/d = 1	20	0.54	0.50	0.46	0.44	0.42	0.38	0.36	0.34	0.30	0.28	0.26	0.24
	r/d = 2	12	0.32	0.30	0.28	0.26	0.25	0.23	0.22	0.20	0.18	0.17	0.16	0.14
	r/d = 3	12	0.32	0.30	0.28	0.26	0.25	0.23	0.22	0.20	0.18	0.17	0.16	0.14
	r/d = 4	14	0.38	0.35	0.32	0.31	0.29	0.27	0.25	0.24	0.21	0.20	0.18	0.17
	r/d = 6	17	0.46	0.43	0.39	0.37	0.36	0.32	0.31	0.29	0.26	0.24	0.22	0.20
	r/d = 8	24	0.65	0.60	0.55	0.53	0.50	0.46	0.43	0.41	0.36	0.34	0.31	0.29
	r/d = 10	30	0.81	0.75	0.69	0.66	0.63	0.57	0.54	0.51	0.45	0.42	0.39	0.36
	r/d = 12	34	0.92	0.85	0.78	0.75	0.71	0.65	0.61	0.58	0.51	0.48	0.44	0.41
	r/d = 14	38	1.03	0.95	0.87	0.84	0.80	0.72	0.68	0.65	0.57	0.53	0.49	0.46
	r/d = 16	42	1.13	1.05	0.97	0.92	0.88	0.80	0.76	0.71	0.63	0.59	0.55	0.50
	r/d = 18	46	1.24	1.15	1.06	1.01	0.97	0.87	0.83	0.78	0.69	0.64	0.60	0.55
	r/d = 20	50	1.35	1.25	1.15	1.10	1.05	0.95	0.90	0.85	0.75	0.70	0.65	0.60
Mitre Bends	α = 0°	2	0.05	0.05	0.05	0.04	0.04	0.04	0.04	0.03	0.03	0.03	0.03	0.02
	α = 15°	4	0.11	0.10	0.09	0.09	0.08	0.08	0.07	0.07	0.06	0.06	0.05	
	α = 30°	8	0.22	0.20	0.18	0.18	0.17	0.15	0.14	0.14	0.12	0.11	0.10	0.10
	α = 45°	15	0.41	0.38	0.35	0.33	0.32	0.29	0.27	0.26	0.23	0.21	0.20	0.18
	α = 60°	25	0.68	0.63	0.58	0.55	0.53	0.48	0.45	0.43	0.38	0.35	0.33	0.30
	α = 75°	40	1.09	1.00	0.92	0.88	0.84	0.76	0.72	0.68	0.60	0.56	0.52	0.48
	α = 90°	60	1.62	1.50	1.38	1.32	1.26	1.14	1.08	1.02	0.90	0.84	0.78	0.72

Calculated from data in Crane Co. Technical Paper No. 410.

Reprinted with permission from *Cameron Hydraulic Data* book.

Friction of Water

Friction Losses in Pipe Fittings

Resistance coefficient K $\left(\text{use in formula } h_f = K\frac{V^2}{2g}\right)$

Note: Fittings are standard with full port openings.

Fitting stop-check valves	L/D	Minimum velocity for full disc lift		Nominal pipe size											
		general ft/sec†	water ft/sec	½	¾	1	1¼	1½	2	2½–3	4	6	8–10	12–16	18–24
				K value*											
	400	$55\sqrt{\bar{V}}$	6.96	10.8	10	9.2	8.8	8.4	7.5	7.2	6.8	6.0	5.6	5.2	4.8
	200	$75\sqrt{\bar{V}}$	9.49	5.4	5	4.6	4.4	4.2	3.8	3.6	3.4	3.0	2.8	2.6	2.4
	350	$60\sqrt{\bar{V}}$	7.59	9.5	8.8	8.1	7.7	7.4	6.7	6.3	6.0	5.3	4.9	4.6	4.2
	300	$60\sqrt{\bar{V}}$	7.59	8.1	7.5	6.9	6.6	6.3	5.7	5.4	5.1	4.5	4.2	3.9	3.6
	55	$140\sqrt{\bar{V}}$	17.7	1.5	1.4	1.3	1.2	1.2	1.1	1.0	.94	.83	.77	.72	.66

Calculated from data in Crane Co. Technical Paper No. 410.
* These K values for flow giving full disc lift. K values are higher for low flows giving partial disc lift.
† In these formulas, $\bar{V}$, is specific volume—ft^3/lb.

Reprinted with permission from *Cameron Hydraulic Data* book.

Friction of Water
Friction Loss in Pipe Fittings

Resistance coefficient K $\left(\text{use in formula } h_f = K\frac{V^2}{2g}\right)$

Note: Fittings are standard with full port openings.

Fitting	L/D	Minimum velocity for full disc lift: general ft/sec†	Minimum velocity for full disc lift: water ft/sec	Nominal pipe size: ½	¾	1	1¼	1½	2	2½–3	4	6	8–10	12–16	18–24
				K value*											
Swing check valve	100	35 $\sqrt{V}$	4.43	2.7	2.5	2.3	2.2	2.1	1.9	1.8	1.7	1.5	1.4	1.3	1.2
	50	48 $\sqrt{V}$	6.08	1.4	1.3	1.2	1.1	1.1	1.0	0.9	0.9	.75	70	.65	.6
Lift check valve	600	40 $\sqrt{V}$	5.06	16.2	15	13.8	13.2	12.6	11.4	10.8	10.2	9.0	8.4	7.8	7.2
	55	140 $\sqrt{V}$	17.7	1.5	1.4	1.3	1.2	1.2	1.1	1.0	.94	.83	.77	.72	.66
Tilting disc check valve	5°	80 $\sqrt{V}$	10.13						.76	.72	.68	.60	.56	.39	.24
	15°	30 $\sqrt{V}$	3.80						2.3	2.2	2.0	1.8	1.7	1.2	.72
Foot valve with strainer poppet disc	420	15 $\sqrt{V}$	1.90	11.3	10.5	9.7	9.3	8.8	8.0	7.6	7.1	6.3	5.9	5.5	5.0
Foot valve with strainer hinged disc	75	35 $\sqrt{V}$	4.43	2.0	1.9	1.7	1.7	1.7	1.4	1.4	1.3	1.1	1.1	1.0	.90

Calculated from data in Crane Co Technical Paper No 410

* These K values for flow giving full disc lift. K values are higher for low flows giving partial disc lift

† In these formulas, V_s is specific volume—ft^3/lb

Reprinted with permission from *Cameron Hydraulic Data* book.

Friction of Water

Friction Loss in Pipe Fittings

Resistance coefficient $\left(\text{use in formula } h_f = K \frac{V^2}{2g}\right)$

Fitting	Description	All pipe sizes
		K value
Pipe exit	projecting sharp edged rounded	1.0
Pipe entrance	inward projecting	0.78
Pipe entrance flush	sharp edged	0.5
	r/d = 0.02	0.28
	r/d = 0.04	0.24
	r/d = 0.06	0.15
	r/d = 0.10	0.09
	r/d = 0.15 & up	0.04

From Crane Co. Technical Paper 410.

Reprinted with permission from *Cameron Hydraulic Data* book.

PART THREE: NET POSITIVE SUCTION HEAD FOR PUMPS

The Net Positive Suction Head is defined as the total suction head measured in feet of liquid (in absolute pressure value at the pump centerline or what is called the impeller eye) from which is deducted the absolute vapor pressure (measured also in feet of liquid) of the liquid being pumped. This value must be positive.

Based on the suction type, there are two formulas:

1. Suction lift or the supply level being below the pump centerline.

$$NPSH = h_a - h_{vpa} - h_{st} - h_{fs}$$

2. Flooded suction or the supply is above the pump centerline.

$$NPSH = h_a - h_{vpa} + h_{st} - h_{fs}$$

where:

h_a = The absolute pressure (measured in feet of liquid) on the surface of the liquid supply level. (This is the atmospheric (atm) pressure for an open tank or sump or the absolute pressure existing in a closed tank.)

h_{vpa} = The head in feet corresponding to the vapor pressure of the liquid at the temperature at which it is being pumped.

h_{st} = The static height in feet that the liquid supply level is above or below the pump centerline.

h_{fs} = All suction line head friction losses (measured in feet) including entrance losses, friction losses through pipe, valves, and fittings, etc.

For an example, see the figure on the next page.

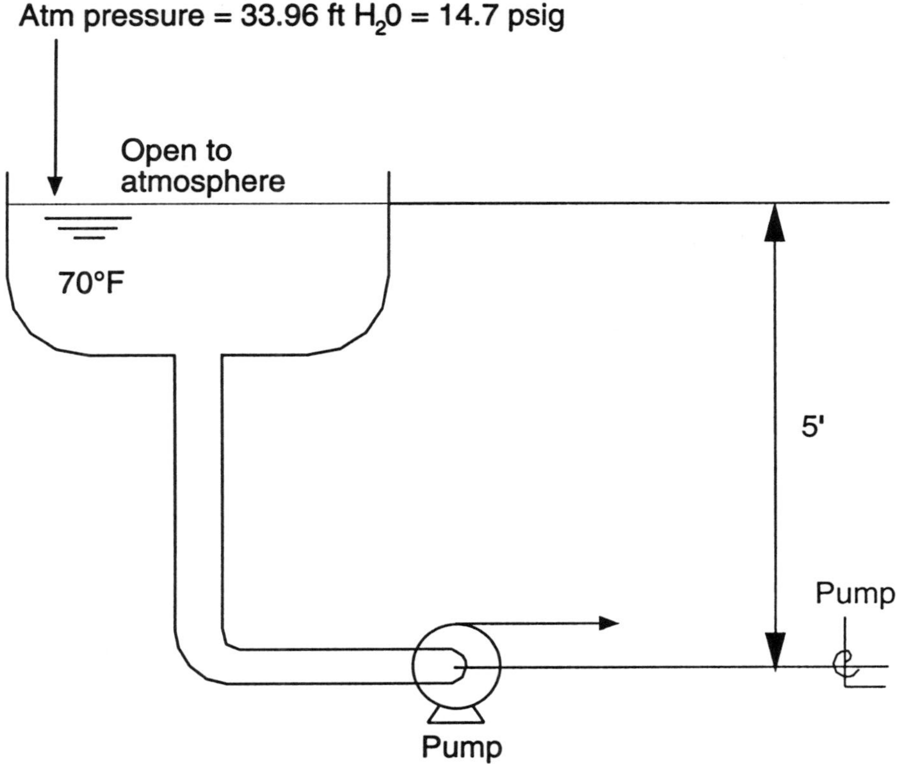

System is flooded (supply above pump centerline).

Liquid water is at 70°F.

Location is at sea level. (It is known that the atm pressure varies with altitude.)

NPSH = 33.96 (ft of water) - 0.79 ft corresponding to vapor pressure + 5 ft static - 2.40 ft head* loss = 35.77 ft

Net Positive Suction Head available is 35.77 feet of H_2O.

*2.40 = Estimated suction line losses including pipe, fittings, and entrance/exit.

PART FOUR: PRESSURE CONVERSION TABLE

Corresponding pressure in pounds per square inch (psi) to corresponding head in feet

psi	0	1	2	3	4	5	6	7	8	9
0		2.3	4.6	6.9	9.2	11.6	13.9	16.2	18.5	20.8
10	23.1	25.4	27.7	30.0	32.3	34.7	37.0	39.3	41.6	43.9
20	46.2	48.5	50.8	53.1	55.4	57.8	60.1	62.4	64.7	67.0
30	69.3	71.6	73.9	76.2	78.5	80.9	83.2	85.5	87.8	90.1
40	92.4	94.7	97.0	99.3	101.6	104.0	106.3	108.6	110.9	113.2
50	115.5	117.8	120.1	122.4	124.7	127.1	129.4	131.7	134.0	136.3
60	138.6	140.9	143.2	145.5	147.8	150.2	152.5	154.8	157.1	159.4
70	161.7	164.0	166.3	168.6	170.9	173.3	175.6	177.9	180.2	182.5
80	184.8	187.1	189.4	191.7	194.0	196.4	198.7	201.0	203.3	205.6
90	207.9	210.2	212.5	214.8	217.7	219.5	221.8	224.1	226.4	228.7
100	231.0	233.3	235.6	237.9	240.2	242.6	244.9	247.2	249.5	251.8
110	254.1	256.4	258.7	261.0	263.3	265.7	268.0	270.3	272.6	274.9
120	277.2	279.5	281.8	284.1	286.4	288.8	291.1	293.4	295.7	298.0
130	300.3	302.6	304.9	307.2	309.5	311.9	314.2	316.5	318.8	321.1
140	323.4	325.7	328.0	330.3	332.6	335.0	337.3	339.6	341.9	344.2
150	346.5	348.8	351.1	353.4	355.7	358.1	360.4	362.7	365.0	367.3
160	369.6	371.9	374.2	376.5	378.8	381.2	383.5	385.8	388.1	390.4
170	392.7	395.0	397.3	399.6	401.9	404.3	406.6	408.9	411.2	413.5
180	415.8	418.1	420.4	422.7	425.0	427.4	429.7	432.0	434.3	436.6
190	438.9	441.2	443.5	445.8	448.1	450.5	452.8	455.1	457.4	459.7
200	462.0	464.3	466.6	468.9	471.2	473.6	475.9	478.2	480.5	482.8
210	485.1	487.4	489.7	492.0	494.3	496.7	499.0	501.3	503.6	505.9
220	508.2	510.5	512.8	515.1	517.4	519.8	522.1	524.4	526.7	529.0
230	531.3	533.6	535.9	538.2	540.5	542.9	545.2	547.5	549.8	552.1
240	554.4	556.7	559.0	561.3	563.6	566.0	568.3	570.6	572.9	575.2
250	577.5	579.8	582.1	584.4	586.7	589.1	591.4	593.7	596.0	598.3

To read the table use the following example as a guide.
For 13 psi, go horizontal from 10 at the left and read under 3 vertically.
Answer is 30 ft (corresponding to 13 psi).

Appendix C

Informative Tables and Charts

This appendix contains the following information:

Part One: Water Characteristic Tables

- Properties of Water at Various Temperatures
- Specific Gravity of Water at Atmospheric Pressure

Part Two: Psychrometric Chart (Courtesy, Carrier Corporation)

Part Three: Air Conditioning Load Estimate (Courtesy, Carrier Corporation)

Part Four: Gas Pipe Sizing (Courtesy, Reznor)

Part Five: Heat Loss Estimate

PART ONE: WATER CHARACTERISTIC TABLES

Properties of Water at Various Temperatures

Temp F	Pressure of saturated vapor lb/in² abs	Specific volume		Density specific wt		Conversion factor ft/lb/in²	Kinematic viscosity centistokes	Temperature	
		ft³/lb	gal/lb	lb/ft³	*g/cm³			F	C
32	0.08859	0.016022	0.1199	62.414	0.9998	2.307	1.79	32	0
33	0.09223	0.016021	0.1198	62.418	0.9999	2.307	1.75	33	0.6
34	0.09600	0.016021	0.1198	62.418	0.9999	2.307	1.72	34	1.1
35	0.09991	0.016020	0.1198	62.420	0.9999	2.307	1.68	35	1.7
36	0.10395	0.016020	0.1198	62.420	0.9999	2.307	1.66	36	2.2
37	0.10815	0.016020	0.1198	62.420	0.9999	2.307	1.63	37	2.8
38	0.11249	0.016019	0.1198	62.425	1.0000	2.307	1.60	38	3.3
39	0.11698	0.016019	0.1198	62.425	1.0000	2.307	1.56	39	3.9
40	0.12163	0.016019	0.1198	62.425	1.0000	2.307	1.54	40	4.4
41	0.12645	0.016019	0.1198	62.426	1.0000	2.307	1.52	41	5
42	0.13143	0.016019	0.1198	62.426	1.0000	2.307	1.49	42	5.6
43	0.13659	0.016019	0.1198	62.426	1.0000	2.307	1.47	43	6.1
44	0.14192	0.016019	0.1198	62.426	1.0000	2.307	1.44	44	6.7
45	0.14744	0.016020	0.1198	62.42	0.9999	2.307	1.42	45	7.2
46	0.15314	0.016020	0.1198	62.42	0.9999	2.307	1.39	46	7.8
47	0.15904	0.016021	0.1198	62.42	0.9999	2.307	1.37	47	8.3
48	0.16514	0.016021	0.1198	62.42	0.9999	2.307	1.35	48	8.9
49	0.17144	0.016022	0.1198	62.41	0.9998	2.307	1.33	49	9.4
50	0.17796	0.016023	0.1199	62.41	0.9998	2.307	1.31	50	10
51	0.18469	0.016023	0.1199	62.41	0.9998	2.307	1.28	51	10.6
52	0.19165	0.016024	0.1199	62.41	0.9997	2.307	1.26	52	11.1
53	0.19883	0.016025	0.1199	62.40	0.9996	2.308	1.24	53	11.7
54	0.20625	0.016026	0.1199	62.40	0.9996	2.308	1.22	54	12.2
55	0.21392	0.016027	0.1199	62.39	0.9995	2.308	1.20	55	12.8
56	0.22183	0.016028	0.1199	62.39	0.9994	2.308	1.19	56	13.3
57	0.23000	0.016029	0.1199	62.39	0.9994	2.308	1.17	57	13.9
58	0.23843	0.016031	0.1199	62.38	0.9993	2.308	1.16	58	14.4
59	0.24713	0.016032	0.1199	62.38	0.9992	2.309	1.14	59	15
60	0.25611	0.016033	0.1199	62.37	0.9991	2.309	1.12	60	15.6
62	0.27494	0.016036	0.1200	62.36	0.9989	2.309	1.09	62	16.7
64	0.29497	0.016039	0.1200	62.35	0.9988	2.310	1.06	64	17.8
66	0.31626	0.016043	0.1200	62.33	0.9985	2.310	1.03	66	18.9
68	0.33889	0.016046	0.1200	62.32	0.9983	2.311	1.00	68	20
70	0.36292	0.016050	0.1201	62.31	0.9981	2.311	0.98	70	21.1
75	0.42964	0.016060	0.1201	62.27	0.9974	2.313	0.90	75	23.9
80	0.50683	0.016072	0.1202	62.22	0.9967	2.314	0.85	80	26.7
85	0.59583	0.016085	0.1203	62.17	0.9959	2.316	0.81	85	29.4
90	0.69813	0.016099	0.1204	62.12	0.9950	2.318	0.76	90	32.2
95	0.81534	0.016114	0.1205	62.06	0.9941	2.320	0.72	95	35
100	0.94924	0.016130	0.1207	62.00	0.9931	2.323	0.69	100	37.8
110	1.2750	0.016165	0.1209	61.98	0.9910	2.328	0.61	110	43.3
120	1.6927	0.016204	0.1212	61.71	0.9886	2.333	0.57	120	48.9
130	2.2230	0.016247	0.1215	61.56	0.9860	2.340	0.51	130	54.4
140	2.8892	0.016293	0.1219	61.38	0.9832	2.346	0.47	140	60
150	3.7184	0.016343	0.1223	61.19	0.9802	2.353	0.44	150	65.6
160	4.7414	0.016395	0.1226	60.99	0.9771	2.361	0.41	160	71.1
170	5.9926	0.016451	0.1231	60.79	0.9737	2.369	0.38	170	76.7
180	7.5110	0.016510	0.1235	60.57	0.9703	2.377	0.36	180	82.2
190	9.340	0.016572	0.1240	60.34	0.9666	2.386	0.33	190	87.8

* Approximately numerically equal to specific gravity basis temperature reference of 39.2 F (4 C)
Calculated from data in ASME Steam Tables

Properties of Water at Various Temperatures (*Continued*)

Temp F	Pressure of saturated vapor lb in² abs	Specific volume ft³/lb	Specific volume gal/lb	Density specific wt. lb ft³	Density specific wt. *g/cm³	Conversion factor ft/lb/in²	Kinematic viscosity centistokes	Temperature °F	Temperature °C
200	11.526	0.016637	0.1245	60.11	0.9628	2.396	0.31	200	93.3
210	14.123	0.016705	0.1250	59.86	0.9589	2.406	0.29	210	98.9
212	14.696	0.016719	0.1251	59.81	0.9580			212	100.0
220	17.186	0.016775	0.1255	59.61	0.9549	2.416		220	104.4
230	20.779	0.016849	0.1260	59.35	0.9507	2.426		230	110
240	24.968	0.016926	0.1266	59.08	0.9464	2.437		240	115.6
250	29.825	0.017006	0.1272	58.80	0.9420	2.449	0.24	250	121.1
260	35.427	0.017089	0.1278	58.52	0.9374	2.461		260	126.7
270	41.856	0.017175	0.1285	58.22	0.9327	2.473		270	132.2
280	49.200	0.017264	0.1291	57.92	0.9279	2.486		280	137.8
290	57.550	0.01736	0.1299	57.60	0.9228	2.500		290	143.3
300	67.005	0.01745	0.1305	57.31	0.9180	2.513	0.20	300	148.9
310	77.667	0.01755	0.1313	56.98	0.9128	2.527		310	154.4
320	89.643	0.01766	0.1321	56.63	0.9071	2.543		320	160
330	103.045	0.01776	0.1329	56.31	0.9020	2.557		330	165.6
340	117.992	0.01787	0.1337	55.96	0.8964	2.573		340	171.1
350	134.604	0.01799	0.1346	55.59	0.8904	2.591	0.17	350	176.7
360	153.010	0.01811	0.1355	55.22	0.8845	2.608		360	182.2
370	173.339	0.01823	0.1364	54.84	0.8787	2.625		370	187.8
380	195.729	0.01836	0.1374	54.47	0.8725	2.644		380	193.3
390	220.321	0.01850	0.1384	54.05	0.8659	2.664		390	198.9
400	247.259	0.01864	0.1394	53.65	0.8594	2.684	0.15	400	204.4
410	276.694	0.01878	0.1404	53.25	0.8530	2.704		410	392.2
420	308.780	0.01894	0.1417	52.80	0.8458	2.727		420	215.6
430	343.674	0.01909	0.1428	52.38	0.8391	2.749		430	221.1
440	381.54	0.01926	0.1441	51.92	0.8317	2.773		440	226.7
450	422.55	0.01943	0.1453	51.47	0.8244	2.798	0.14	450	232.2
460	466.87	0.01961	0.1467	50.99	0.8169	2.824		460	237.8
470	514.67	0.01980	0.1481	50.51	0.8090	2.851		470	243.3
480	566.15	0.02000	0.1496	50.00	0.8010	2.880		480	248.9
490	621.48	0.02021	0.1512	49.48	0.7926	2.910		490	254.4
500	680.86	0.02043	0.1528	48.95	0.7841	2.942	0.13	500	260
510	744.47	0.02067	0.1546	48.38	0.7750	2.976		510	265.6
520	812.53	0.02091	0.1564	47.82	0.7661	3.011		520	271.1
530	885.23	0.02118	0.1584	47.21	0.7563	3.050		530	276.7
540	962.79	0.02146	0.1605	46.60	0.7465	3.090		540	282.2
550	1045.43	0.02176	0.1628	45.96	0.7362	3.133	0.12	550	287.8
560	1133.38	0.02207	0.1651	45.31	0.7258	3.178		560	293.3
570	1226.88	0.02242	0.1677	44.60	0.7145	3.228		570	298.9
580	1326.17	0.02279	0.1705	43.88	0.7029	3.281		580	304.4
590	1431.5	0.02319	0.1735	43.12	0.6908	3.339		590	310
600	1543.2	0.02364	0.1768	42.30	0.6776	3.404	0.12	600	315.6
610	1661.6	0.02412	0.1804	41.46	0.6641	3.473		610	321.1
620	1786.9	0.02466	0.1845	40.55	0.6496	3.551		620	326.6
630	1919.5	0.02526	0.1890	39.59	0.6342	3.637		630	332.2
640	2059.9	0.02595	0.1941	38.54	0.6173	3.737		640	337.8
650	2203.4	0.02674	0.2000	37.40	0.5991	3.851		650	343.3
670	2532.2	0.02884	0.2157	34.67	0.5554	4.153		670	354.4
690	2895.7	0.03256	0.2436	30.71	0.4920	4.689		690	365.6
700	3094.3	0.03662	0.2739	27.31	0.4374	5.273		700	371.1
705.47	3208.2	0.05078	0.3799	19.69	0.3155	7.312		705.47	374.15

* Approximately numerically equal to specific gravity basis temperature reference of 39.2°F (4°C).
Calculated from data in ASME Steam Tables

Reprinted with permission from *Cameron Hydraulic Data* book.

Specific Gravity of Water at Atmospheric Pressure

Degrees			Degrees		
Fahrenheit	**Celsius**	**Specific Gravity**	**Fahrenheit**	**Celsius**	**Specific Gravity**
32	0	0.9998	158	70	0.9778
39.5	4	1.0000	165.2	74	0.9755
41	5	0.9999	172.4	78	0.9731
50	10	0.9997	176	80	0.9719
57.2	14	0.9993	185	85	0.9689
64.4	18	0.9986	194	90	0.9657
68	20	0.9983	203	95	0.9623
71.6	22	0.9978	212	100	0.9584
78.8	26	0.9968	230	110	0.9514
82.4	28	0.9963	248	120	0.9433
86	30	0.9957	266	130	0.9348
96.8	36	0.9937	284	140	0.9263
104	40	0.9922	302	150	0.9171
111.2	44	0.9907	338	170	0.8976
118.4	48	0.9889	356	180	0.8872
122	50	0.9881	374	190	0.8761
129.2	54	0.9862	392	200	0.8647
136.4	58	0.9842	410	210	0.8531
140	60	0.9832	428	220	0.8404
147.2	64	0.9811	446	230	0.8275
154.4	68	0.9789			

Specific gravity is the ratio of the density of one substance to the density of a second (or reference) substance. Water is used as the reference substance.

Specific volume is the volume per unit of mass, and it is the reciprocal of specific density.

Part Two: Psychrometric Chart

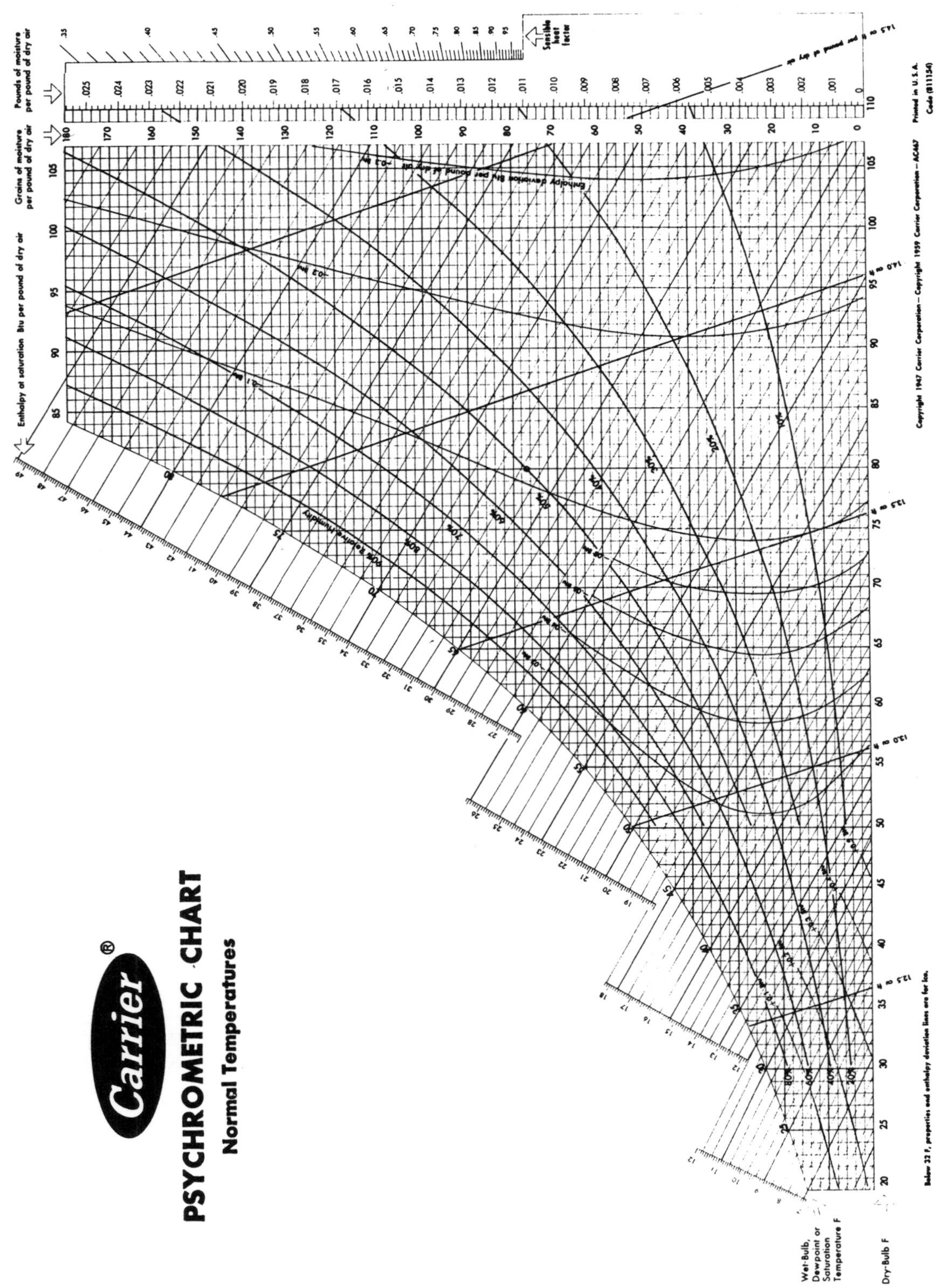

Courtesy, Carrier Corporation

PART THREE: AIR CONDITIONING LOAD ESTIMATE

SHEET ____ Carrier Air Conditioning Company DATE ____

PREPARED BY ____ OFFICE ____ PROP NO. ____ JOB NO. ____

NAME OF JOB ____

LOCATION ____ APPROVED ____

SPACE USED FOR ____

SIZE ____ × ____ = ____ SQ FT × ____ = ____ CU FT

ITEM	AREA OR QUANTITY	SUN GAIN OR TEMP. DIFF.	FACTOR	BTU/HOUR
SOLAR GAIN—GLASS				
GLASS	SQ FT ×		×	
GLASS	SQ FT ×		×	
GLASS	SQ FT ×		×	
GLASS	SQ FT ×		×	
SKYLIGHT	SQ FT ×		×	
SOLAR & TRANS. GAIN—WALLS & ROOF				
WALL	SQ FT ×		×	
WALL	SQ FT ×		×	
WALL	SQ FT ×		×	
WALL	SQ FT ×		×	
ROOF—SUN	SQ FT ×		×	
ROOF—SHADED	SQ FT ×		×	
TRANS. GAIN—EXCEPT WALLS & ROOF				
ALL GLASS	SQ FT ×		×	
PARTITION	SQ FT ×		×	
CEILING	SQ FT ×		×	
FLOOR	SQ FT ×		×	
INFILTRATION	CFM ×		× 1.08	
INTERNAL HEAT				
PEOPLE		PEOPLE ×		
POWER		HP OR KW ×		
LIGHTS		WATTS × 3.4 ×		
APPLIANCES, ETC.		×		
ADDITIONAL HEAT GAINS		×		
			SUB TOTAL	
STORAGE	SQ FT ×	× (−)		
			SUB TOTAL	
SAFETY FACTOR	%			
		ROOM SENSIBLE HEAT ■		
SUPPLY DUCT HEAT GAIN % + SUPPLY DUCT LEAK. LOSS % + FAN H. P. %				
OUTDOOR AIR	CFM ×	F ×	BF × 1.08	
		EFFECTIVE ROOM SENSIBLE HEAT ■		
LATENT HEAT				
INFILTRATION	CFM ×		GR/LB × 0.68	
PEOPLE	PEOPLE ×			
STEAM	LB/HR × 1050			
APPLIANCES, ETC.				
ADDITIONAL HEAT GAINS				
VAPOR TRANS.	SQ FT × 1/100 ×	GR/LB ×		
			SUB TOTAL	
SAFETY FACTOR	%			
		ROOM LATENT HEAT		
SUPPLY DUCT LEAKAGE LOSS		%		
OUTDOOR AIR	CFM ×	GR/LB ×	BF × 0.68	
		EFFECTIVE ROOM LATENT HEAT		
		EFFECTIVE ROOM TOTAL HEAT ■		
OUTDOOR AIR HEAT				
SENSIBLE:	CFM ×	F × (1 —	BF) × 1.08	
LATENT:	CFM ×	GR/LB × (1 —	BF) × 0.68	
RETURN DUCT HEAT GAIN % + RETURN DUCT LEAK. GAIN % + HP PUMP % + DEHUM. & PIPE LOSS %			SUB TOTAL	
		GRAND TOTAL HEAT ■		

ESTIMATE FOR	LOCAL TIME SUN TIME	PEAK LOAD	LOCAL TIME SUN TIME
HOURS OF OPERATION			

CONDITIONS	DB	WB	% RH	DP	GR/LB
OUTDOOR (OA)					
ROOM (RM)					
DIFFERENCE		X X X	X X X	X X X	

OUTDOOR AIR

VENTILATION: ____ PEOPLE × ____ CFM/PERSON = ____; ____ SQ FT × ____ CFM/SQ FT = ____; CFM VENTILATION ■

INFILTRATION: SWINGING REVOLVING DOORS ____ PEOPLE × ____ CFM/PERSON = ____; OPEN DOORS ____ DOORS × ____ CFM/DOOR = ____; EXHAUST FAN ____; CRACK ____ FEET × ____ CFM/FT = ____; CFM INFILTRATION ■

CFM OUTDOOR AIR THRU APPARATUS ■ ____ CFM_{OA}

APPARATUS DEWPOINT

ESHF: EFFECTIVE SENS HEAT FACTOR = EFFECTIVE ROOM SENS. HEAT / EFFECTIVE ROOM TOTAL HEAT = ____

ADP: INDICATED ADP = ____ F SELECTED ADP = ____ F

DEHUMIDIFIED AIR QUANTITY

TEMP. RISE: $(1 - \_\_ BF) \times (T_{RM} \_\_ F - T_{ADP} \_\_ F) = \_\_ F$

DEHUM. CFM: EFFECTIVE ROOM SENS. HEAT / (1.08 × ____ F TEMP. RISE) = ____ CFM_{DA}

OUTLET TEMP. DIFF.: ROOM SENS. HEAT / (1.08 × ____ CFM_{DA}) = ____ F (RM—OUTLET AIR)*

SUPPLY AIR QUANTITY

SUPPLY CFM: ROOM SENS. HEAT / (1.08 × ____ F DESIRED DIFF) = ____ CFM_{SA}

BYPASS CFM: ____ CFM_{SA} − ____ CFM_{DA} = ____ CFM_{BA}

RESULTING ENT & LVG CONDITIONS AT APPARATUS

EDB: $T_{RM} \_\_ F + \frac{CFM_{OA}}{CFM\dagger} \times (T_{OA} \_\_ F - T_{RM} \_\_ F) = T_{EDB} \_\_ F$

LDB: $T_{ADP} \_\_ F + \_\_ BF \times (T_{EDB} \_\_ F - T_{ADP} \_\_ F) = T_{LDB} \_\_ F$

FROM PSYCH. CHART: T_{EWB} ____ F, T_{LWB} ____ F

NOTES

*IF THIS ΔT IS TOO HIGH, DETERMINE SUPPLY CFM FOR DESIRED DIFFERENCE BY SUPPLY AIR QUANTITY FORMULA.

†WHEN BYPASSING A MIXTURE OF OUTDOOR AND RETURN AIR, USE SUPPLY CFM. WHEN BYPASSING RETURN AIR ONLY, USE DEHUMIDIFIED CFM.

AIR CONDITIONING LOAD ESTIMATE — FORM E-20 (2-68)

Courtesy, Carrier Corporation

PART FOUR: GAS PIPE SIZING

NOTE:

Specifications subject to change without notice.

ALL MODELS

PIPE SIZING, GAS METER FLOW TIME, HEAT LOSS FORM 600D
EFFECTIVE MAY, 1989/OBSOLETES FORM 600C

REZNOR® MERCER, PA. 16137

APPLIES TO: **Technical Data**

SHORT CUT HEAT LOSS ESTIMATING AND SIZING GAS PIPE SUPPLY LINES

The purpose of this sheet is to provide a quick method of estimating heat loss of buildings or of individual rooms accurate within 5% of the results obtained by longer methods. The method is such that it can be used by a service engineer where an estimate of radiation capacity is desired but where highly accurate calculations are not necessary at the time. It can also be used by other engineers for preliminary or checking calculations. The method was developed by the engineers of a large utility company and reported by Alford G. Canar.

HOW TO USE THE METHOD

1. Measure length, width and ceiling height, taking account of any shape of floor area other than one simple square or rectangle. Calculate floor area and multiply by ceiling height to obtain cubical contents.
2. Make a rapid scouting survey to observe what the prevailing conditions are so that the proper factor may be selected. See whether there are any party walls, heated space above, flat or eaved roof, storm windows, or skylights.
3. Select proper factor from Table 1 and Table 2.
4. Multiply cubical contents by the factor from Table 1 (and from Table 2 where necessary) and thus obtain the heat loss in Btu per hour.

For conditions in parts of the country where design temperature differentials other than 0°F and 70°F are customary, the factors from Table 2 are applicable by multiplying the heat loss obtained by using the heat factors in Table 1.

TABLE 2

CORRECTION FACTORS	
Minimum Design Temperature	Correction Factor
+50°F	0.29
+40°F	0.43
+30°F	0.57
+20°F	0.72
+10°F	0.86
0°F	1.00
-10°F	1.14
-20°F	1.28
-30°F	1.43

TABLE 1

TYPE OF BUILDING	MAJOR QUALIFICATIONS	MINOR QUALIFICATIONS	MULTIPLY CUBIC CONTENTS BY†
Factories or Warehouses 65°F Inside	One Story	Skylight in roof	5.8
		No skylight in roof	5.3
	No Skylights	Two story	4.3
		Three story	4.0
		Four story	3.8
		Five story	3.6
		Six story	3.4
Public Garages 60°F Inside	All Walls Exposed	Skylight in roof	5.5
		No skylight in roof	5.1
		Heated space above	4.0
	Warm Party Walls on Both Long Sides	Skylight in roof	4.7
		No skylight in roof	4.4
		Heated space above	3.0
	One Long Warm Party Wall	Skylight in roof	5.0
		No skylight in roof	4.9
		Heated space above	3.4
Stores[1] 70°F Inside	All Walls Exposed	Flat roof	6.9
		Heated space above	5.2
	Warm Party Walls on Both Long Sides	Flat roof	5.8
		Heated space above	4.1
	One Long Warm Party Wall	Flat roof	6.3
		Heated space above	4.7

† Figures are in all cases BTU heat loss per cu. ft. contents. Factors apply only to inside temperatures listed in first column and to 0°F outside temperature.
[1] If building has bad north and west exposures, increase heat loss by 10 percent.

Courtesy, Reznor

DETERMINING PIPE SIZES

To determine proper pipe size for your heating system, add up the longest length of pipe required, i.e., the distance between the main pressure regulator and the farthest heating unit in the system. To this, add the equivalent pipe length of the fittings, as determined from Table 4. Divide this total length of pipe into the total available pressure drop.

$$\frac{P_a}{L} \times 100 = P_b$$

where

P_a = Total available pressure drop (Pressure from utility minus manufacturer's recommended minimum supply pressure);

L = Total equivalent pipe length in feet;

P_b = Allowable pressure drop per 100 feet.

Determine the required CFH of gas in that branch line and find this CFH in left hand column of Table 3. Read across to the appropriate total allowable pressure drop per 100 feet from the formula. Select the proper pipe size from top of Table. NOTE: Many complex systems will show an advantage by breaking the system up into major trunk lines for closer calculations.

TABLE 3 — PRESSURE DROP PER 100 FEET

(1000 BTU gas at .60 specific gravity)

	PIPE SIZE							
CFH	½"	¾"	1"	1¼"	1½"	2"	3"	4"
25	0.125	0.029	0.008	0.002				
50	0.500	0.118	0.033	0.008	0.003			
75	1.125	0.265	0.074	0.017	0.007	0.002		
100	2.000	0.471	0.132	0.031	0.013	0.004		
125	3.125	0.736	0.206	0.048	0.020	0.006		
150	4.500	1.060	0.297	0.070	0.029	0.009	0.001	
175	6.125	1.442	0.404	0.095	0.040	0.012	0.001	
200	4.62 oz.	1.884	0.528	0.124	0.052	0.015	0.002	
250	7.23 oz.	2.944	0.825	0.194	0.081	0.024	0.003	0.001
300	10.40 oz.	4.239	1.188	0.279	0.117	0.034	0.004	0.001
350	14.16 oz.	5.770	1.617	0.380	0.159	0.046	0.006	0.001
400	1.16 lb.	7.536	2.112	0.496	0.208	0.060	0.008	0.002
500	1.81 lb.	6.81 oz.	3.300	0.775	0.325	0.095	0.012	0.003
600	2.60 lb.	9.80 oz.	4.752	1.116	0.468	0.136	0.017	0.004
800	4.62 lb.	1.09 lb.	4.88 oz.	1.984	0.833	0.242	0.030	0.007
1000	7.22 lb.	1.70 lb.	7.63 oz.	3.100	1.301	0.378	0.047	0.011
1250	11.28 lb.	2.66 lb.	11.92 oz.	4.844	2.033	0.591	0.073	0.017
1500	16.25 lb.	3.83 lb.	1.07 lb.	6.975	2.927	0.851	0.106	0.025
2000	28.88 lb.	6.80 lb.	1.91 lb.	7.17 oz.	5.204	1.512	0.188	0.044
2500	45.13 lb.	10.63 lb.	2.98 lb.	11.20 oz.	4.70 oz.	2.363	0.294	0.069
3000	64.98 lb.	15.30 lb.	4.29 lb.	1.01 lb.	6.77 oz.	3.402	0.423	0.099

NOTE: Pressure drop data above not designated in pounds or ounces is expressed in inches water-column.

TABLE 4

EQUIVALENT LENGTHS OF PIPE FITTINGS AND VALVES (IN FEET)									
TYPE OF FITTING OR VALVE	NOMINAL PIPE SIZE — INCHES								
	½	¾	1	1¼	1½	2	2½	3	4
Standard tee with entry discharge through side	3.10	4.12	5.24	6.90	8.04	10.30	12.30	15.30	20.20
90° elbow	1.55	2.06	2.62	3.45	4.02	5.17	6.16	7.67	10.10
Medium sweep elbow	1.3	1.8	2.3	3.0	3.7	4.6	5.4	6.8	9.0
Long sweep elbow or run of a standard tee or butterfly valve	1.0	1.3	1.7	2.3	2.7	3.5	4.2	5.3	7.0
45° elbow	0.73	0.96	1.22	1.61	1.88	2.41	2.88	3.58	4.70
180° close return bend	3.47	4.6	5.82	7.66	8.95	11.5	13.7	17.1	22.4
Gate valve, wide open or slight bushing reduction	0.36	0.48	0.61	0.81	0.94	1.21	1.44	1.79	2.35

Courtesy, Reznor

PART FIVE: HEAT LOSS ESTIMATE

Design Conditions **Prepared by** ______________________ **Date** __________

Outside Temperature _______°F **Desired Inside Temperature** _______°F

Temperature Difference (TD) _______°F **Direction of Prevailing Cold Winds** _________________

Heat Loss General Formula

Heat loss = Area x U x TD where: TD = Temperature difference outside-inside **Estimate Btuh**

U = Transmission coefficient

1. Windows________________________________ sq ft x ____ x ____ °F = ______________
2. Wall_______ sq ft - _______ sq ft windows = _______ sq ft x ____ x ____ °F = ______________
3. Wall_______ sq ft - _______ sq ft windows = _______ sq ft x ____ x ____ °F = ______________
4. Wall_______ sq ft - _______ sq ft windows = _______ sq ft x ____ x ____ °F = ______________
5. Roof or ceiling ____________________________ sq ft x ____ x ____ °F = ______________
6. Floor ____________________________________ sq ft x ____ x ____ °F = ______________
7. Partition__________________________________ sq ft x ____ x ____ °F = ______________
8. Ventilation and/or infiltration _______cfm x _____°F x 1.08_______________ = ______________
9. Humidification________ Space Volume in ft^3 x 0.5 Btu/ft^3 _______________ = ______________
10. Heat loss ____ = ______________
11. Safety factor____% of (item 10) _________________________________ = ______________
12. Total heat loss (add items 10 and 11) ____________________________ **Total** = _________

All heat loads are in Btu per hour (Btuh).

Transmission coefficient U shall be based on construction (R factors).

Informative U Factors**	**Btu/sq ft/hr/°F**
Windows — single	1.13
Windows — double	0.57
Windows — glass block	0.46
Walls — heavy masonry	0.30
Walls — average masonry	0.40
Walls — insulated masonry or frame	0.15
Walls — average frame	0.30
Partition — double thickness	0.35
Floor	0.30
Ceiling and pitched roof combined	0.35*
Ceiling and flat roof combined	0.43*
Ceiling under occupied floor	0.30*

*No insulation. Multiply coefficient given by 0.4 for 1" insulation, 0.3 for 2" insulation, 0.2 for 4" insulation.

** These U factors are to be used for overall calculations only. For specific calculations use ASHRAE *Handbook -- Fundamentals.*

UNIT AND CONVERSION FACTORS

The following are conversions used specifically for hydronic applications.

Multiply	By	To obtain
Watt	3.41	Btuh
Kilowatt	3413	Btuh
Btuh	0.000293	Kilowatts
Btuh	0.2985×10^4	Boiler horsepower
Boiler horsepower	33,475	Btuh
Inches	2.54	Centimeters
Feet of water	2.24	Centimeters of mercury
Cubic feet	28.32	Liters
Cubic feet of water	62.3	Pound of water at 70°F
Feet (length)	30.48	Centimeters
Feet of water	0.433	Pounds per square inch (psi)
Psi	2.31	Feet of water
Gallons	3.78	Liters
Gallons of water	8.34	Pound of water at 70°F
Btuh	0.252	Kilocalories per hour
Kilocalories per hour	3.97	Btuh
Kilograms per cm^2	14.22	Psi
Meter	3.28	Feet
Tons of refrigeration	12,000	Btuh
Pound of ice	144	Btu latent heat
Ton (short)	2000	Pound
Horsepower	2546	Btuh
Psig + 14.7	—	Psia

To convert from other systems of measurement to SI values, the following conversion factors are to be used. (Note: For additional conversion equivalents not shown herein, refer to ANSI Z210.1. Also issued as ASTM E380.)

UNIT AND CONVERSION FACTORS

a. Linear acceleration

$ft/s^2 = 0.3048\ m/s^2$	$m/s^2 = 3.28\ ft/s^2$
$in/s^2 = 0.0254\ m/s^2$	$m/s^2 = 9.37\ in/s^2$

b. Area

$acre = 4046.9\ m^2$	$m^2 = 0.0000247\ acre$
$ft^2 = 0.0929\ m^2$	$m^2 = 10.76\ ft^2$
$in^2 = 0.000645\ m^2 = 645.16\ mm^2$	$m^2 = 1550.39\ in^2$
$mi^2 = 2{,}589{,}988\ m^2 = 1.59$	$km^2 = 0.39\ mi^2$
$yd^2 = 0.836\ m^2$	$m^2 = 1.2\ yd^2$

c. Bending moment (torque)

pound-force·inch (lbf·in) = 0.113 Newton meter (N·m)	N·m = 8.85 lbf·in
lbf·ft = 1.356 N·m	N·m = 0.74 lbf·in

d. Bending moment (torque) per unit length

lbf·in/in = 4.448 N·m/m	N·m/m = 0.225 lbf·in/in
lbf·ft/in = 53.379 N·m/m	N·m/m = 0.019 lbf·ft/in

e. Electricity and magnetism

ampere	=	1 A
ampere-hour	=	3600 Ah
coulomb	=	1 C
farad	=	1 F
henry	=	1 H
ohm	=	1 W
volt	=	1 V

f. Energy (work)

British thermal unit (Btu) = 1055 Joule (J)	J = 0.000948 Btu
ft·lbf = 1.356 J	J = 0.074 ft·lbf
kWh = 3,600,000 J	J = 0.000000278 kWh

g. Energy per unit area per unit time

Btu/(ft^2·s) = 11,349 W/m^2	W/m^2 = 0.000088 Btu/(ft^2·s)

h. Force

ounce-force (ozf) = 0.287 N	N = 3.48 ozf
pound-force (lbf) = 4.448 N	N = 0.23 lbf
kilogram-force (kgf) = 9.807 N	N = 0.1 kgf

i. Force per unit length

lbf/in = 175.1 N/m	N/m = 0.0057 lbf/in
lbf/ft = 14.594 N/m	N/m = 0.069 lbf/ft

j. Heat

Btu·in/(s·ft^2·°F) = 519.2 W/(m·K)	W/(m·K) = 0.002 Btu·in/(s·ft)
Btu·in/(h·ft^2·°F) = 0.144 W/(m·K)	W/(m·K) = 6.94 Btu·in/(h·ft^2)
Btu/ft^2 = 11,357 J/m^2	J/m^2 = 0.000088 Btu/ft^2
Btu/(h·ft^2·°F) = 5.678 W/(m^2·K)	W/(m^2·K) = 0.176 Btu/(h·ft^2·°F)
Btu/lbm = 2326 J/kg	J/kg = 0.00043 Btu/lbm
Btu/(lbm·°F) = 4186.8 J/(kg·K)	J/(kg·K) = 0.000239 Btu/(lbm·°F)
(°F·h·ft^2)/Btu = 0.176 (K·m^2)/W	(K·m^2)/W = 5.68 (°F·h·ft^2)/Btu

k. Length

in = 0.0254 m	m = 39.37 in
ft = 0.3048 m	m = 3.28 ft
yd = 0.914 m	m = 1.1 yd
mi = 1609.3 m	m = 0.000621 mi

l. Light (illuminance)

footcandle (fc) = 10.764 1x	1x = 0.093 fc

m. Mass

ounce-mass (ozm) = 0.028 kg	kg = 35.7 ozm
pound-mass (lbm) = 0.454 kg	kg = 2.2 lbm

n. Mass per unit area

lbm/ft^2 = 4.882 kg/m^2	kg/m^2 = 0.205 lbm/ft^2

o. Mass per unit length

lbm/ft = 1.488 kg/m	kg/m = 0.67 lbm/ft

p. Mass per unit time (flow)

lbm/h = 0.0076 kg/s	kg/s = 131.58 lbm/h

q. Mass per unit volume (density)

lbm/ft^3 = 16.019 kg/m^3	kg/m^3 = 0.062 lbm/ft^3
lbm/in^3 = 27,680 kg/m^3	kg/m^3 = 0.000036 lbm/in^3
lbm/gal = 119.8 kg/m^3	kg/m^3 = 0.008347 lbm/gal

r. Moment of inertia

lb/ft^2 = 0.042 kg·m^2	kg·m^2 = 23.8 lb/ft^2

s. Plane angle

degree = 17.453 mrad	mrad = 0.057 deg
minute = 290.89 μrad	μrad = 0.00344 min
second = 4.848 μrad	μrad = 0.206 s

t. Power

Btu/h = 0.293 W	W = 3.41 Btu/h
(ft·lbf)/h = 0.38 mW	mW = 2.63 (ft·lbf)/h
horsepower (hp) = 745.7 W	W = 0.00134 hp

u. **Pressure (stress) force per unit area**

atmosphere = 101.325 kiloPascal (kPa)	kPa = 0.009869 atm
inch of mercury (at 60°F) = 3.3769 kPa	kPa = 0.296 in of Hg
inch of water (at 60°F) = 248.8 Pa	Pa = 0.004 in of water
lbf/ft^2 = 47.88 Pa	Pa = 0.02 lbf/ft^2
lbf/in^2 = 6.8948 kPa	kPa = 0.145 lbf/in^2

v. **Temperature equivalent**

$t_k = (t_f + 459.67)/1.8$	$t_f = 1.8\ t_k - 459.67$
$t_c = (t_f - 32)/1.8$	$t_f = 1.8\ t_c + 32$

w. **Velocity (length per unit time)**

ft/h = 0.085 mm/s	mm/s = 11.76 ft/h
ft/min = 5.08 mm/s	mm/s = 0.197 ft/min
ft/s = 0.3048 m/s	m/s = 3.28 ft/s
in/s = 0.0254 m/s	m/s = 39.37 in/s
mi/h = 0.447 m/s	m/s = 2.24 mi/h

x. **Volume**

ft^3 = 0.028 m^3 = 28.317 L	m^3 = 35.71 ft^3
in^3 = 16,378 mL	mL = 0.061 m^3
gal = 3.785L	L = 0.264 gal
oz = 29.574 mL	mL = 0.034 oz
pt = 473.18 mL	mL = 0.002 pt
qt = 946.35 mL	mL = 0.001 qt
acre/ft = 1233.49 m^3	m^3 = 0.00081 acre/ft

y. **Volume per unit time (flow)**

ft^3/min = 0.472 L/s	L/s = 2.12 ft^3/m
in^3/min = 0.273 mL/s	mL/s = 3.66 in^3/m
gal/min = 0.063 L/s	L/s = 15.87 gal/min

TEMPERATURE CONVERSION FACTORS

The numbers in the center column refer to the known temperature, either in °F or °C, to be converted to the other scale. If converting from °F to °C, the number in the center column represents the known temperature, in °F, and its equivalent temperature, in °C, will be found in the left column. If converting from °C to °F, the number in the center represents the known temperature, in °C, and its equivalent temperature, in °F, will be found in the right column.

°C	Known temp. °F or °C	°F	°C	Known temp. °F or °C	°F
- 59	- 74	- 101	- 32.2	- 26	- 14.8
- 58	- 73	- 99	- 31.6	- 25	- 13.0
- 58	- 72	- 98	- 31.1	- 24	- 11.2
- 57	- 71	- 96	- 30.5	- 23	- 9.4
- 57	- 70	- 94	- 30.0	- 22	- 7.6
- 56	- 69	- 92	- 29.4	- 21	- 5.8
- 56	- 68	- 90	- 28.9	- 20	- 4.0
- 55	- 67	- 89	- 28.3	- 19	- 2.2
- 54	- 66	- 87	- 27.7	- 18	- 0.4
- 54	- 65	- 85	- 27.2	- 17	1.4
- 53	- 64	- 83	- 26.6	- 16	3.2
- 53	- 63	- 81	- 26.1	- 15	5.0
- 52	- 62	- 80	- 25.5	- 14	6.8
- 52	- 61	- 78	- 25.0	- 13	8.6
- 51	- 60	- 76	- 24.4	- 12	10.4
- 51	- 59	- 74	- 23.8	- 11	12.2
- 50	- 58	- 72	- 23.3	- 10	14.0
- 49	- 57	- 71	- 22.7	- 9	15.8
- 49	- 56	- 69	- 22.2	- 8	17.6
- 48	- 55	- 67	- 21.6	- 7	19.4
- 48	- 54	- 65	- 21.1	- 6	21.2
- 47	- 53	- 63	- 20.5	- 5	23.0
- 47	- 52	- 62	- 20.0	- 4	24.8
- 46	- 51	- 60	- 19.4	- 3	26.6
- 45.6	- 50	- 58.0	- 18.8	- 2	28.4
- 45.0	- 49	- 56.2	- 18.3	- 1	30.2
- 44.4	- 48	- 54.4	- 17.8	0	32.0
- 43.9	- 47	- 52.6	- 17.2	1	33.8
- 43.3	- 46	- 50.8	- 16.7	2	35.6
- 42.8	- 45	- 49.0	- 16.1	3	37.4
- 42.2	- 44	- 47.2	- 15.6	4	39.2
- 41.7	- 43	- 45.4	- 15.0	5	41.0
- 41.1	- 42	- 43.6	- 14.4	6	42.8
- 40.6	- 41	- 41.8	- 13.9	7	44.6
- 40.0	- 40	- 40.0	- 13.3	8	46.4
- 39.4	- 39	- 38.2	- 12.8	9	48.2
- 38.9	- 38	- 36.4	- 12.2	10	50.0
- 38.3	- 37	- 34.6	- 11.7	11	51.8
- 37.8	- 36	- 32.8	- 11.1	12	53.6
- 37.2	- 35	- 31.0	- 10.6	13	55.4
- 36.7	- 34	- 29.2	- 10.0	14	57.2
- 36.1	- 33	- 27.4	- 9.4	15	59.0
- 35.5	- 32	- 25.6	- 8.9	16	60.8
- 35.0	- 31	- 23.8	- 8.3	17	62.6
- 34.4	- 30	- 22.0	- 7.8	18	64.4
- 33.9	- 29	- 20.2	- 7.2	19	66.2
- 33.3	- 28	- 18.4	- 6.7	20	68.0
- 32.8	- 27	- 16.6	- 6.1	21	69.8

°C	Known temp. °F or °C	°F	°C	Known temp. °F or °C	°F
- 5.6	22	71.6	25.6	78	172.4
- 5.0	23	73.4	26.1	79	174.2
- 4.4	24	75.2	26.7	80	176.0
- 3.9	25	77.0	27.2	81	177.8
- 3.3	26	78.8	27.8	82	179.6
- 2.8	27	80.6	28.3	83	181.4
- 2.2	28	82.4	28.9	84	183.2
- 1.7	29	84.2	29.4	85	185.0
- 1.1	30	86.0	30.0	86	186.8
- 0.6	31	87.8	30.6	87	188.6
0	32	89.6	31.1	88	190.4
0.6	33	91.4	31.7	89	192.2
1.1	34	93.2	32.2	90	194.0
1.7	35	95.0	32.8	91	195.8
2.2	36	96.8	33.3	92	197.6
2.8	37	98.6	33.9	93	199.4
3.3	38	100.4	34.4	94	201.2
3.9	39	102.2	35.0	95	203.0
4.4	40	104.0	35.6	96	204.8
5.0	41	105.8	36.1	97	206.6
5.6	42	107.6	36.7	98	208.4
6.1	43	109.4	37.2	99	210.2
6.7	44	111.2	37.8	100	212.0
7.2	45	113.0	38.3	101	212
7.8	46	114.8	43.3	110	230
8.3	47	116.6	48.9	120	248
8.9	48	118.4	54.4	130	266
9.4	49	120.2	60	140	284
10.0	50	122.0	65.6	150	302
10.6	51	123.8	71.1	160	320
11.1	52	125.6	76.7	170	338
11.7	53	127.4	82.2	180	356
12.2	54	129.2	87.8	190	374
12.8	55	131.0	93.3	200	392
13.3	56	132.8	98.9	210	410
13.9	57	134.6	100	212	414
14.4	58	136.4	104	220	428
15.0	59	138.2	110	230	446
15.6	60	140.0	116	240	464
16.1	61	141.8	121	250	482
16.7	62	143.6	127	260	500
17.2	63	145.4	132	270	518
17.8	64	147.2	138	280	536
18.3	65	149.0	143	290	554
18.9	66	150.8	149	300	572
19.4	67	152.6	154	310	590
20.0	68	154.4	160	320	608
20.6	69	156.2	166	330	626
21.1	70	158.0	171	340	644
21.7	71	159.8	177	350	662
22.2	72	161.6	182	360	680
22.8	73	163.4	188	370	698
23.3	74	165.2	193	380	716
23.9	75	167.0	199	390	734
24.4	76	168.8	204	400	752
25.0	77	170.6	210	410	770

Heat Losses from Pipes (Bare and Insulated)

The tables contained in this appendix are reprinted with permission from the *Handbook of Air Conditioning, Heating, and Ventilating, Third Edition*, E. Stamper and R. Koral, Industrial Press, Inc., 1979.

HEAT LOSSES FROM BARE STEEL PIPE

HORIZONTAL PIPES

Diameter of Pipe, Inches	Temperature of Pipe, Deg. F.										
	100	120	150	180	210	240	270	300	330	360	390
	Temperature Difference, Pipe to Air, Deg. F.										
	30	50	80	110	140	170	200	230	260	290	320
	Heat Loss per Lineal Foot of Pipe, Btu per Hour										
½	13	22	40	60	82	106	133	162	193	227	265
¾	15	27	50	74	100	131	163	199	238	280	325
1	19	34	61	90	123	160	199	243	292	343	399
1¼	23	42	75	111	152	198	248	302	362	427	496
1½	27	48	85	126	173	224	280	343	410	483	563
2	33	59	104	154	212	275	344	420	503	594	692
2½	39	70	123	184	252	327	410	502	600	709	827
3	46	84	148	221	303	393	493	601	721	852	994
3½	52	95	168	250	342	444	556	680	816	964	1125
4	59	106	187	278	381	496	621	759	911	1076	1257
5	71	129	227	339	464	603	755	924	1109	1311	1532
6	84	151	267	398	546	709	890	1088	1306	1544	1806
8	107	194	341	509	697	906	1137	1391	1671	1977	2312
10	132	238	420	626	857	1114	1399	1714	2060	2437	2852
12	154	279	491	732	1003	1305	1640	2009	2415	2860	3346
14	181	326	575	856	1173	1527	1918	2350	2826	3347	3918
16	203	366	644	960	1314	1711	2149	2634	3168	3753	4395
18	214	385	678	1011	1385	1802	2266	2777	3339	3958	4635
20	236	426	748	1115	1529	1990	2501	3066	3690	4373	5123

VERTICAL PIPES

Diameter of Pipe, Inches	Temperature of Pipe, Deg. F.										
	100	120	150	180	210	240	270	300	330	360	390
	Temperature Difference, Pipe to Air, Deg. F.										
	30	50	80	110	140	170	200	230	260	290	320
	Heat Loss per Lineal Foot of Pipe, Btu per Hour										
½	11	20	35	52	71	93	116	142	170	201	235
¾	14	25	44	65	89	116	145	177	213	252	294
1	17	31	55	81	111	145	181	222	266	315	368
1¼	22	39	69	103	141	183	230	281	337	398	465
1½	25	45	79	118	161	210	263	321	386	456	532
2	31	56	99	147	201	262	328	401	481	569	665
2½	37	68	120	178	244	317	397	486	583	687	805
3	46	83	146	217	297	386	484	592	710	839	980
3½	52	94	166	248	339	440	552	676	810	958	1119
4	59	106	187	279	382	496	622	760	912	1078	1259
5	72	131	231	344	472	612	768	939	1126	1331	1555
6	86	156	275	410	562	729	915	1119	1342	1587	1853
8	112	203	358	534	731	950	1191	1456	1747	2065	2412
10	140	254	447	667	913	1186	1487	1818	2181	2578	3012
12	166	301	530	790	1081	1404	1761	2154	2584	3054	3567
14	195	354	624	930	1273	1653	2073	2536	3042	3596	4200
16	221	400	705	1051	1438	1868	2343	2865	3437	4063	4745
18	234	425	748	1115	1526	1982	2486	3040	3648	4311	5036
20	260	472	831	1239	1696	2203	2763	3378	4053	4791	5596

HEAT LOSSES FROM BARE BRIGHT COPPER TUBE

HORIZONTAL TUBES

Nominal Diameter of Tube, Inches	Temperature of Tube, Deg. F.										
	100	120	150	180	210	240	270	300	330	360	390
	Temperature Difference, Tube to Air, Deg. F.										
	30	50	80	110	140	170	200	230	260	290	320
	Heat Loss per Lineal Foot of Tube, Btu per Hr.										
1/4	3	6	11	16	22	28	34	40	47	54	61
3/8	4	8	13	20	27	35	43	51	60	69	78
1/2	5	9	16	24	33	42	51	61	72	82	94
5/8	6	10	19	28	38	49	60	71	83	95	108
3/4	7	12	21	32	43	55	68	79	95	109	123
1	8	15	26	39	53	67	83	98	115	133	151
1 1/4	9	17	31	46	63	80	99	117	137	158	179
1 1/2	10	20	36	53	72	92	113	135	158	181	206
2	13	25	44	66	90	115	141	168	196	226	257
2 1/2	15	29	52	78	107	136	167	200	233	268	305
3	18	34	61	90	123	157	192	229	268	309	352
3 1/2	20	38	68	102	139	178	218	260	305	351	400
4	23	43	77	113	154	198	243	289	339	391	445
5	27	51	91	137	185	237	292	347	407	469	533
6	31	59	106	157	213	270	336	400	464	541	616
8	40	75	134	198	271	347	426	507	594	686	790
10	47	89	159	239	323	413	509	607	710	822	935
12	54	104	184	276	377	482	593	707	827	957	1090

VERTICAL TUBES

Nominal Diameter of Tube, Inches	Temperature of Tube, Deg. F.										
	100	120	150	180	210	240	270	300	330	360	390
	Temperature Difference, Tube to Air, Deg. F.										
	30	50	80	110	140	170	200	230	260	290	320
	Heat Loss per Lineal Foot of Tube, Btu per Hr.										
1/4	2	4	7	10	14	17	21	25	30	34	39
3/8	3	5	9	13	18	23	28	34	40	46	52
1/2	3	6	11	17	23	29	36	43	50	57	65
5/8	4	7	13	20	27	35	43	51	60	69	78
3/4	5	9	16	23	32	40	50	59	70	80	91
1	6	11	20	30	41	52	64	76	89	103	117
1 1/4	7	14	25	37	50	64	78	93	110	126	143
1 1/2	9	16	29	43	59	75	92	110	129	149	169
2	11	21	38	57	77	98	121	144	169	195	221
2 1/2	14	26	47	70	95	121	150	178	209	241	274
3	17	31	56	84	113	144	178	212	249	286	325
3 1/2	19	36	65	97	131	167	206	246	289	332	377
4	22	41	74	110	149	191	235	280	329	378	429
5	27	51	92	137	185	237	291	348	408	469	533
6	32	61	110	163	221	282	348	415	487	560	636
8	43	81	146	217	294	376	463	553	648	746	847
10	54	101	182	271	366	468	576	687	806	928	1054
12	64	121	217	324	438	560	689	822	965	1110	1260

HEAT LOSSES FROM BARE TARNISHED COPPER TUBE

HORIZONTAL TUBES

Nominal Diameter of Tube, Inches	Temperature of Tube, Deg. F.										
	100	120	150	180	210	240	270	300	330	360	390
	Temperature Difference, Tube to Air, Deg. F.										
	30	50	80	110	140	170	200	230	260	290	320
	Heat Loss per Lineal Foot of Tube, Btu per Hr.										
1/4	4	8	14	21	29	37	46	56	66	77	88
3/8	6	10	18	28	37	48	60	72	85	99	114
1/2	7	13	22	33	45	59	72	88	104	121	139
5/8	8	15	26	39	53	68	85	102	121	141	163
3/4	9	17	30	45	61	79	97	117	139	162	187
1	11	21	37	55	75	97	120	146	173	201	232
1 1/4	14	25	45	66	90	117	145	175	207	242	279
1 1/2	16	29	52	77	105	135	167	203	241	281	324
2	20	37	66	97	132	171	212	257	305	356	411
2 1/2	24	44	78	117	160	206	255	310	367	429	496
3	28	51	92	136	186	240	297	360	428	501	578
3 1/2	32	59	104	156	212	274	340	412	490	573	662
4	36	66	118	174	238	307	381	462	550	644	744
5	43	80	142	212	288	373	464	561	669	783	905
6	51	93	166	246	336	432	541	656	776	915	1059
8	66	120	215	317	435	562	699	848	1010	1184	1372
10	80	146	260	387	527	681	848	1031	1227	1442	1670
12	94	172	304	447	621	802	999	1214	1446	1699	1969

VERTICAL TUBES

Nominal Diameter of Tube, Inches	Temperature of Tube, Deg. F.										
	100	120	150	180	210	240	270	300	330	360	390
	Temperature Difference, Tube to Air, Deg. F.										
	30	50	80	110	140	170	200	230	260	290	320
	Heat Loss per Lineal Foot of Tube, Btu per Hr.										
1/4	3	6	10	15	21	27	34	41	49	57	66
3/8	4	8	14	21	28	36	45	55	65	77	88
1/2	5	10	17	26	35	46	57	69	82	96	111
5/8	6	12	21	31	42	54	68	82	98	114	132
3/4	7	14	24	36	49	64	79	96	114	134	155
1	10	18	31	46	63	82	102	123	147	172	198
1 1/4	12	21	38	57	77	100	125	151	180	210	243
1 1/2	14	25	45	67	91	118	147	178	212	248	287
2	18	33	59	88	120	155	192	233	277	325	375
2 1/2	22	41	73	109	148	191	238	288	343	402	464
3	27	49	87	129	176	227	283	343	408	478	552
3 1/2	31	57	101	150	204	264	328	398	474	554	641
4	35	64	114	171	232	300	374	453	539	631	729
5	43	80	142	212	288	373	464	561	669	783	905
6	52	96	170	253	344	445	554	670	798	934	1080
8	69	127	226	337	458	592	737	892	1063	1244	1438
10	86	158	281	419	570	737	917	1110	1322	1548	1789
12	103	189	336	501	682	881	1097	1328	1582	1851	2140

HEAT LOSS THROUGH PIPE INSULATION — ¾ INCH STEEL PIPE

Insulation Conductivity k	Temperature Difference, Pipe to Air, Deg. F										
	30	50	80	110	140	170	200	230	260	290	320
	Heat Loss per Lineal Foot of Bare Pipe, Btu per Hour										
	15	27	50	74	100	131	163	199	238	280	325
	Heat Loss per Lineal Foot of Insulated Pipe, Btu per Hour										
1 Inch Thick Insulation — ¾ Inch Pipe											
0.20	2	6	7	10	13	16	18	21	24	27	29
0.25	3	6	9	12	16	19	23	26	29	33	36
0.30	4	7	11	15	19	23	27	31	35	39	43
0.35	5	8	12	17	21	26	30	35	40	44	49
0.40	5	9	14	19	24	29	34	39	44	50	55
0.45	6	10	15	21	27	32	38	44	49	55	61
0.50	6	10	17	23	29	35	42	48	54	60	67
0.55	7	11	18	25	32	38	45	52	59	65	72
0.60	7	12	19	27	34	41	48	56	63	70	77
1½ Inch Thick Insulation — ¾ Inch Pipe											
0.20	2	4	6	8	10	12	14	16	19	21	23
0.25	3	5	7	10	13	16	18	21	24	27	29
0.30	3	5	9	12	15	19	22	25	28	32	35
0.35	4	6	10	14	18	21	25	29	33	37	40
0.40	4	7	11	16	20	24	29	33	37	41	46
0.45	5	8	13	17	22	27	32	37	41	46	51
0.50	5	9	14	19	25	30	35	40	46	51	56
0.55	6	10	15	21	27	32	38	44	49	55	61
0.60	6	10	16	23	29	35	41	47	53	59	66
2 Inch Thick Insulation — ¾ Inch Pipe											
0.20	2	3	5	7	9	11	13	15	17	19	21
0.25	2	4	6	9	11	14	16	18	21	23	26
0.30	3	5	8	11	13	16	19	22	25	28	31
0.35	3	6	9	12	16	19	22	26	29	32	36
0.40	4	6	10	14	18	21	25	29	33	37	40
0.45	4	7	11	16	20	24	28	32	37	41	45
0.50	5	8	12	17	22	26	31	36	40	45	50
0.55	5	8	14	19	24	29	34	39	44	49	54
0.60	5	9	15	20	26	31	37	42	48	53	59
2½ Inch Thick Insulation — ¾ Inch Pipe											
0.20	2	3	5	6	8	10	12	13	15	17	19
0.25	2	3	5	7	9	11	13	15	17	19	21
0.30	3	4	7	10	12	15	17	20	23	25	28
0.35	3	5	8	11	14	17	20	23	26	29	32
0.40	3	6	9	13	16	20	23	26	30	33	37
0.45	3	6	10	14	18	22	26	29	33	37	41
0.50	4	7	11	16	20	24	28	33	37	41	45
0.55	5	8	12	17	22	26	31	36	40	45	50
0.60	5	8	13	18	24	29	34	39	44	49	54

HEAT LOSS THROUGH PIPE INSULATION—1 INCH STEEL PIPE

Insulation Conductivity *k*	Temperature Difference, Pipe to Air, Deg. F										
	30	50	80	110	140	170	200	230	260	290	320
	Heat Loss per Lineal Foot of Bare Pipe, Btu per Hour										
	19	34	61	90	123	160	199	243	292	343	399
	Heat Loss per Lineal Foot of Insulated Pipe, Btu per Hc										
1 Inch Thick Insulation — 1 Inch Pipe											
0.20	3	5	8	12	15	18	21	24	27	30	34
0.25	4	6	10	14	18	23	26	30	34	37	41
0.30	5	8	12	17	21	26	30	35	40	44	49
0.35	5	9	14	19	24	30	35	40	45	50	56
0.40	6	10	16	22	27	33	39	45	51	57	63
0.45	7	11	17	24	30	37	43	50	56	63	69
0.50	7	12	19	26	33	40	47	55	62	69	76
0.55	8	13	20	28	36	44	51	59	67	74	82
0.60	8	14	22	30	39	47	55	63	72	80	88
1½ Inch Thick Insulation — 1 Inch Pipe											
0.20	2	4	7	9	12	14	17	19	22	24	27
0.25	3	5	8	11	15	18	21	24	27	30	33
0.30	4	6	10	14	17	21	25	29	32	36	40
0.35	4	7	11	16	20	24	29	33	37	41	46
0.40	5	8	13	18	23	27	32	37	42	47	52
0.45	5	9	14	20	25	31	36	41	47	52	58
0.50	6	10	16	22	28	33	39	45	51	57	63
0.55	6	11	17	24	30	37	43	49	56	62	69
0.60	7	12	19	26	32	39	46	53	60	67	74
2 Inch Thick Insulation — 1 Inch Pipe											
0.20	2	4	6	8	10	13	15	17	19	21	24
0.25	3	5	7	10	13	15	18	21	23	26	29
0.30	3	5	9	12	15	18	21	25	28	31	34
0.35	4	6	10	14	17	21	25	29	32	36	40
0.40	4	7	11	15	20	24	28	32	37	41	45
0.45	5	8	13	17	22	27	31	36	41	46	50
0.50	5	9	14	19	24	29	35	40	45	50	55
0.55	6	9	15	21	26	32	38	43	49	55	60
0.60	6	10	16	23	29	35	41	47	53	59	66
2½ Inch Thick Insulation — 1 Inch Pipe											
0.20	2	3	5	7	9	11	13	15	17	19	21
0.25	3	4	7	10	12	15	17	20	23	25	28
0.30	3	5	8	11	14	16	19	22	25	28	31
0.35	3	6	9	12	16	19	22	26	29	32	36
0.40	4	6	10	14	18	22	26	29	33	37	41
0.45	4	7	11	16	20	24	29	33	37	41	46
0.50	5	8	13	17	22	27	32	36	41	46	51
0.55	5	9	14	19	24	29	34	40	45	50	55
0.60	6	9	15	21	26	32	37	43	49	54	60

HEAT LOSS THROUGH PIPE INSULATION — 1¼ INCH STEEL PIPE

Insulation Conductivity k	Temperature Difference, Pipe to Air, Deg. F										
	30	50	80	110	140	170	200	230	260	290	320
	Heat Loss per Lineal Foot of Bare Pipe, Btu per Hour										
	23	42	75	111	152	198	248	302	362	427	496
	Heat Loss per Lineal Foot of Insulated Pipe, Btu per Hour										
1 Inch Thick Insulation — 1¼ Inch Pipe											
0.20	4	6	10	14	17	21	25	28	32	36	39
0.25	5	8	12	17	21	26	30	35	39	44	48
0.30	5	9	14	19	25	30	35	41	46	51	57
0.35	6	10	16	22	28	35	41	47	53	59	65
0.40	7	11	18	25	32	39	46	52	59	66	73
0.45	8	13	20	28	35	43	50	58	66	73	81
0.50	8	14	22	30	39	47	55	63	72	80	88
0.55	9	15	24	33	42	50	59	68	77	86	95
0.60	10	16	26	35	45	54	64	73	83	92	102
1½ Inch Thick Insulation — 1¼ Inch Pipe											
0.20	3	5	8	11	14	16	19	22	25	28	31
0.25	4	6	10	13	17	20	24	27	31	35	38
0.30	4	7	11	16	20	24	28	32	37	41	45
0.35	5	8	13	18	23	28	33	37	42	47	52
0.40	6	9	15	20	26	31	37	42	48	53	59
0.45	6	10	16	23	29	35	41	47	53	59	66
0.50	7	11	18	25	32	38	45	52	59	65	72
0.55	7	12	20	27	34	42	49	56	64	71	78
0.60	8	13	21	29	37	45	53	61	69	77	84
2 Inch Thick Insulation — 1¼ Inch Pipe											
0.20	2	4	7	9	12	14	17	19	21	24	26
0.25	3	5	8	11	14	17	20	23	27	30	33
0.30	4	6	10	13	17	21	24	28	32	35	39
0.35	4	7	11	16	19	24	28	32	37	41	45
0.40	5	8	13	18	22	27	32	37	42	47	51
0.45	5	9	14	20	25	30	36	41	46	52	57
0.50	6	10	16	22	27	33	39	45	51	57	63
0.55	6	11	17	24	30	36	43	49	56	62	68
0.60	7	12	19	26	32	39	46	53	60	67	74
2½ Inch Thick Insulation — 1¼ Inch Pipe											
0.20	2	4	6	8	10	12	15	17	19	21	23
0.25	3	5	7	10	13	16	18	21	24	26	29
0.30	3	5	9	12	15	19	22	25	28	32	35
0.35	4	6	10	14	18	21	25	29	33	37	40
0.40	4	7	11	16	20	24	29	33	37	41	46
0.45	5	8	13	18	22	27	32	37	42	46	51
0.50	5	9	14	19	25	30	35	41	46	51	57
0.55	6	10	15	21	27	33	39	44	50	56	62
0.60	6	11	17	23	29	36	42	48	55	61	67

HEAT LOSS THROUGH PIPE INSULATION—1½ INCH STEEL PIPE

Insulation Conductivity k	Temperature Difference, Pipe to Air, Deg. F										
	30	50	80	110	140	170	200	230	260	290	320
	Heat Loss per Lineal Foot of Bare Pipe, Btu per Hour										
	27	48	85	126	173	224	280	343	410	483	563
	Heat Loss per Lineal Foot of Insulated Pipe, Btu per Hour										
1 Inch Thick Insulation — 1½ Inch Pipe											
0.20	4	7	11	15	19	23	27	31	35	39	43
0.25	5	8	13	18	23	28	32	37	42	47	52
0.30	6	10	16	21	27	33	39	45	50	56	62
0.35	7	11	18	24	31	38	44	51	58	64	71
0.40	7	12	20	27	35	42	50	57	65	72	80
0.45	8	14	22	30	39	47	55	63	72	80	88
0.50	9	15	24	33	42	51	60	69	78	87	96
0.55	10	16	26	36	46	55	65	75	85	94	104
0.60	10	17	28	38	49	59	70	80	90	100	111
1½ Inch Thick Insulation — 1½ Inch Pipe											
0.20	3	5	8	12	15	18	21	24	27	30	34
0.25	4	7	10	14	18	22	26	30	34	38	42
0.30	5	8	12	17	22	26	31	35	40	45	49
0.35	5	9	14	19	25	30	35	41	46	51	57
0.40	6	10	16	22	28	34	40	46	52	58	64
0.45	7	11	18	25	31	38	45	51	58	65	71
0.50	7	12	20	27	34	41	49	56	63	71	78
0.55	8	13	21	29	37	45	53	61	69	77	85
0.60	9	14	23	32	40	49	57	66	75	83	92
2 Inch Thick Insulation — 1½ Inch Pipe											
0.20	3	4	7	10	12	15	18	20	23	25	28
0.25	3	6	9	12	15	19	22	25	29	32	35
0.30	4	7	10	14	18	22	26	30	34	38	42
0.35	5	8	12	17	21	26	30	35	44	49	53
0.40	5	9	14	19	24	29	34	40	45	50	55
0.45	6	10	15	21	27	33	38	44	50	56	61
0.50	6	11	17	23	30	36	43	49	55	62	68
0.55	7	12	18	25	32	39	46	53	60	67	74
0.60	8	13	20	28	35	43	50	58	65	73	80
2½ Inch Thick Insulation — 1½ Inch Pipe											
0.20	2	4	6	9	11	13	16	18	21	23	25
0.25	3	5	8	11	14	17	20	23	25	28	31
0.30	4	6	9	13	16	20	23	27	30	34	37
0.35	4	7	11	15	19	23	27	31	35	39	44
0.40	5	8	12	17	22	26	31	35	40	45	49
0.45	5	9	14	19	24	29	34	40	45	50	55
0.50	6	10	15	21	27	32	38	44	49	55	61
0.55	6	10	17	23	29	35	42	48	54	60	67
0.60	7	11	18	25	32	38	45	52	59	65	72

HEAT LOSS THROUGH PIPE INSULATION — 2 INCH STEEL PIPE

Insulation Conductivity k	Temperature Difference, Pipe to Air, Deg. F										
	30	50	80	110	140	170	200	230	260	290	320
	Heat Loss per Lineal Foot of Bare Pipe, Btu per Hour										
	33	59	104	154	212	275	344	420	503	594	692
	Heat Loss per Lineal Foot of Insulated Pipe, Btu per Hour										
1 Inch Thick Insulation — 2 Inch Pipe											
0.20	5	8	13	17	22	27	32	36	41	46	51
0.25	6	10	15	21	27	32	39	44	50	56	62
0.30	7	11	18	25	32	39	45	52	59	66	73
0.35	8	13	21	29	36	44	52	60	68	75	83
0.40	9	15	23	32	41	49	58	67	76	84	93
0.45	10	16	26	35	45	55	64	74	83	93	103
0.50	10	18	28	39	49	60	70	81	91	102	112
0.55	11	19	30	42	53	64	76	87	99	110	121
0.60	12	20	32	45	57	69	81	93	106	118	130
1½ Inch Thick Insulation — 2 Inch Pipe											
0.20	4	6	10	13	17	21	24	28	32	35	39
0.25	5	8	12	17	21	26	30	35	39	44	48
0.30	5	9	14	20	25	30	36	41	46	52	57
0.35	6	10	16	23	29	35	41	47	53	59	66
0.40	7	12	18	25	32	39	46	53	60	67	74
0.45	8	13	21	28	36	44	51	59	67	75	82
0.50	9	15	23	32	41	49	58	67	75	84	93
0.55	9	15	25	34	43	52	61	71	80	89	98
0.60	10	17	26	36	46	55	66	76	83	96	106
2 Inch Thick Insulation — 2 Inch Pipe											
0.20	3	5	8	11	14	17	20	23	27	30	33
0.25	4	6	10	14	18	22	25	29	33	37	41
0.30	4	8	12	17	21	26	30	35	39	44	48
0.35	5	9	14	19	24	30	35	40	45	50	56
0.40	6	10	16	22	28	33	39	45	51	57	63
0.45	7	11	18	24	31	37	44	51	57	64	70
0.50	7	12	19	27	34	41	48	56	63	70	77
0.55	8	13	21	29	37	45	53	61	69	77	84
0.60	9	14	23	31	40	49	57	66	74	83	92
2½ Inch Thick Insulation — 2 Inch Pipe											
0.20	3	5	7	10	13	15	18	21	23	26	29
0.25	3	6	9	12	16	19	22	26	29	32	36
0.30	4	7	11	15	19	23	27	31	35	39	43
0.35	5	8	12	17	22	26	31	35	40	45	49
0.40	5	9	14	19	25	30	35	40	46	51	56
0.45	6	10	16	22	27	33	39	45	51	57	63
0.50	6	11	17	24	30	37	43	50	56	63	69
0.55	7	12	19	26	33	40	48	54	61	68	76
0.60	8	13	21	29	37	45	53	61	69	77	85

HEAT LOSS THROUGH PIPE INSULATION — 2½ INCH STEEL PIPE

Insulation Conductivity *k*	Temperature Difference, Pipe to Air, Deg. F										
	30	50	80	110	140	170	200	230	260	290	320
	Heat Loss per Lineal Foot of Bare Pipe, Btu per Hour										
	39	70	123	184	252	327	410	502	600	709	827
	Heat Loss per Lineal Foot of Insulated Pipe, Btu per Hour										
1 Inch Thick Insulation — 2½ Inch Pipe											
0.20	5	9	14	20	25	31	36	42	47	52	58
0.25	7	11	18	24	31	38	44	51	57	64	71
0.30	8	13	21	29	36	44	52	60	68	75	83
0.35	9	15	24	33	42	50	59	68	77	86	95
0.40	10	17	27	37	47	57	67	77	87	97	107
0.45	11	18	29	40	52	63	74	85	96	107	118
0.50	12	21	33	45	57	70	82	94	107	119	131
0.55	13	22	35	48	61	74	87	100	113	126	139
0.60	14	23	37	51	65	79	93	106	120	134	148
1½ Inch Thick Insulation — 2½ Inch Pipe											
0.20	4	7	11	15	19	24	28	32	36	40	44
0.25	5	9	14	19	24	29	34	39	44	50	55
0.30	6	10	16	22	28	35	41	47	53	59	65
0.35	7	12	19	26	33	40	47	54	61	68	75
0.40	8	13	21	29	37	45	53	61	69	77	84
0.45	9	15	23	32	41	50	59	67	76	85	94
0.50	10	16	26	35	45	55	64	74	83	93	103
0.55	10	17	28	38	49	59	70	80	91	101	112
0.60	11	19	30	41	53	64	75	86	98	109	120
2 Inch Thick Insulation — 2½ Inch Pipe											
0.20	3	6	9	13	16	20	23	26	30	33	37
0.25	4	7	11	16	20	24	29	33	37	41	46
0.30	5	9	14	19	24	29	34	39	44	49	54
0.35	6	10	16	22	27	33	39	45	51	57	63
0.40	7	11	18	24	31	38	44	51	58	64	71
0.45	7	12	20	27	35	42	50	57	64	72	79
0.50	8	14	22	30	38	46	55	63	71	79	87
0.55	9	15	24	33	42	50	59	68	77	86	95
0.60	10	16	26	35	45	55	64	74	84	93	103
2½ Inch Thick Insulation — 2½ Inch Pipe											
0.20	3	5	8	11	14	17	20	23	26	29	32
0.25	4	6	10	14	18	21	25	29	33	36	40
0.30	4	7	12	16	21	25	30	34	39	43	48
0.35	5	9	14	19	24	29	35	40	45	50	55
0.40	6	10	16	22	27	33	39	45	51	57	63
0.45	7	11	18	24	31	37	44	50	57	64	70
0.50	7	12	19	27	34	41	48	55	63	70	77
0.55	8	13	21	29	37	45	53	61	69	77	84
0.60	9	14	23	31	40	49	57	66	74	83	92

HEAT LOSS THROUGH PIPE INSULATION — 3 INCH STEEL PIPE

Insulation Conductivity *k*	Temperature Difference, Pipe to Air, Deg. F										
	30	50	80	110	140	170	200	230	260	290	320
	Heat Loss per Lineal Foot of Bare Pipe, Btu per Hour										
	46	84	148	221	303	393	493	601	721	852	994
	Heat Loss per Lineal Foot of Insulated Pipe, Btu per Hour										
	1 Inch Thick Insulation — 3 Inch Pipe										
0.20	6	11	17	23	30	36	42	49	55	61	68
0.25	8	13	21	28	36	44	52	60	67	75	83
0.30	9	15	24	33	43	52	61	70	79	88	97
0.35	10	17	28	38	49	59	69	80	90	101	111
0.40	12	19	31	43	54	66	78	89	101	113	124
0.45	13	21	34	47	60	73	86	99	111	124	137
0.50	14	23	37	51	65	79	93	107	121	135	149
0.55	15	25	40	55	71	86	101	116	131	146	161
0.60	16	27	43	59	76	92	108	124	140	156	173
	1½ Inch Thick Insulation — 3 Inch Pipe										
0.20	5	8	13	18	22	27	32	37	41	46	51
0.25	6	10	16	22	27	33	39	45	51	57	63
0.30	7	12	19	26	33	39	46	53	60	67	74
0.35	8	13	22	30	39	46	54	62	70	78	86
0.40	9	15	24	33	42	51	60	69	78	87	96
0.45	10	17	27	37	47	57	67	77	87	97	107
0.50	11	18	29	40	51	62	73	84	95	106	117
0.55	12	20	32	44	56	68	80	92	104	116	128
0.60	13	21	34	47	60	73	86	99	112	124	137
	2 Inch Thick Insulation — 3 Inch Pipe										
0.20	4	7	11	15	18	22	26	30	34	38	42
0.25	5	8	13	18	23	28	33	38	43	47	52
0.30	6	10	15	21	27	33	39	45	51	56	62
0.35	7	11	18	25	31	38	45	52	58	65	72
0.40	8	13	20	28	36	43	51	58	66	74	81
0.45	8	14	23	31	40	48	57	65	74	82	91
0.50	9	16	25	34	44	53	62	72	81	90	100
0.55	10	17	27	37	48	58	68	78	88	98	109
0.60	11	18	29	40	51	62	73	84	95	106	117
	2½ Inch Thick Insulation — 3 Inch Pipe										
0.20	3	6	9	12	16	19	23	26	29	33	36
0.25	4	7	11	16	20	24	28	33	37	41	45
0.30	5	8	13	19	24	29	34	39	44	49	54
0.35	6	10	16	22	27	33	39	45	51	57	63
0.40	7	11	18	24	31	38	44	51	58	64	71
0.45	7	12	20	27	35	42	50	57	64	72	79
0.50	8	14	22	30	38	46	55	63	71	79	87
0.55	9	15	24	33	42	51	60	69	78	87	96
0.60	10	16	26	36	45	55	65	74	84	94	103

HEAT LOSS THROUGH PIPE INSULATION — 3½ INCH STEEL PIPE

Insulation Conductivity k	Temperature Difference, Pipe to Air, Deg. F										
	30	50	80	110	140	170	200	230	260	290	320
	Heat Loss per Lineal Foot of Bare Pipe, Btu per Hour										
	52	95	168	250	342	444	556	680	816	964	1125
	Heat Loss per Lineal Foot of Insulated Pipe, Btu per Hour										
1 Inch Thick Insulation — 3½ Inch Pipe											
0.20	7	12	19	26	33	40	47	54	61	68	75
0.25	9	14	23	32	40	49	57	66	75	83	92
0.30	10	17	27	37	47	57	67	78	88	98	108
0.35	12	19	31	42	54	65	77	89	100	112	123
0.40	13	22	34	47	60	73	86	99	112	125	138
0.45	14	24	38	52	67	81	95	109	124	138	152
0.50	16	26	41	57	72	88	104	119	135	150	166
0.55	17	28	45	61	78	95	112	128	145	162	179
0.60	18	30	48	66	84	102	119	137	155	173	191
1½ Inch Thick Insulation — 3½ Inch Pipe											
0.20	5	9	14	19	25	30	35	41	46	51	56
0.25	7	11	17	24	30	37	43	50	56	63	70
0.30	8	13	21	28	36	44	51	59	67	75	82
0.35	9	15	24	33	41	50	59	68	77	86	95
0.40	10	17	27	37	47	57	67	77	87	97	107
0.45	11	19	30	41	52	63	74	85	96	107	118
0.50	12	20	32	45	57	69	81	93	105	118	130
0.55	13	22	35	48	62	75	88	101	114	128	141
0.60	14	24	38	52	66	81	95	109	123	138	152
2 Inch Thick Insulation — 3½ Inch Pipe											
0.20	4	7	12	16	20	25	29	33	38	42	46
0.25	5	9	14	20	25	30	36	41	47	52	57
0.30	6	11	17	23	30	36	43	49	55	62	68
0.35	7	12	20	27	34	42	49	56	64	71	79
0.40	8	14	22	31	39	47	56	64	72	81	89
0.45	9	15	25	34	43	53	62	71	80	90	99
0.50	10	17	27	37	48	58	68	78	89	99	109
0.55	11	19	30	41	52	63	74	85	96	108	119
0.60	12	20	32	44	56	68	80	92	104	116	128
2½ Inch Thick Insulation — 3½ Inch Pipe											
0.20	4	6	10	14	18	21	25	29	33	36	40
0.25	5	8	12	17	22	26	31	36	40	45	50
0.30	6	9	15	20	26	31	37	43	48	54	59
0.35	6	11	17	24	30	36	43	49	56	62	68
0.40	7	12	19	27	34	41	49	56	63	70	78
0.45	8	14	22	30	38	46	54	62	70	79	87
0.50	9	15	24	33	42	51	60	69	78	87	96
0.55	10	16	26	36	46	55	65	75	85	95	104
0.60	11	18	28	39	49	60	71	81	92	102	113

HEAT LOSS THROUGH PIPE INSULATION — 4 INCH STEEL PIPE

Insulation Conductivity k	Temperature Difference, Pipe to Air, Deg. F										
	30	50	80	110	140	170	200	230	270	300	320
	Heat Loss per Lineal Foot of Bare Pipe, Btu per Hour										
	59	106	187	278	381	496	621	759	911	1076	1257
	Heat Loss per Lineal Foot of Insulated Pipe, Btu per Hour										
	1 Inch Thick Insulation — 4 Inch Pipe										
0.20	8	13	20	28	33	43	51	59	66	74	82
0.25	10	16	25	35	44	54	63	73	82	92	101
0.30	11	19	30	41	52	63	74	86	97	108	119
0.35	13	21	34	47	59	72	85	98	110	123	136
0.40	14	24	38	52	66	81	95	109	123	137	152
0.45	16	26	42	58	73	89	105	121	136	152	168
0.50	17	28	45	62	80	97	114	131	148	165	182
0.55	18	31	49	67	86	104	123	141	159	178	196
0.60	20	33	53	72	92	112	132	151	171	191	211
	1½ Inch Thick Insulation — 4 Inch Pipe										
0.20	6	10	15	21	27	33	38	44	50	56	61
0.25	7	12	19	26	33	40	47	55	62	69	76
0.30	8	14	22	28	39	48	56	64	73	81	90
0.35	10	16	26	36	45	55	65	74	84	94	103
0.40	11	18	29	40	51	62	73	84	95	106	116
0.45	12	20	32	44	56	69	81	93	105	117	129
0.50	13	22	35	49	62	75	88	101	115	128	141
0.55	14	24	38	53	67	81	96	110	124	139	153
0.60	15	26	41	57	72	88	103	118	134	149	165
	2 Inch Thick Insulation — 4 Inch Pipe										
0.20	5	8	13	17	22	27	31	36	41	46	50
0.25	6	10	16	21	27	33	39	45	51	57	62
0.30	7	12	19	26	32	39	46	53	60	67	74
0.35	8	13	21	29	37	45	53	61	69	77	85
0.40	9	15	24	33	42	51	60	69	79	88	97
0.45	10	17	27	37	47	57	67	78	88	98	108
0.50	11	19	30	41	52	63	74	85	96	107	118
0.55	12	20	32	44	56	69	81	93	105	117	129
0.60	13	22	35	48	61	74	87	101	114	127	140
	2½ Inch Thick Insulation — 4 Inch Pipe										
0.20	4	7	11	15	19	23	27	31	35	39	44
0.25	5	8	13	18	24	29	34	39	44	49	54
0.30	6	10	16	22	28	34	40	46	52	58	64
0.35	7	12	19	26	32	39	46	53	60	67	74
0.40	8	13	21	29	37	45	52	60	68	76	84
0.45	9	15	24	32	41	50	59	68	76	85	94
0.50	10	16	26	36	45	55	65	75	84	94	104
0.55	11	18	28	39	49	60	71	81	92	102	113
0.60	11	19	31	42	53	65	76	88	99	111	122

HEAT LOSS THROUGH PIPE INSULATION — 5 INCH STEEL PIPE

Insulation Conductivity *k*	Temperature Difference, Pipe to Air, Deg. F										
	30	50	80	110	140	170	200	230	260	290	320
	Heat Loss per Lineal Foot of Bare Pipe, Btu per Hour										
	71	129	227	339	464	603	755	924	1109	1311	1532
	Heat Loss per Lineal Foot of Insulated Pipe, Btu per Hour										
1 Inch Thick Insulation — 5 Inch Pipe											
0.20	8	14	23	31	40	48	57	65	74	82	90
0.25	10	17	28	38	48	59	69	79	90	100	110
0.30	12	20	32	45	57	69	81	93	105	117	130
0.35	14	23	37	51	65	79	93	106	120	134	148
0.40	15	26	41	57	73	88	104	119	135	150	166
0.45	17	29	46	63	80	97	114	131	148	165	182
0.50	19	31	50	68	87	106	124	143	180	199	217
0.55	20	34	54	74	94	114	134	154	174	194	214
0.60	21	36	57	79	100	122	143	165	186	208	229
1½ Inch Thick Insulation — 5 Inch Pipe											
0.20	6	11	17	23	30	36	42	49	55	61	68
0.25	8	13	21	29	37	44	52	60	68	76	84
0.30	9	15	25	34	43	53	62	71	80	90	99
0.35	11	18	28	39	50	60	71	82	92	103	114
0.40	12	20	32	44	56	68	80	92	104	116	128
0.45	13	22	36	49	62	75	89	102	115	129	142
0.50	15	24	39	53	68	83	97	112	126	141	156
0.55	16	26	42	58	74	90	106	121	137	153	169
0.60	17	28	45	62	80	97	114	131	148	165	182
2 Inch Thick Insulation — 5 Inch Pipe											
0.20	5	9	14	19	24	29	34	40	45	50	55
0.25	6	11	17	23	30	36	43	49	55	62	68
0.30	8	13	20	28	35	43	51	58	66	73	81
0.35	9	15	23	32	41	50	58	67	76	85	93
0.40	10	17	26	36	46	56	66	76	86	96	106
0.45	11	18	29	40	51	62	73	84	95	106	117
0.50	12	20	32	44	56	68	80	92	105	117	129
0.55	13	22	35	48	61	75	88	101	114	128	141
0.60	14	24	38	52	67	81	95	109	124	138	152
2½ Inch Thick Insulation — 5 Inch Pipe											
0.20	4	7	12	16	21	25	29	34	38	43	47
0.25	5	9	15	20	25	31	36	42	47	53	58
0.30	7	11	17	24	30	37	43	50	56	63	69
0.35	8	13	20	28	35	43	50	58	65	73	80
0.40	9	14	23	31	40	48	57	66	74	83	91
0.45	10	16	25	35	45	54	64	73	83	92	102
0.50	11	18	28	39	49	60	70	81	91	102	112
0.55	11	19	31	42	53	65	76	88	99	111	122
0.60	12	21	33	45	58	70	83	95	107	120	132

HEAT LOSS THROUGH PIPE INSULATION — 6 INCH STEEL PIPE

Insulation Conductivity *k*	Temperature Difference, Pipe to Air, Deg. F										
	30	50	80	110	140	170	200	230	260	290	320
	Heat Loss per Lineal Foot of Bare Pipe, Btu per Hour										
	84	151	267	398	546	709	890	1088	1306	1544	1806
	Heat Loss per Lineal Foot of Insulated Pipe, Btu per Hour										
1 Inch Thick Insulation — 6 Inch Pipe											
0.20	11	18	29	40	51	61	72	83	94	105	115
0.25	13	22	35	48	62	75	88	101	114	128	141
0.30	15	26	41	57	72	88	103	119	134	150	165
0.35	18	29	47	65	82	100	118	135	153	171	188
0.40	20	33	53	72	92	112	132	151	171	191	211
0.45	22	36	58	80	102	123	145	167	189	210	232
0.50	24	39	63	87	110	134	158	181	205	229	252
0.55	25	42	68	93	119	144	170	195	221	246	272
0.60	27	45	73	100	127	154	182	209	236	263	291
1½ Inch Thick Insulation — 6 Inch Pipe											
0.20	8	13	21	29	37	45	53	61	69	77	85
0.25	10	16	26	36	45	55	65	75	84	94	104
0.30	12	19	31	42	54	65	77	88	100	111	123
0.35	13	22	35	49	62	75	88	101	115	128	141
0.40	15	25	40	55	70	84	99	114	129	144	159
0.45	17	28	44	61	77	94	110	127	143	160	176
0.50	18	30	48	66	84	103	121	139	157	175	193
0.55	20	33	52	72	92	111	131	151	170	190	209
0.60	21	35	56	77	99	120	141	162	183	204	225
2 Inch Thick Insulation — 6 Inch Pipe											
0.20	6	11	17	23	30	36	43	49	55	62	68
0.25	8	13	21	29	37	45	53	60	68	76	84
0.30	9	16	25	34	44	53	62	72	81	90	100
0.35	11	18	29	40	50	61	72	83	94	104	115
0.40	12	20	33	45	57	69	81	94	106	118	130
0.45	14	23	36	50	63	77	91	104	118	131	145
0.50	15	25	40	55	70	85	100	114	129	144	159
0.55	16	27	43	60	76	92	108	125	141	157	173
0.60	18	29	47	64	82	99	117	134	152	170	187
2½ Inch Thick Insulation — 6 Inch Pipe											
0.20	5	9	14	20	25	31	36	41	47	52	58
0.25	7	11	18	25	31	38	45	51	58	65	72
0.30	8	13	21	29	37	45	53	61	69	77	85
0.35	9	15	25	34	43	52	62	71	80	89	98
0.40	10	17	28	38	49	59	70	80	91	101	111
0.45	12	19	31	43	54	66	78	89	101	113	124
0.50	13	21	34	47	60	73	86	99	111	124	137
0.55	14	23	37	51	65	79	93	107	121	135	149
0.60	15	25	40	56	71	86	101	116	131	147	162

HEAT LOSS THROUGH PIPE INSULATION — 8 INCH STEEL PIPE

Insulation Conductivity *k*	Temperature Difference, Pipe to Air, Deg. F										
	30	50	80	110	140	170	200	230	260	290	320
	Heat Loss per Lineal Foot of Bare Pipe, Btu per Hour										
	107	194	341	509	697	906	1137	1391	1671	1977	2312
	Heat Loss per Lineal Foot of Insulated Pipe, Btu per Hour										
1 Inch Thick Insulation — 8 Inch Pipe											
0.20	14	23	36	50	64	78	91	105	119	132	146
0.25	17	28	44	61	78	95	111	128	145	161	178
0.30	20	33	52	72	91	111	130	150	169	189	209
0.35	22	37	59	82	104	126	149	171	193	215	238
0.40	25	42	66	91	116	141	166	191	216	241	266
0.45	27	46	73	101	128	155	183	210	238	265	293
0.50	30	50	79	109	139	169	199	228	258	288	318
0.55	32	53	86	118	150	182	214	246	278	310	342
0.60	34	57	91	126	160	194	229	263	297	331	366
1½ Inch Thick Insulation — 8 Inch Pipe											
0.20	10	16	26	36	46	56	65	75	85	95	105
0.25	12	20	32	44	56	68	80	92	104	117	129
0.30	14	24	38	52	66	81	95	109	123	138	152
0.35	16	27	44	60	76	93	109	125	142	158	174
0.40	18	31	49	68	86	104	123	141	160	178	196
0.45	20	34	54	75	95	116	136	156	177	197	218
0.50	22	37	60	82	104	127	149	171	194	216	238
0.55	24	40	65	89	113	137	161	186	210	234	258
0.60	26	43	69	95	122	148	174	200	226	252	278
2 Inch Thick Insulation — 8 Inch Pipe											
0.20	8	13	21	29	37	45	53	61	68	76	84
0.25	10	16	26	36	46	55	65	75	85	94	104
0.30	12	19	31	42	54	66	77	89	100	112	124
0.35	13	22	36	49	62	76	89	102	116	129	142
0.40	15	25	40	55	70	85	101	116	131	146	161
0.45	17	28	45	62	78	95	112	129	145	162	179
0.50	18	31	49	68	86	104	123	141	160	178	197
0.55	20	33	53	74	94	114	134	154	174	194	214
0.60	22	36	58	79	101	123	144	166	187	209	231
2½ Inch Thick Insulation — 8 Inch Pipe											
0.20	7	11	18	24	31	37	44	51	57	64	71
0.25	8	14	22	30	38	46	55	63	71	79	87
0.30	10	16	26	36	45	55	65	75	84	94	104
0.35	11	19	30	41	53	64	75	86	98	109	120
0.40	13	21	34	47	60	72	85	98	111	123	136
0.45	14	24	38	52	66	81	95	109	123	138	152
0.50	16	26	42	57	73	89	104	120	136	151	167
0.55	17	28	46	63	80	97	114	131	148	165	182
0.60	18	31	49	68	86	105	123	142	160	179	197

HEAT LOSS THROUGH PIPE INSULATION — 10 INCH STEEL PIPE

Insulation Conductivity *k*	Temperature Difference, Pipe to Air, Deg. F										
	30	50	80	110	140	170	200	230	260	290	320
	Heat Loss per Lineal Foot of Bare Pipe, Btu per Hour										
	132	238	420	626	857	1114	1399	1714	2060	2437	2852
	Heat Loss per Lineal Foot of Insulated Pipe, Btu per Hour										
1 Inch Thick Insulation — 10 Inch Pipe											
0.20	16	27	43	60	76	92	109	125	141	158	174
0.25	20	33	53	73	93	113	133	152	172	192	212
0.30	23	39	62	85	109	132	155	179	202	225	249
0.35	27	44	71	97	124	151	177	204	230	257	283
0.40	30	49	79	109	138	168	198	227	257	287	317
0.45	33	54	87	120	152	185	218	250	283	316	348
0.50	36	59	95	130	166	201	237	272	308	343	379
0.55	38	64	102	140	179	217	255	293	332	370	408
0.60	41	68	109	150	191	232	273	313	354	395	436
1½ Inch Thick Insulation — 10 Inch Pipe											
0.20	12	20	32	44	55	67	79	91	103	115	127
0.25	15	24	39	54	68	83	97	112	127	141	156
0.30	17	29	46	63	81	98	115	132	150	167	184
0.35	20	33	53	73	93	112	132	152	172	192	211
0.40	22	37	59	82	104	126	149	171	193	216	238
0.45	25	41	66	91	115	140	165	189	214	239	264
0.50	27	45	72	99	126	153	180	207	234	261	288
0.55	29	49	78	107	137	166	195	225	254	283	313
0.60	32	53	84	116	147	179	210	242	273	305	336
2 Inch Thick Insulation — 10 Inch Pipe											
0.20	9	16	25	35	44	54	63	73	82	92	101
0.25	12	20	31	43	55	66	78	90	102	113	125
0.30	14	23	37	51	65	79	93	107	120	134	148
0.35	16	27	43	59	75	91	107	123	139	155	171
0.40	18	30	48	66	84	103	121	139	157	175	193
0.45	20	34	54	74	94	114	134	154	174	195	215
0.50	22	37	59	81	103	125	147	169	192	214	236
0.55	24	40	64	88	112	136	160	184	208	232	256
0.60	26	43	69	95	121	147	173	199	225	251	277
2½ Inch Thick Insulation — 10 Inch Pipe											
0.20	8	13	21	29	37	44	52	60	68	76	84
0.25	10	16	26	36	45	55	65	74	84	94	104
0.30	12	19	31	42	54	65	77	89	100	112	123
0.35	13	22	36	49	62	76	89	102	116	129	142
0.40	15	25	40	55	71	86	101	116	131	146	161
0.45	17	28	45	62	79	96	112	129	146	163	180
0.50	19	31	49	68	87	105	124	142	161	179	198
0.55	20	34	54	74	94	115	135	155	175	196	216
0.60	22	36	58	80	102	124	146	168	190	211	233

HEAT LOSS THROUGH PIPE INSULATION—12 INCH STEEL PIPE

Insulation Conductivity k	Temperature Difference, Pipe to Air, Deg. F										
	30	50	80	110	140	170	200	230	260	290	320
	Heat Loss per Lineal Foot of Bare Pipe, Btu per Hour										
	154	279	491	732	1003	1305	1640	2009	2415	2860	3346
	Heat Loss per Lineal Foot of Insulated Pipe, Btu per Hour										
	1 Inch Thick Insulation — 12 Inch Pipe										
0.20	19	32	51	70	89	108	127	146	165	184	204
0.25	23	39	62	85	109	132	155	179	202	225	248
0.30	27	45	73	100	127	155	182	209	237	264	291
0.35	31	52	83	114	145	176	207	239	270	301	332
0.40	35	58	93	127	162	197	232	266	301	336	371
0.45	38	64	102	140	178	217	255	293	331	369	408
0.50	42	69	111	152	194	235	277	319	360	402	443
0.55	45	75	119	164	209	254	298	343	388	433	477
0.60	48	80	127	175	223	271	319	366	414	462	510
	1½ Inch Thick Insulation — 12 Inch Pipe										
0.20	14	23	36	50	64	77	91	105	118	132	146
0.25	17	28	45	61	78	95	112	129	145	162	179
0.30	20	33	53	73	92	112	132	152	172	191	211
0.35	23	38	61	83	106	129	152	174	197	220	243
0.40	26	43	68	94	119	145	171	196	222	247	273
0.45	28	47	76	104	132	161	189	217	246	274	302
0.50	31	52	83	114	145	176	207	238	269	300	331
0.55	34	56	90	123	157	191	224	258	291	325	359
0.60	36	60	96	132	169	205	241	277	313	349	385
	2 Inch Thick Insulation — 12 Inch Pipe										
0.20	11	18	29	41	52	63	74	85	96	107	118
0.25	14	23	36	50	64	77	91	105	118	132	146
0.30	16	27	43	59	75	92	108	124	140	156	173
0.35	19	31	50	68	87	106	124	143	161	180	199
0.40	21	35	56	77	98	119	140	161	182	203	224
0.45	23	39	62	86	109	133	156	179	203	226	250
0.50	26	43	68	94	120	146	171	197	223	248	274
0.55	28	47	74	102	130	158	186	214	242	270	298
0.60	31	52	83	114	145	176	207	238	269	300	332
	2½ Inch Thick Insulation — 12 Inch Pipe										
0.20	9	15	24	34	43	52	61	70	79	89	98
0.25	11	19	30	42	53	64	76	87	98	110	121
0.30	13	22	36	49	63	76	90	103	117	130	144
0.35	16	26	42	57	73	88	104	119	135	151	166
0.40	18	29	47	65	82	100	118	135	153	170	188
0.45	20	33	52	72	92	111	131	151	170	190	210
0.50	22	36	58	79	101	123	144	166	187	209	231
0.55	24	39	63	86	110	134	157	181	204	228	251
0.60	25	42	68	93	119	144	170	195	221	246	272

HEAT LOSS THROUGH PIPE INSULATION — 14 INCH STEEL PIPE

Insulation Conductivity k	Temperature Difference, Pipe to Air, Deg. F										
	30	50	80	110	140	170	200	230	260	290	320
	Heat Loss per Lineal Foot of Bare Pipe, Btu per Hour										
	181	326	575	856	1173	1527	1918	2350	2826	3347	3918
	Heat Loss per Lineal Foot of Insulated Pipe, Btu per Hour										
1 Inch Thick Insulation — 14 Inch Pipe											
0.20	22	36	57	79	100	122	143	165	187	208	230
0.25	26	44	70	96	122	149	175	201	227	253	280
0.30	31	51	82	113	143	174	205	235	266	297	327
0.35	35	58	93	128	163	198	233	268	303	338	373
0.40	39	65	104	143	182	221	260	299	338	377	416
0.45	43	72	114	157	200	243	286	329	372	415	458
0.50	47	78	124	171	218	264	311	357	404	451	497
0.55	50	84	134	184	234	284	335	385	435	485	535
0.60	54	89	143	196	250	303	357	411	464	518	571
1½ Inch Thick Insulation — 14 Inch Pipe											
0.20	15	26	41	56	72	87	102	118	133	148	164
0.25	19	31	50	69	88	107	126	145	163	182	201
0.30	22	37	59	82	104	126	148	171	193	215	237
0.35	26	43	68	94	119	145	170	196	221	247	272
0.40	29	48	77	105	134	163	191	220	249	278	306
0.45	32	53	85	117	148	180	212	244	276	307	339
0.50	35	58	93	127	162	197	232	267	301	336	371
0.55	38	63	100	138	176	213	251	289	326	364	402
0.60	40	67	108	148	189	229	270	310	350	391	431
2 Inch Thick Insulation — 14 Inch Pipe											
0.20	12	20	31	43	55	66	78	90	102	113	125
0.25	14	24	39	53	68	82	97	111	125	140	154
0.30	17	29	46	63	80	97	114	132	149	166	183
0.35	20	33	53	73	92	112	132	152	171	191	211
0.40	22	37	60	82	104	127	149	171	194	216	238
0.45	25	41	66	91	116	141	166	190	215	240	265
0.50	27	45	73	100	127	155	182	209	236	264	291
0.55	30	49	79	109	138	168	198	227	257	287	317
0.60	32	53	85	117	149	181	213	245	277	309	341
2½ Inch Thick Insulation — 14 Inch Pipe											
0.20	10	16	26	36	46	56	65	75	85	95	105
0.25	12	20	32	45	57	69	81	93	105	117	130
0.30	14	24	38	53	67	82	96	111	125	140	154
0.35	17	28	44	61	78	95	111	128	145	161	178
0.40	19	31	50	69	88	107	126	145	164	182	201
0.45	21	35	56	77	98	119	140	161	182	203	225
0.50	23	39	62	85	108	131	154	178	201	224	247
0.55	25	42	67	93	118	143	168	194	219	244	269
0.60	27	45	73	100	127	155	189	209	237	264	291

Appendix F

Useful Installation Details

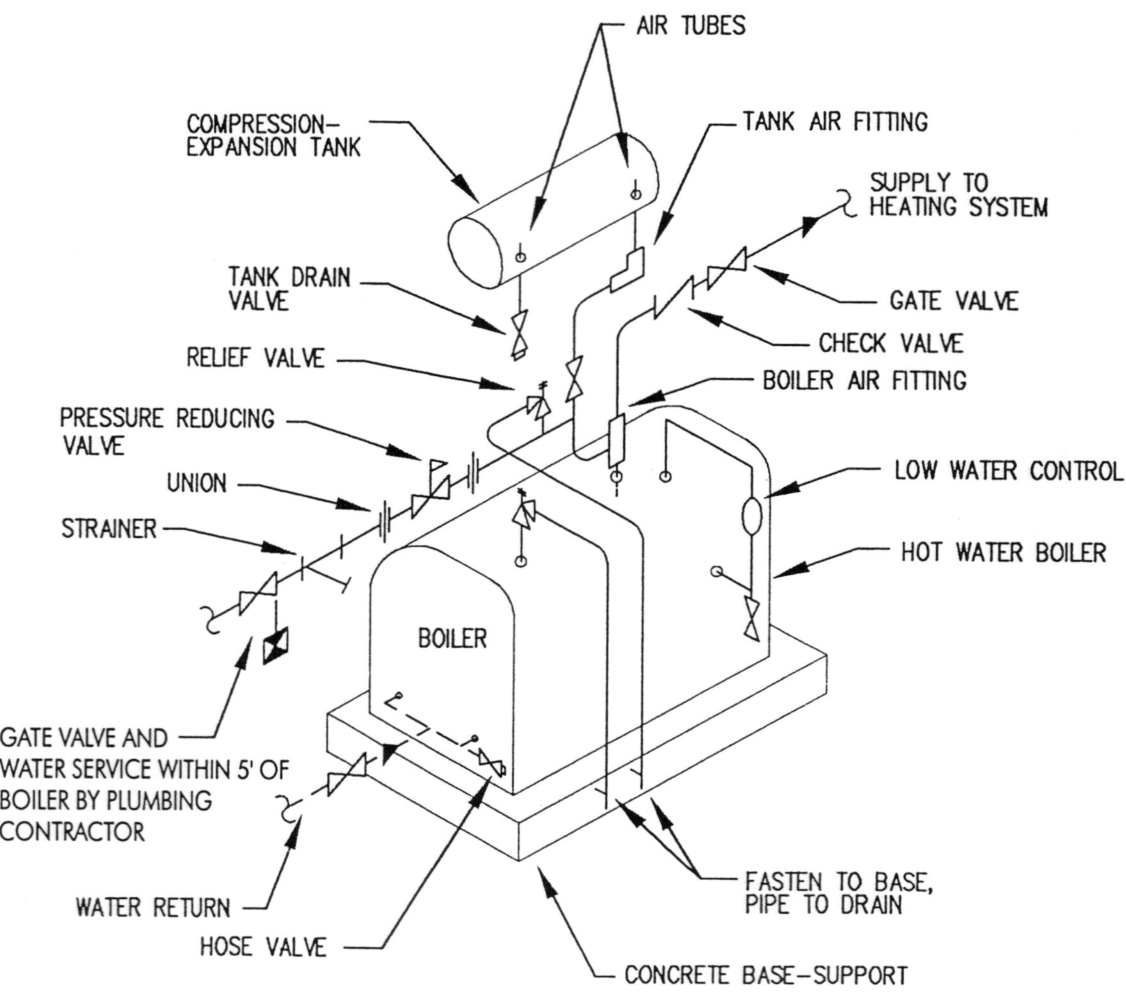

HOT WATER BOILER PIPING AND APPURTENANCES ARRANGEMENT DETAIL

NOT TO SCALE

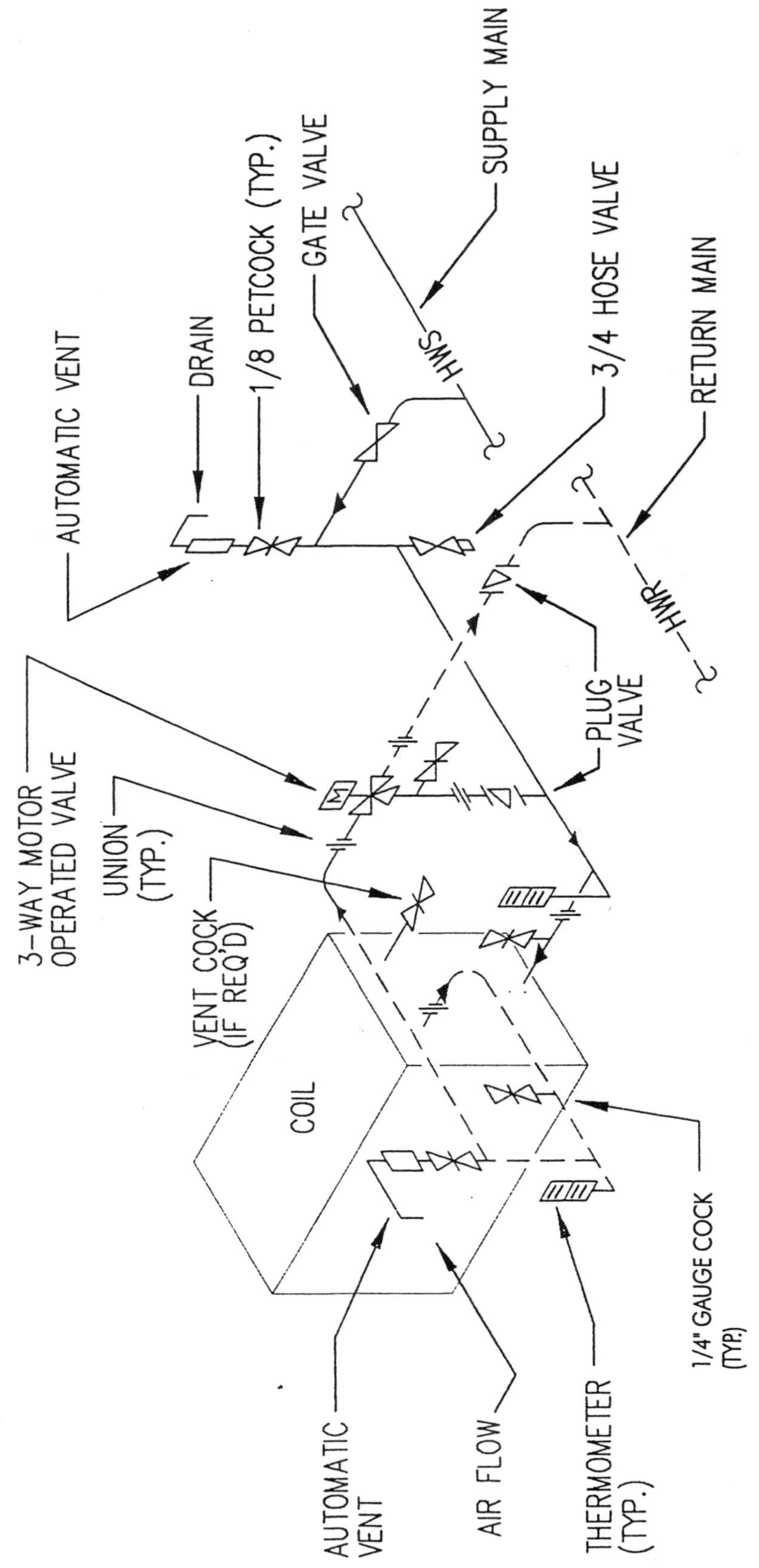

HOT WATER COIL PIPING - DETAIL

NOT TO SCALE

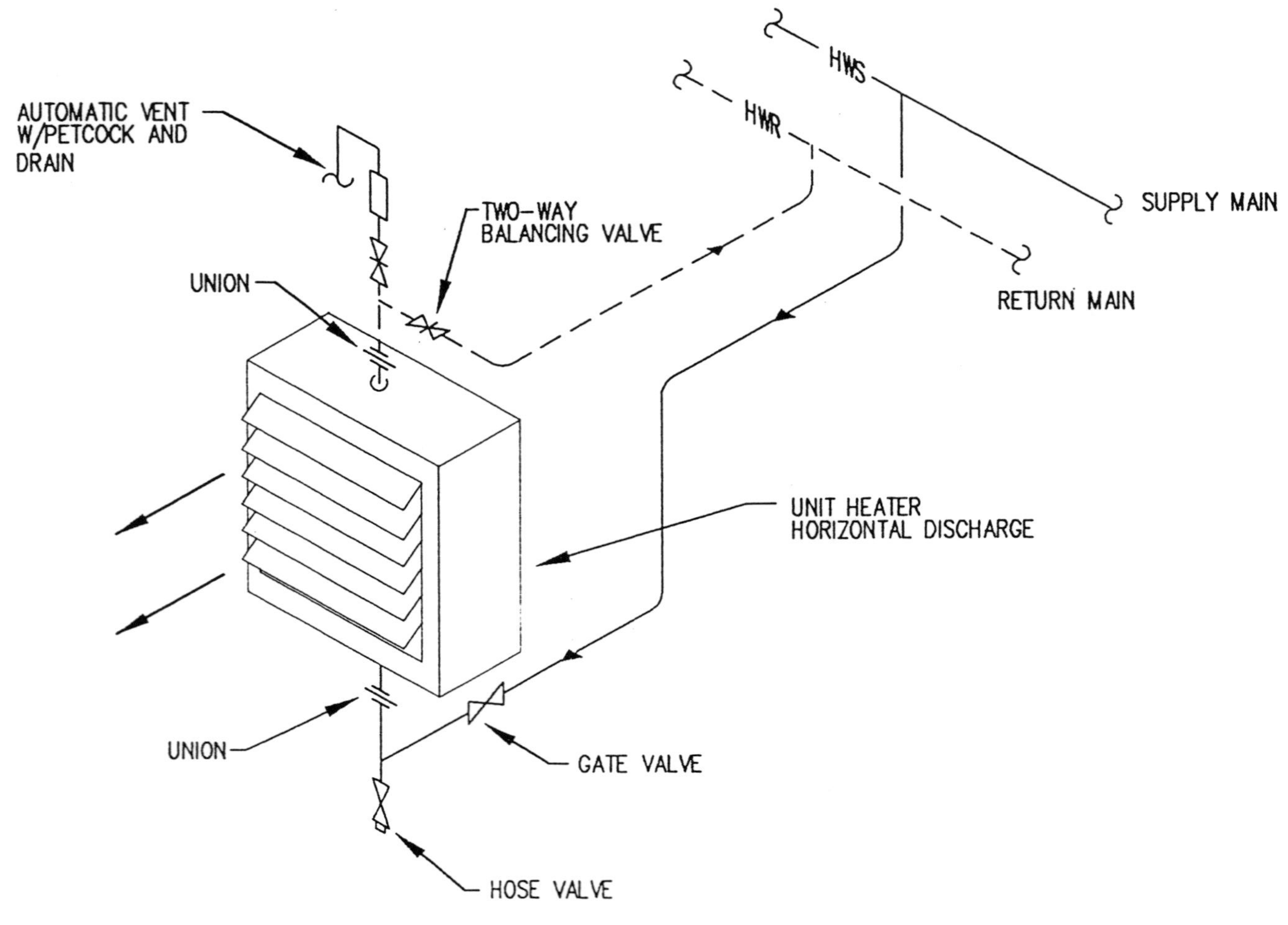

HOT WATER-UNIT HEATER PIPING ARRANGEMENT DETAIL

NOT TO SCALE

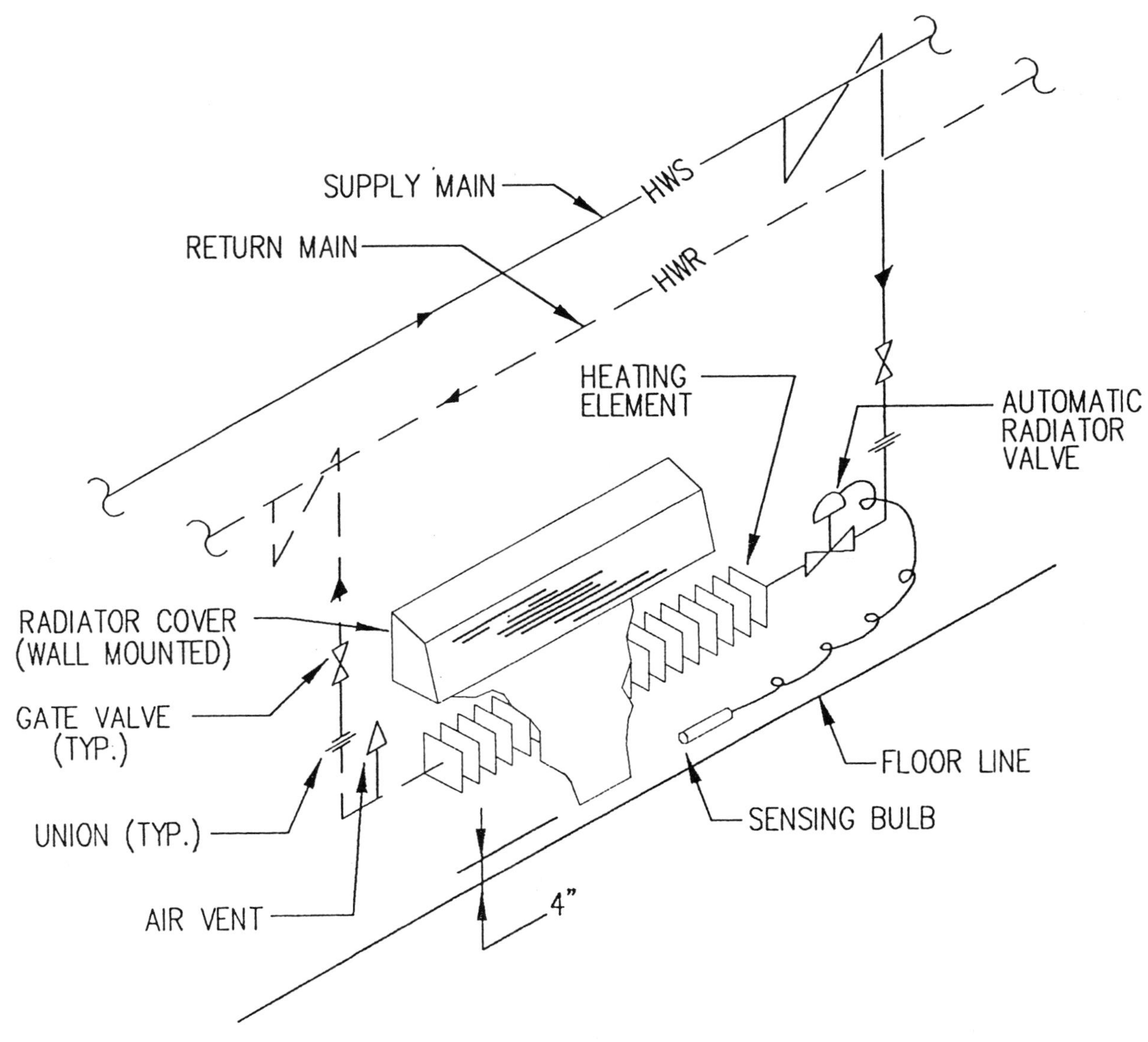

DOWN FED

HOT WATER FINNED TUBE CONVECTOR

DETAIL

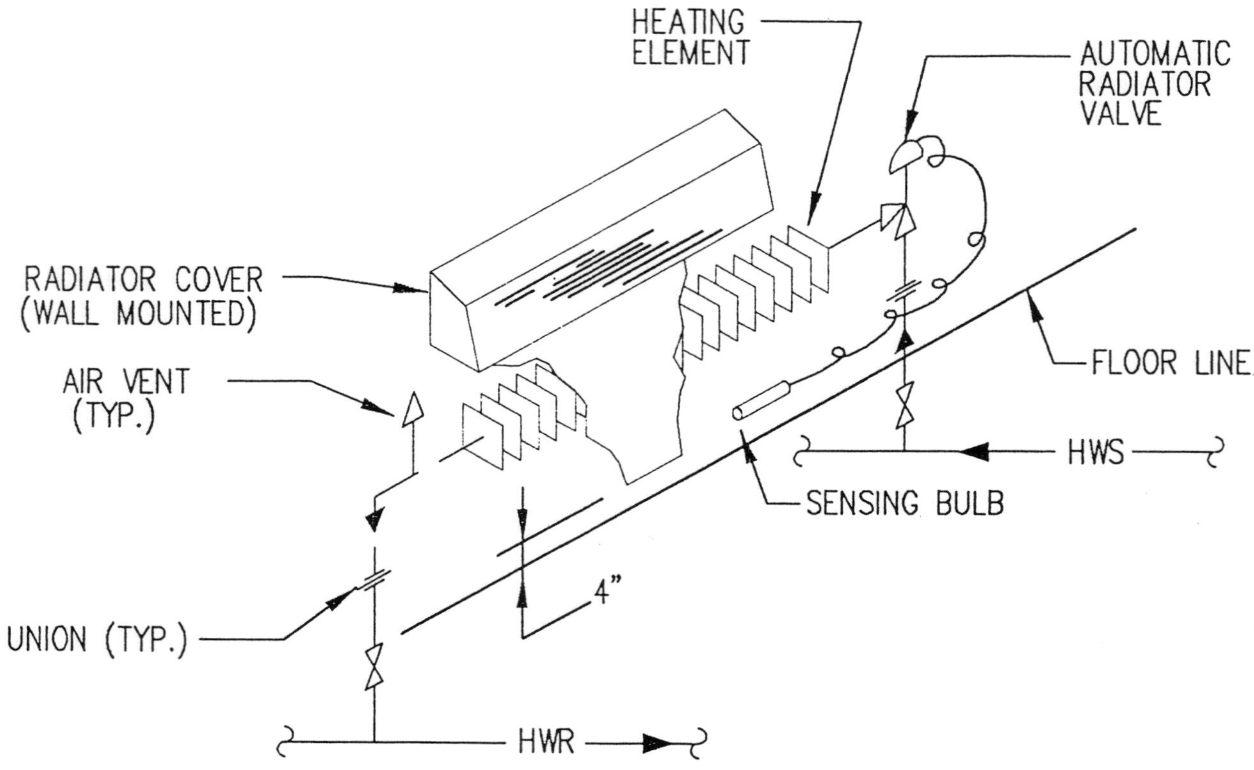

UP FED

HOT WATER FINNED TUBE CONVECTOR

DETAIL

NOT TO SCALE

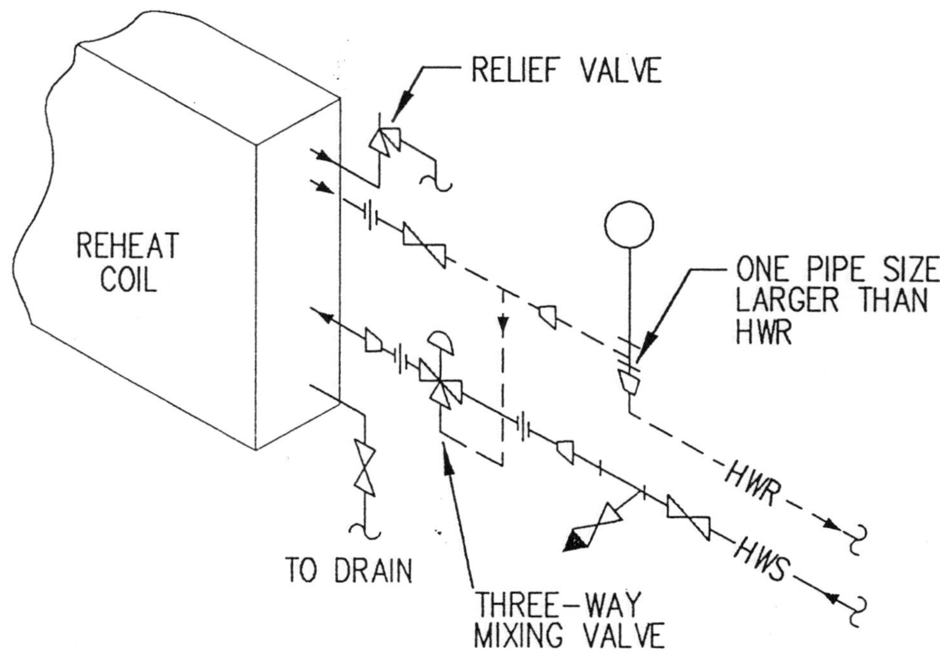

HOT WATER REHEAT COIL

DETAIL

NOT TO SCALE

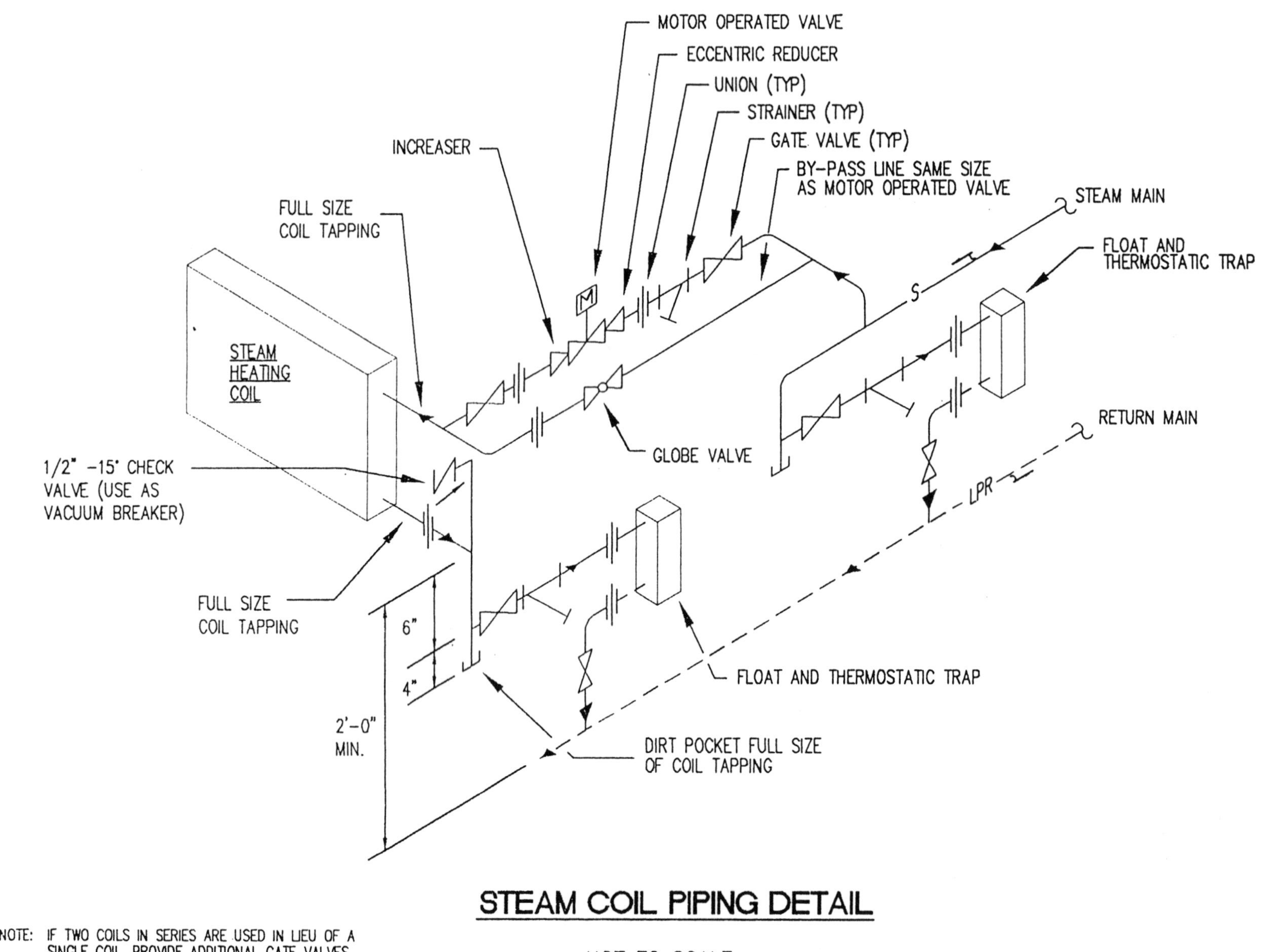

STEAM COIL PIPING DETAIL

NOT TO SCALE

NOTE: IF TWO COILS IN SERIES ARE USED IN LIEU OF A SINGLE COIL, PROVIDE ADDITIONAL GATE VALVES IN STEAM AND RETURN PIPING TO ISOLATE EACH COIL, AND TRAP EACH COIL SEPARATELY WITH AN F & T TRAP.

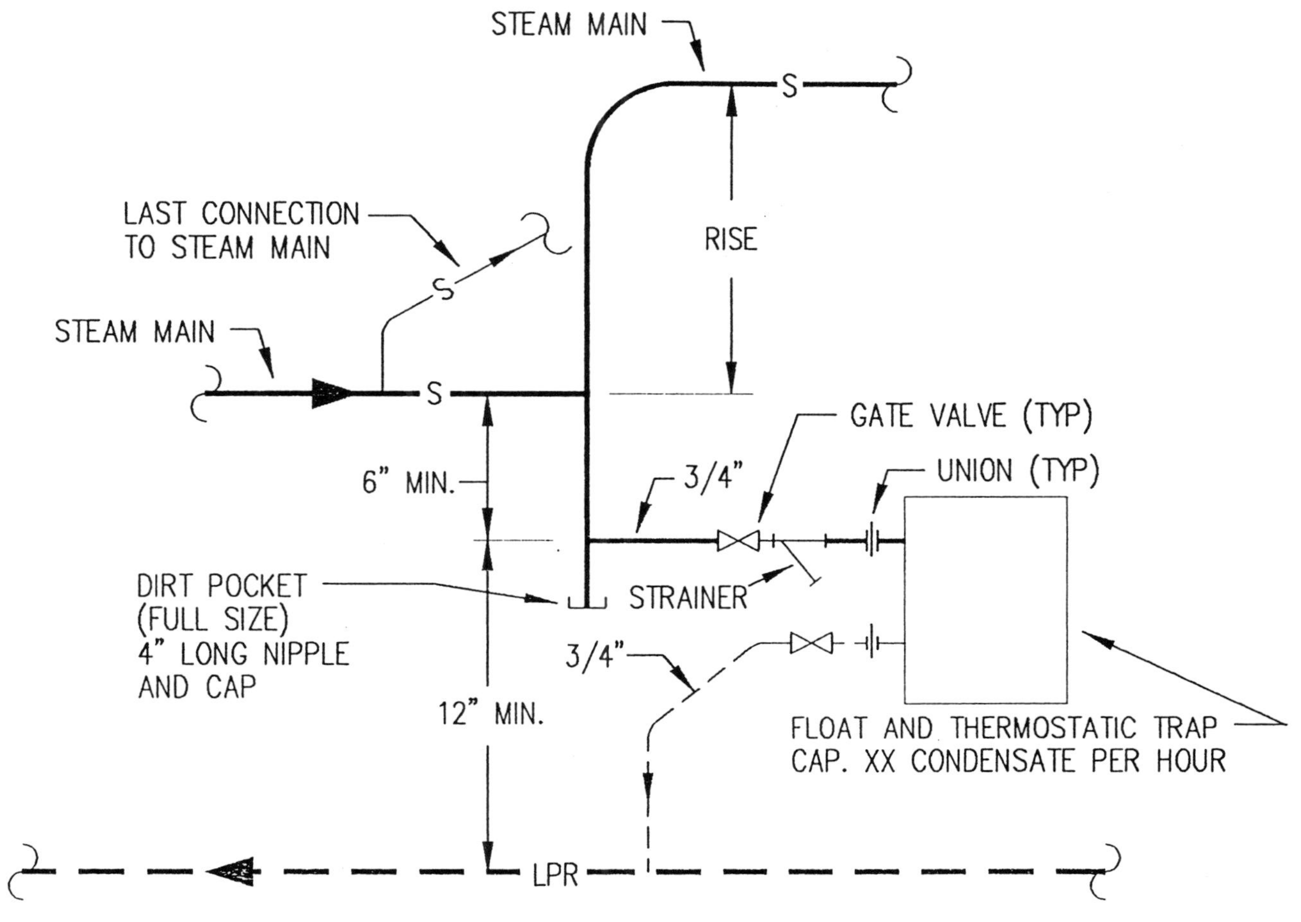

END OF STEAM MAIN DRIP
OR
DRIP AND RISE
DETAIL

NOT TO SCALE

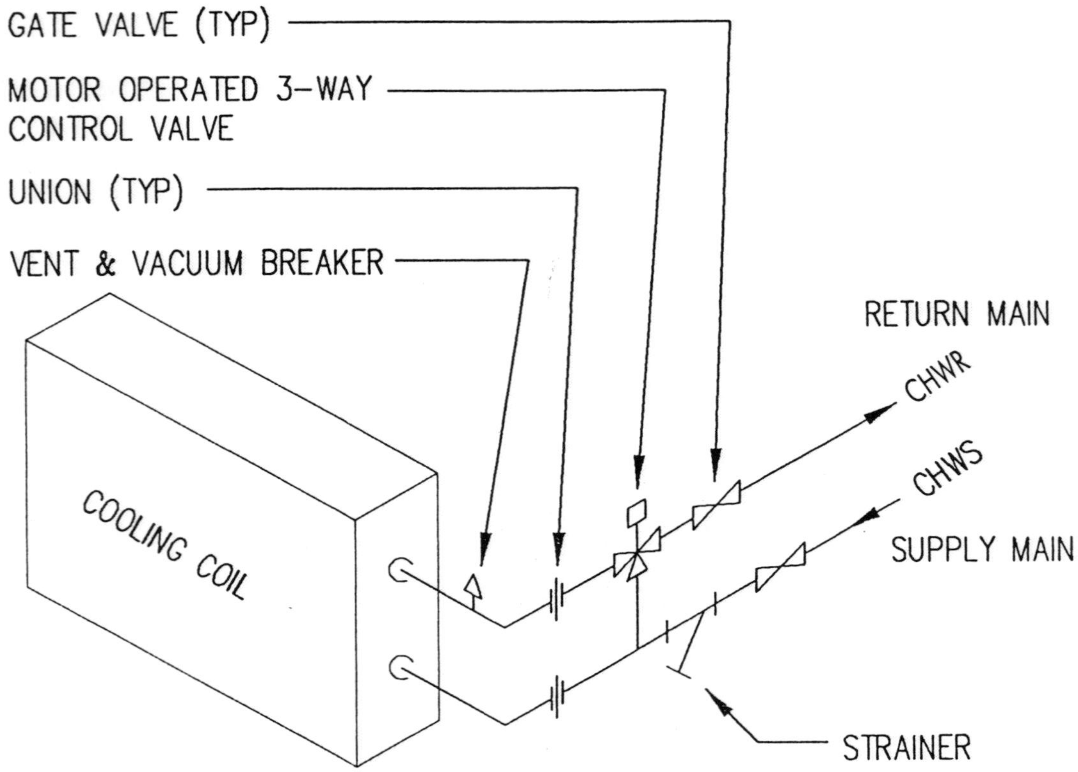

TYPICAL COOLING COIL DETAIL

NOT TO SCALE

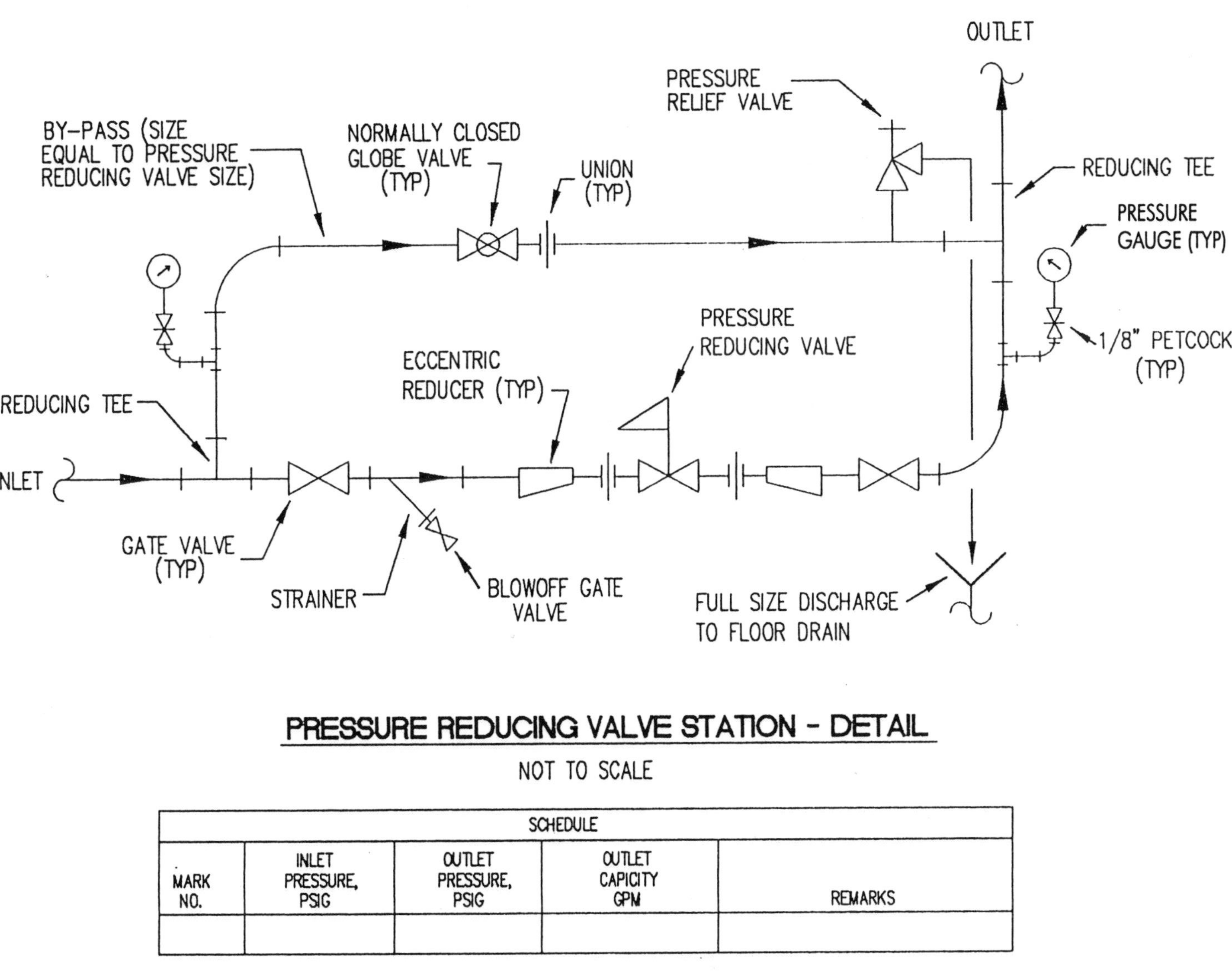

PRESSURE REDUCING VALVE STATION - DETAIL

NOT TO SCALE

SCHEDULE				
MARK NO.	INLET PRESSURE, PSIG	OUTLET PRESSURE, PSIG	OUTLET CAPICITY GPM	REMARKS

Sample Hydronic Heating/ Cooling Design

This sample specification consists of two sections:

Part One: Hydronic Heating Project A
Part Two: Hydronic Heating and Cooling Project B

HYDRONIC HEATING PROJECT A

I. PROJECT DATA:

Type of Building:	Residential, One Story
Location:	Boston, Massachusetts
Type of Heating:	Hot Water: Supply Temp. 200°F Return Temp. 180°F Average Temp. 190°F
Type of Heating System:	Two Pipe, Reverse-Return
Fuel:	No. 2 Heating Oil
Type of Pipes:	Copper Tubing, Type K
Number of Zones:	Four
Type of Controls:	One Circulating Pump, Four Zone Control Valves

II. CALCULATION OF THE HEATING SYSTEM

Based on the architectural drawing and upon Owner's approval, it has been decided that the four heating zones will be most suitable for the occupants of this dwelling and will provide required flexibility in selecting the temperatures for various areas.

The individual rooms will be grouped to create four heating zones as follows:

ZONE A
- Dining Room
- Foyer
- Living Room

ZONE B
- Closet 1
- Master Bath
- Master Bedroom
- Bath
- Closet 2

ZONE C
- Study

ZONE D
- Family Room
- Mud Room
- Breakfast Room
- Kitchen

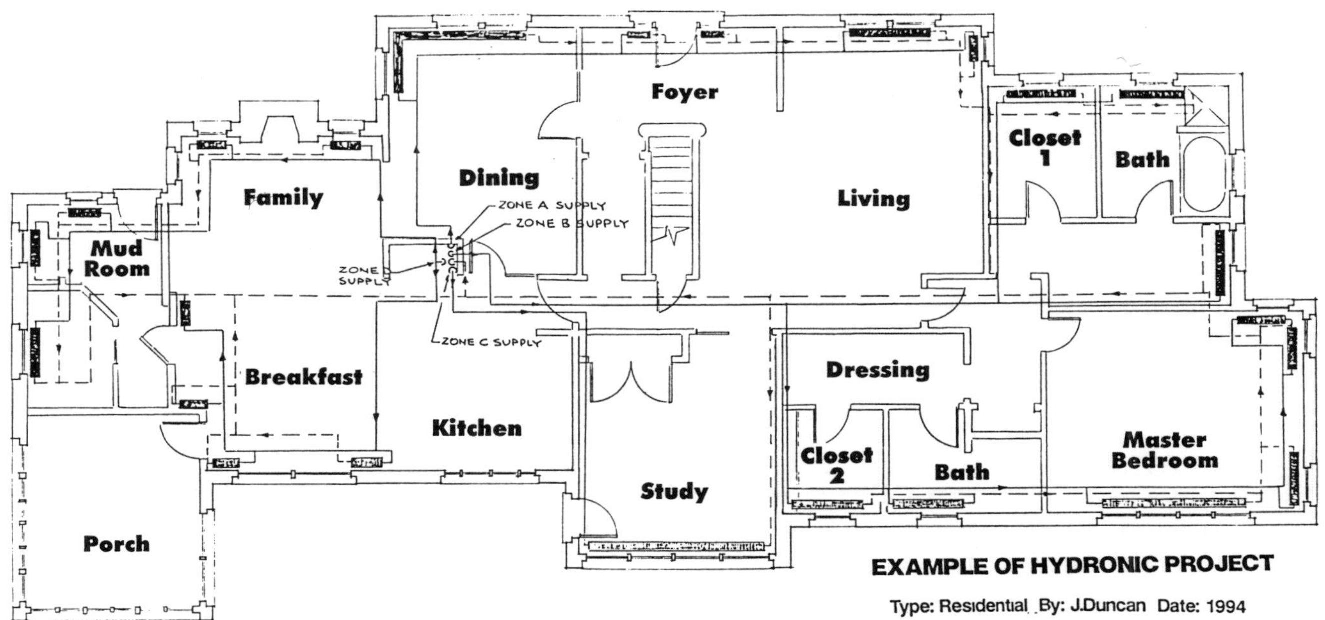

Residential Building Plan

Good engineering practice requires that the baseboard radiators be located under the windows to prevent cold drafts. After consultations with the architect and review of the architectural plans, location of the baseboard units and routing of the hot water piping have been selected as indicated on Sketch No. 1.

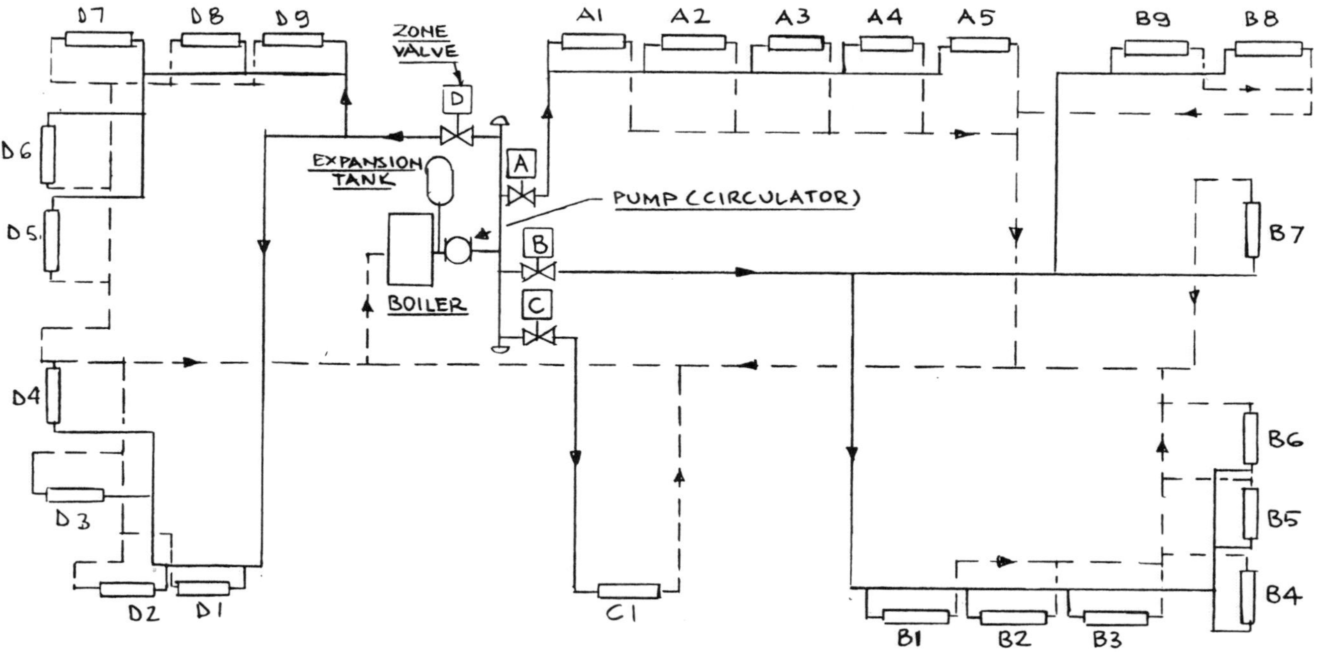

Sketch No. 1

The piping arrangement, superimposed on the architectural drawing, will be used later to determine the length of the piping runs.

III. REQUIRED CAPACITY OF THE HOT WATER BOILER

Based on the heat loss calculations, the capacity of the boiler and hot water flow (based on 20°F temperature difference) has been determined (not included) for each zone.

Table A-1 summarizes heating loads and hot water requirements for each heating zone.

TABLE A-1

ZONE	HEAT LOSS[1] (Btu/hr)	WATER FLOW (gpm[2])
A	34,000	3.4
B	27,000	2.7
C	11,000	1.1
D	22,000	2.2
TOTAL	94,000	9.4

(1) Calculations for the heat loss are not included.
(2) gpm = (Btu/hr)/(500 x 20°F)

Table A-2 is an excerpt from the manufacturer, I=B=R certified, performance data representing the typical capacity rating of the oil-fired hot water boiler, using No. 2 oil, with a heating value of 140,000 Btu/gal, and code required piping pick-up allowance of 1.15. The net rating of the boiler must be equal, or exceed, the total estimated heat loss.

TABLE A-2

BOILER SIZE	BURNER CAPACITY	GPH HEATING CAPACITY (MBH)	I=B=R NET RATING (MBH)	CHIMNEY SIZE (INxINxFT)	BOILER EFFICIENCY (%)
1.0	0.85	98	85.2	8x8x20	83.3
2.0	0.90	103	89.6	8x8x20	83.1
3.0	1.05	120	104.3	8x8x20	82.5
4.0	1.25	142	123.5	8x8x20	81.7
5.0	1.45	164	142.6	8x8x20	81.5

Boiler size 3.0 is selected because its net I=B=R rating capacity of 104.3 MBH closely matches the calculated heat loss.

IV. SELECTION OF THE BASEBOARD RADIATORS

The total length of baseboard radiation is determined based on heat loss calculations and manufacturers' rated capacities.

Table A-3 represents the typical capacity of the baseboard radiators provided by the equipment manufacturer.

TABLE A-3

ELEMENT 3/4" nominal copper tubing, 2-5/8" x 2-1/8", 0.009" aluminum fins, 55 fins/ft	**WATER FLOW (gpm)**	I=B=R approved hot water rating, Btu/hr at average water temperature:					
		160°	170°	180°	190°	200°	210°
	1.0	450	510	580	640	710	770

Average water temperature of 190°F, selected for this application, provides the baseboard radiator capacity of 640 (Btu/hr)/ln ft.

V. DETERMINATION OF THE REQUIRED LENGTH OF THE BASEBOARD RADIATORS

For the purpose of this calculation the detailed design will be performed for Zone B only. In order to determine the length of all radiators, similar calculations must be performed for Zones A, C, and D.

Sketch No. 2 represents an isometric of Zone B piping. From this sketch, the number of fittings will be established and, in conjunction with the pipe lengths indicated on the architectural drawings, will be used for sizing the hot water circulating pump.

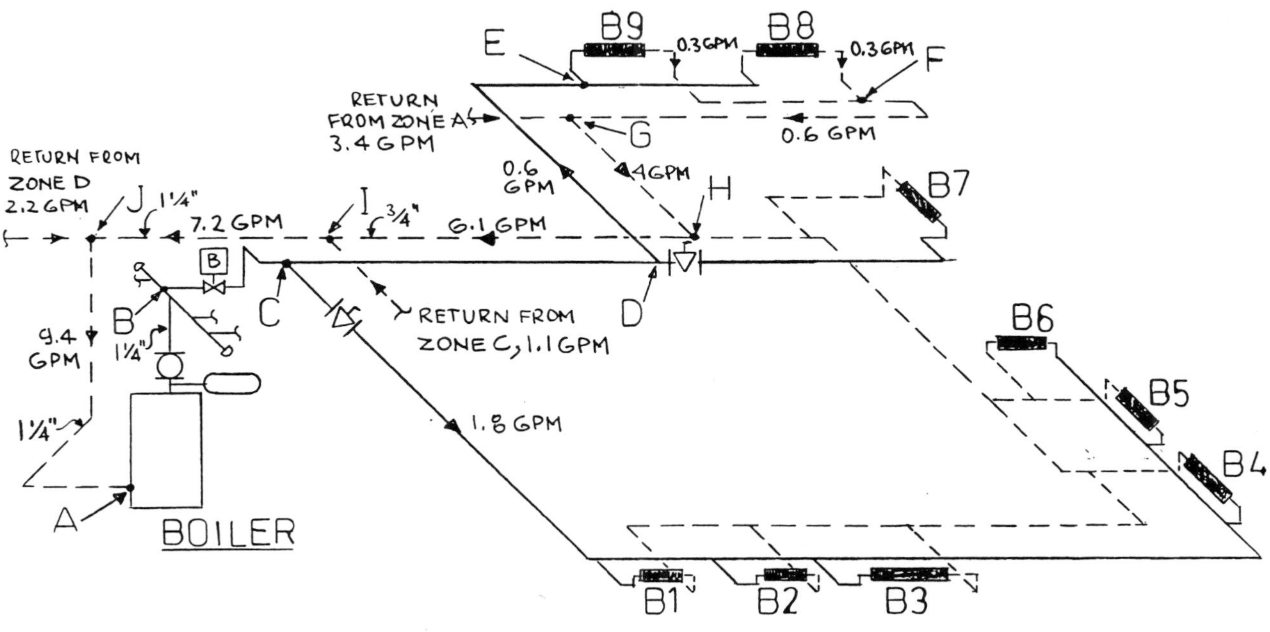

Sketch No. 2

Table A-4 summarizes the Zone B individual room heat losses and the lengths of baseboard radiators (see Sketch No. 2).

TABLE A-4

ROOM	HEAT LOSS (Btu/hr)	CALCULATED TOTAL LENGTH OF BASEBOARDS(1) (ft)	INSTALLED LENGTH OF INDIVIDUAL BASEBOARDS (ft)
CLOSET 1	2,800	4.4	5 (B9)
BATH	2,800	4.4	5 (B8)
CORRIDOR	2,800	4.4	5(B7)
MASTER BEDROOM	13,400	22.0	10.0 (B6) 4.0 (B5) 4.0 (B4) 4.0 (B3)
MASTER BATH	3,200	5.0	5.0 (B2)
CLOSET 2	2,000	3.2	4.0 (B1)

(1) LENGTH (ft) = { HEAT LOSS (Btu/hr)}/{ 640 (Btu/hr)/ln ft}
The calculated length of installed radiators will be rounded-up to the next foot.

VI. DETERMINATION OF THE REQUIRED HOT WATER PUMP (CIRCULATOR) HEAD

To eliminate the need for balancing valves at each baseboard radiator, and subsequent system balancing, the reverse-return piping arrangement has been selected as it results in the combined length of the supply and return piping, and number of fittings, approximately equal for all baseboard radiators. Therefore, for the purpose of calculating the pump head, equal flow through all baseboard radiators will be assumed.

Evaluation of Sketch No. 1 indicates that the Zone B, baseboard radiator B8, has the longest pipe length and, in conjunction with the pressure loss in elbows and tees, will determine the selection of the circulating pump. The equivalent length method (refer to Chapter 2) will be utilized to determine the longest pipe run pressure loss.

Table A-5 summarizes the pressure loss associated with the baseboard radiator B8 loop.

TABLE A-5

Section	Flow gpm	Pipe Size in	Velocity fps	Loss ft/ln ft	Length of Pipe and Equivalent Length of Fittings, ft		Total Loss ft
A-B	9.4	1 1/4	2.4	0.021	Pipe 90^0Elbow (1) Boiler Total	9.50 3.45 12.00 24.95	0.52
B-C	2.7	3/4	1.8	0.023	Pipe 90^0Elbow (3) Zone Valve Total	30.5 10.35 23.3 64.15	1.41
C-D	0.9	3/4	0.6	0.003	Pipe	14.00	0.04
D-E	0.6	3/4	0.4	0.002	Pipe 90^0Elbow (1) Tee-Branch (1) Tee-Through (1) Total	14.00 2.06 4.10 1.40 21.56	0.04
E-F	0.3	3/4	0.2	0.001	Pipe 90^0Elbow (5) Total	19.00 10.30 29.30	0.03
F-G	0.6	3/4	0.4	0.002	Pipe 90^0Elbow (2) Tee-Branch (1) Total	15.00 4.12 4.10 23.22	0.05
G-H	4.0	3/4	2.7	0.045	Pipe Tee-Branch (1) Total	13.00 4.10 17.10	0.77
H-I	6.1	3/4	4.1	0.1	Pipe Tee-Through (1) Total	14.00 1.40 15.40	1.54
I-J	7.2	1 1/4	1.85	0.014	Pipe Tee-Branch (1) Total	27.00 4.10 31.10	0.44
J-A	9.4	1 1/4	2.4	0.021	Pipe 90^0Elbow (2) Total	7.00 4.12 11.12	0.23
							5.03

VII. SELECTION OF THE HOT WATER PUMP (CIRCULATOR)

As determined in Sections III and VI, the pump must be capable of developing at least 5.03 ft of total head while circulating 9.4 gpm of water through the system. For the pump selection, the total head will be rounded up to 5.1 ft.

Figure A-1 represents performance curves, developed by the pump (circulator) manufacturer, designed for application in the hot water heating systems. Pump size 005 will deliver the required flow and head.

FIGURE A-1

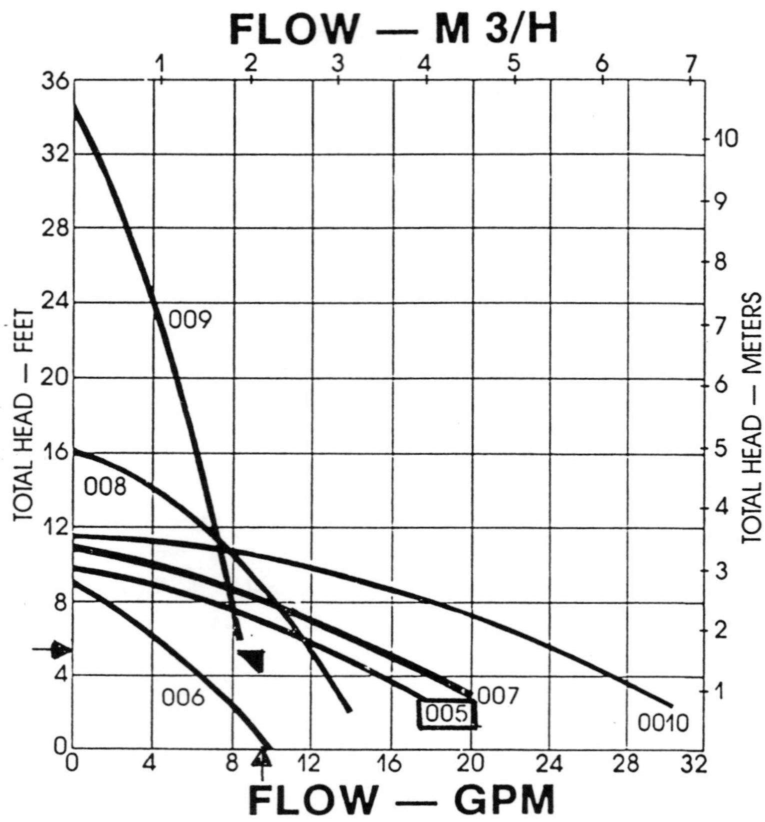

(Courtesy, Taco, Inc.)

VIII. SIZING OF THE HOT WATER EXPANSION TANK

The following formula will be used to determine the minimum volume of the tank. Please note that the tank is located on the suction side of the pump.

From the piping layout it is estimated that the system contains 60 gallons of water:

$v_s = 60$ gallons

Other system parameters are:

The high point of the system is 5 ft above the boiler.
The average water temperature is t = 190°F.
The system relief valve is set at 15 psig. We will add 5 psig safety margin to prevent inadvertent opening of the valve.

$p_o = 15 + 5 = 20$ psig

System fill pressure is 5 psig.

p_f = (5 x 2.3) + 5 + 34 = 50.5 ft abs

p_o = (20 x 2.3) + 34 = 80 ft abs

p_a = 34 ft abs

V_t = [(0.00041 x 190) - (0.0466)60]/[(34/50.5)-(34/80)] = 7.6 gal

A standard tank of 10 gallon capacity will be selected for this system.

HYDRONIC HEATING AND COOLING PROJECT B

HOT WATER SYSTEM

I. PROJECT DATA FOR HEATING SYSTEM:

Type of Building:	Commercial
Location:	Boston, Massachusetts
Type of Heating:	Hot Water: Supply Temp. 200°F Return Temp. 180°F Average Temp. 190°F
Type of Heating System:	Two Pipe, Reverse-Return
Fuel:	No. 2 Heating Oil
Type of Pipes:	Copper Tubing, Type K
Number of Zones:	Nine
Type of Controls:	One Circulating Pump, Nine Zone Control Valves

II. DETERMINATION OF ZONING FOR THE HEATING SYSTEM

Based on the architectural drawing and upon Owner's approval, it has been decided that nine heating zones will be most suitable for this application.

The rooms will be grouped based on the exposure and the internal load profile, resulting in nine heating zones, as follows:

ZONE A
Mechanical Room

ZONE B
Standby Generator
Sprinkler Room

ZONE C
Office 1

ZONE D
Lobby

ZONE E
Warehouse

ZONE F
Conference Room

ZONE G
Staff Entry
Director
Supervisor
Foreman
Clerk
Accounting

ZONE H
Office 2
Receiving

ZONE I
Loading Dock

Notes: 1. Refrigerator Space does not require heating.
2. Electrical Room will be provided with electric heat to avoid potential water leaks on electrical equipment.

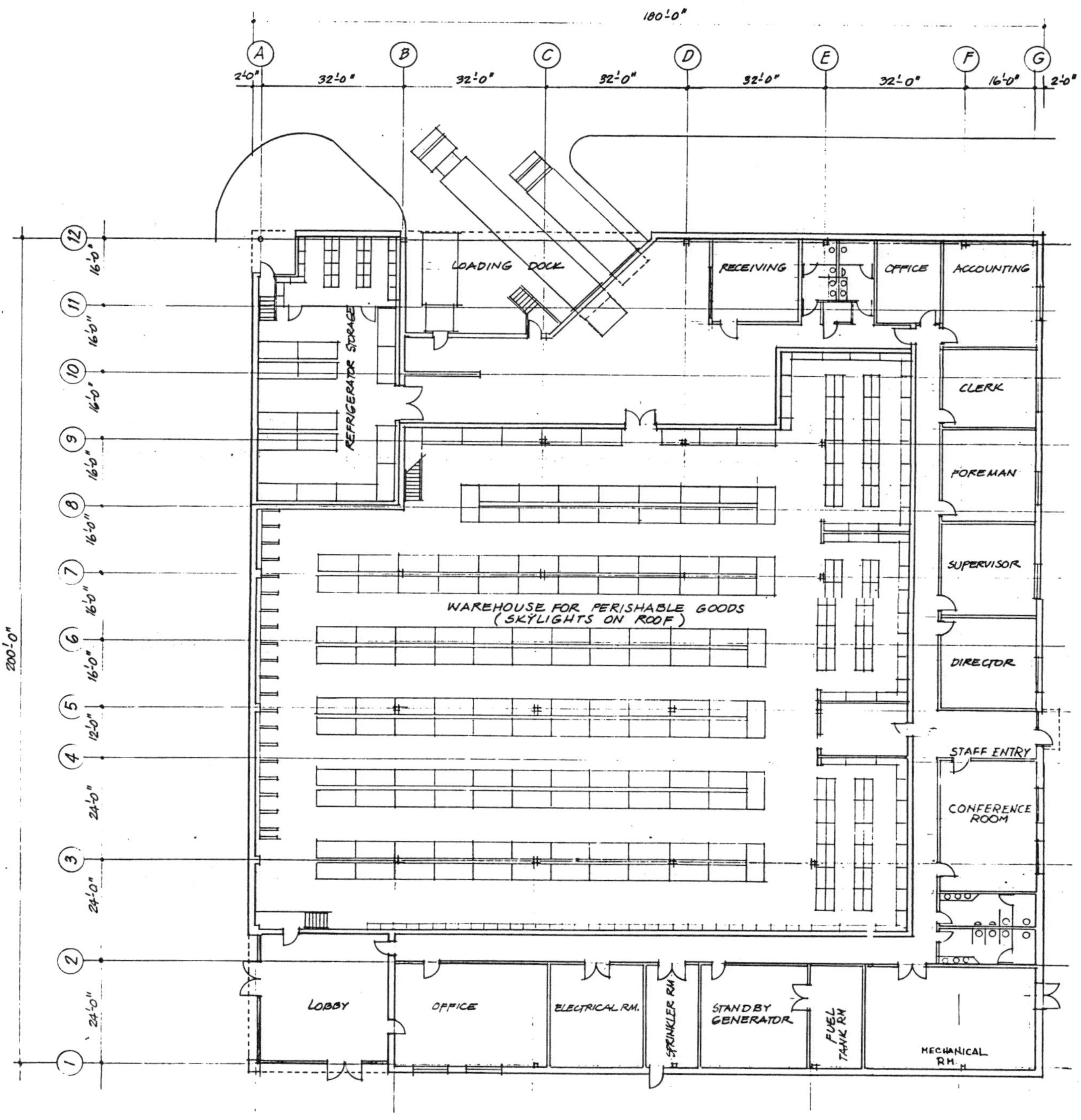

Commercial Building Plan

Good engineering practice requires that the baseboard radiators be located under the windows to prevent cold drafts.

Where possible, the warm air discharge of the wall-mounted unit heaters should be directed along the walls with an outside exposure, to minimize cold downdrafts.

Since the Warehouse Area is provided with the heating/cooling units, the units will be provided with the hot water heating coils.

The Loading Dock Area will be heated with down-blast unit heaters delivering heat at the truck loading doors and at the same time providing unobstructed headroom for loading activities.

Sketch No. 1 indicates location of the heating equipment and the piping arrangement, superimposed on the architectural drawing, which will be used later to determine the length of the piping runs.

III. REQUIRED CAPACITY OF THE HOT WATER BOILER

Based on the heat loss calculations, the capacity of the boiler and hot water flow (based on 20°F temperature difference) has been determined (not included) for each zone.

Table B-1 summarizes heating loads and hot water flow requirements for each heating zone.

TABLE B-1

ZONE	HEAT LOSS(1) (Btu/hr)	WATER FLOW (gpm(2))
A	6,500	0.50(3)
B	18,900	1.00(3)
C	12,000	1.20
D	12,000	1.20
E	75,000	7.50(4)
F	11,000	1.10
G	43,000	4.30
H	14,000	1.40
I	39,000	1.50(3)
TOTAL	231,000	19.70

(1) Calculations for the heat loss are not included.
(2) gpm = (Btu/hr)/(500 x 20°F)
(3) Water flow of 0.5 gpm per unit heater based on unit heater rating listed in Table B-5.
(4) Total water flow of 7.5 gpm for units E1 and E2 (3.75 gpm each) based on heating coil rating in Table B-7.

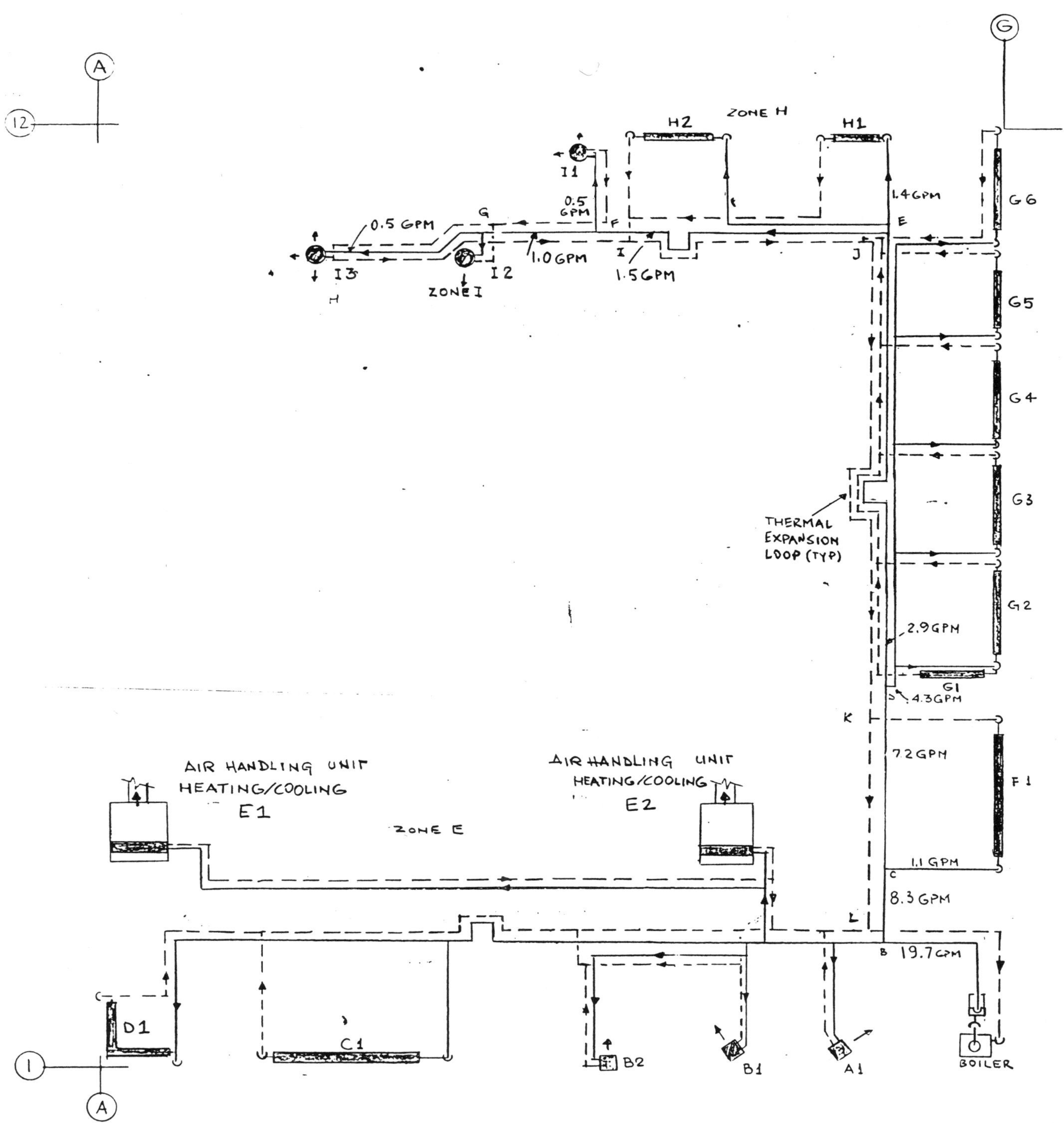
ZONE H
H2
H1
I1
0.5 GPM
G
0.5 GPM
F
1.4GPM
E
I
J
1.0GPM
1.5GPM
I3
I2
ZONE I
G6
G5
G4
G3
THERMAL EXPANSION LOOP (TYP)
G2
2.9GPM
G1
4.3GPM
K
7.2GPM
F1
AIR HANDLING UNIT HEATING/COOLING
E1
AIR HANDLING UNIT HEATING/COOLING
E2
ZONE E
1.1 GPM
8.3 GPM
19.7GPM
D1
C1
B2
B1
A1
BOILER

Sketch No. 1

Table B-2 is an excerpt from the manufacturer, I=B=R certified, performance data representing the typical capacity rating of the oil-fired, hot water boiler, using No. 2 oil, with a heating value of 140,000 Btu/gal, and code required piping pick-up allowance of 1.15. The net rating of the boiler must be equal to, or exceed, the total calculated heat loss.

TABLE B-2

BOILER SIZE	BURNER CAPACITY (GPH)	HEATING CAPACITY (MBH)	I=B=R NET RATING (MBH)	CHIMNEY SIZE (INxINxFT)	BOILER EFFICIENCY (%)
8.0	2.05	230	123.5	8x8x20	81.7
9.0	2.25	252	142.6	8x8x20	81.5
10.0	2.45	287	234.5	8x8x20	81.7

Boiler size 10.0 is selected because its net I=B=R rating capacity of 234.5 MBH closely matches the calculated heat loss.

IV. SELECTION OF THE HEATING EQUIPMENT

a. Selection of Baseboard Radiators

The total length of baseboard radiation will be determined based on heat loss calculations and manufacturers' rated capacities.

Table B-3 represents the typical capacity of the baseboard radiators provided by the equipment manufacturer.

TABLE B-3

ELEMENT	WATER FLOW (gpm)	I=B=R approved hot water rating, Btu/hr at average water temperature:					
		160°	170°	180°	190°	200°	210°
3/4" nominal copper tubing, 2-5/8" x 2-1/8", 0.009" aluminum fins, 55 fins/ft	1.0	450	510	580	640	710	770

Average water temperature of 190°F, selected for this application, provides the baseboard radiator capacity of 640 (Btu/hr)/ln ft.

b. Determination of the Required Length of the Baseboard Radiators

For the purpose of this calculation, the detailed design will be performed for Zone G only. In order to determine the length of all radiators, similar calculations must be performed for Zones C, D, F, and H. Table B-4 summarizes the Zone G individual room heat losses and the lengths of baseboard radiators (see Sketch No. 2).

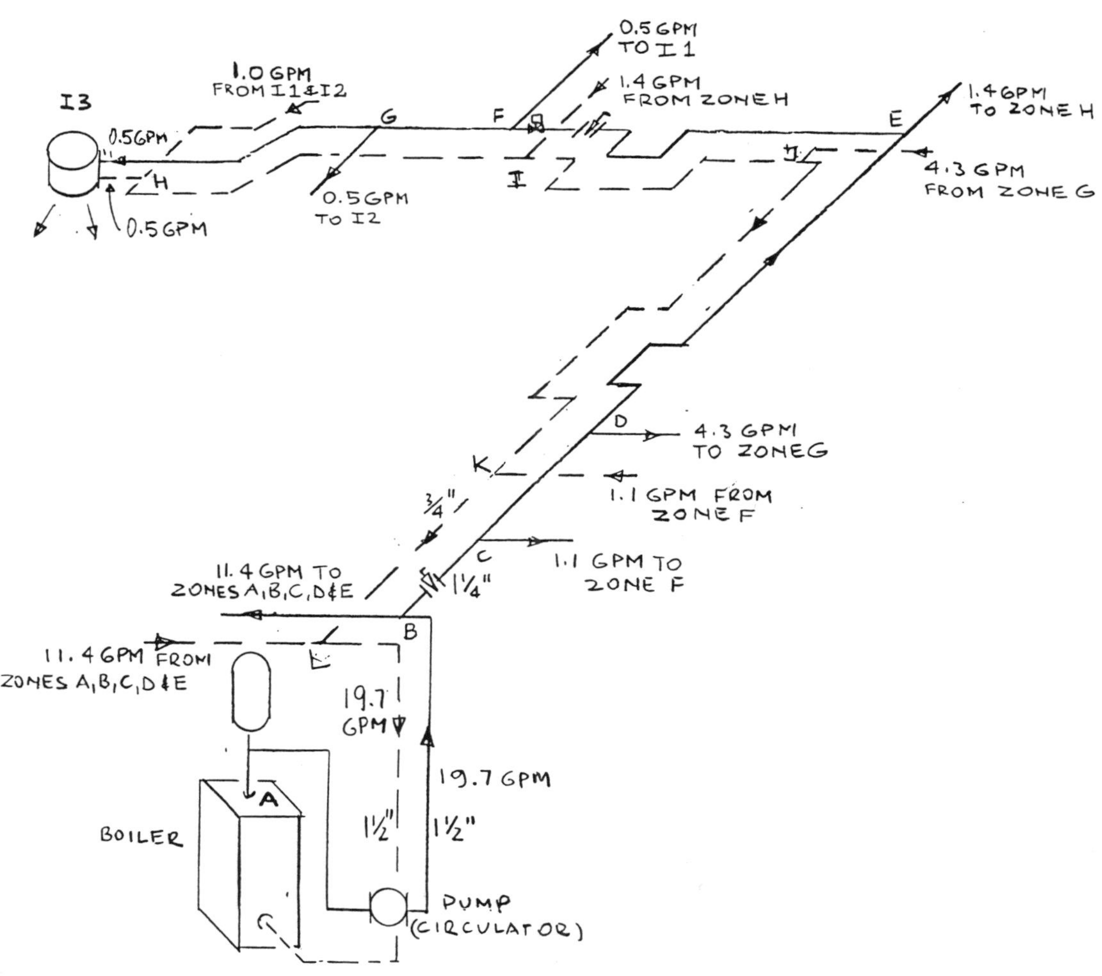

Sketch No. 2

TABLE B-4

ROOM	HEAT LOSS (Btu/hr)	CALCULATED LENGTH OF BASEBOARD(1) (ft)	INSTALLED LENGTH OF BASEBOARDS (ft)
STAFF ENTRY	3,000	4.7	5(G1)
DIRECTOR	7,000	10.9	12(G2)
SUPERVISOR	7,000	10.9	12(G3)
FOREMAN	7,000	10.9	12(G4)
CLERK	6,000	9.4	10(G5)
ACCOUNTING	13,000	20.3	21(G6)

(1) LENGTH (ft) = { HEAT LOSS (Btu/hr)}/{ 640 (Btu/hr)/ln ft}

c. Selection of the Unit Heaters

Table B-5 represents the typical rating of the unit heaters provided by the equipment manufacturer.

TABLE B-5

TYPE/UNIT SIZE	WATER FLOW (gpm)	CAPACITY (Btu/hr)	WATER PRESSURE LOSS (ft)
1.0 (WALL)	0.15	6,800	0.09
380 cfm 1,550	0.50	11,400	0.70
rpm 1/60 hp	1.00	13,300	2.30
	2.00	14,500	7.50
2.0			
(DOWNBLAST)	0.15	8,200	0.85
550 cfm 1,550	0.50	14,700	1.01
rpm 1/20 hp	1.00	17,800	1.85
	2.00	19,900	6.90

Table B-6 lists the selected unit heater sizes for individual rooms.

TABLE B-6

ROOM	HEAT LOSS (Btu/Hr)	LISTED UNIT HEATER CAPACITY/ SIZE (Btu/hr)	HOT WATER FLOW (gpm)	WATER PRESSURE DROP (ft)
MECHANICAL ROOM (A1)	9,500	11,400/1.0	0.50	0.70
STANDBY GENERATOR (B1)	6,300	6,800/1.0	0.15	0.09
SPRINKLER ROOM (B2)	6,000	6,800/1.0	0.15	0.09
LOADING DOCK (I1)	13,000	14,700/2.0	0.50	1.01
LOADING DOCK (I2)	13,000	14,700/2.0	0.50	1.01
LOADING DOCK (I3)	13,000	14,700/2.0	0.50	1.01

d. Selection of the Zone E Air Handling Unit Heating Capacity

Table B-7 represents the typical capacity of the heating coil, installed in the air handling unit, provided by the equipment manufacturer.

TABLE B-7

UNIT SIZE	WATER FLOW (gpm)	CAPACITY (Btu/hr)	WATER TEMP. DROP (°F)	FINAL AIR TEMP. (°F)	WATER PRESSURE DROP (ft)
1.0	3.75	39,200	20.9	100.0	0.04
900 cfm	4.25	41,900	19.7	103.0	0.07
1,550 rpm	8.00	49,400	12.4	110.0	0.22
1/20 hp	12.50	52,800	8.4	114.0	0.43
	16.00	55,000	6.9	116.0	0.71

Zone E heat loss equals 75,000 Btu/hr. Since two units are provided, each unit capacity must be at least 37,500 Btu/hr.
Unit size 1.0, having the heating capacity of 39,200 Btu/hr and the hot water flow of 3.75 gpm, will be selected, as it matches closely the required capacity of 37,500 Btu/hr.

V. DETERMINATION OF THE REQUIRED HOT WATER PUMP (CIRCULATOR) HEAD

To eliminate the need for balancing valves at each baseboard radiator or the unit heater, and subsequent system balancing, the reverse-return piping arrangement has been selected as it results in the combined length of the supply and return piping and number of fittings, approximately equal for all heating equipment.

Evaluation of Sketch No. 1 indicates that the Zone I unit heater (I3) has the longest pipe length and, in conjunction with the pressure loss in elbows and tees, will determine the selection of the circulating pump. The equivalent length method (refer to Chapter 2) will be utilized to determine the longest pipe run pressure loss.

Table B-8 summarizes the pressure loss associated with the unit heater I3 loop.

Section	Flow gpm	Pipe Size in	Velocity fps	Loss ft/ln ft	Length of Pipe and Equivalent Length of Fittings, ft		Total Loss ft
A-B	19.7	1 1/2	3.7	0.036	Pipe 90°Elbow (4) Tee-Branch(1) Total	52.5 8.24 4.10 64.84	2.33
B-C	8.3	1 1/4	2.1	0.017	Pipe Tee-Through(1) Gate Valve Total	15.5 1.40 0.55 17.45	0.30
C-D	7.2	3/4	1.8	0.013	Pipe Tee-Through(1) Total	38.0 1.40 39.40	0.51
D-E	2.9	3/4	2.1	0.029	Pipe 90°Elbow (4) Tee-Branch (1) Total	102.0 8.24 4.10 114.34	3.32
E-F	1.5	3/4	1.1	0.009	Pipe 90°Elbow (4) Gate Valve(1) Zone Valve(1) Total	64.5 8.24 0.55 23.3 96.59	0.90
F-G	1.0	3/4	0.74	0.004	Pipe Tee-Through(1) Total	23.5 1.40 24.9	0.10
G-H	0.5	3/4	0.37	0.002	Pipe 45°Elbow(2) Tee-Branch (1) Unit Heater Total	34.0 2.20 4.10 1.01 41.31	0.08
H-I	1.5	3/4	1.1	0.009	Pipe 45°Elbow(2) Tee-Through (1) Total	59.0 2.20 1.40 62.6	0.56
I-J	2.9	3/4	2.1	0.017	Pipe 90°Elbow (4) Tee-Branch(1) Total	62.5 8.24 4.10 74.84	1.27
J-K	7.2	3/4	1.8	0.013	Pipe 90°Elbow (5) Tee-Through(1) Total	108.0 10.35 1.40 119.75	1.57
K-L	8.3	1 1/4	2.1	0.017	Pipe Tee-Branch(1) Total	43.5 4.10 47.60	0.81
L-A	19.7	1 1/2	3.7	0.036	Pipe 90°Elbow (3) Boiler Total	70.0 10.35 12.00 92.35	3.32
							15.07

VI. SELECTION OF THE HOT WATER PUMP (CIRCULATOR)

As determined in Sections III and V, the pump must be capable of developing at least 15.07 ft of total head while circulating 19.7 gpm of water through the system. For the pump selection, the rounded up total flow of 20 gpm and the total head of 15.5 ft will be used.

Figure B-1 represents performance curves for Pump Model 1507, developed by the pump (circulator) manufacturer. This pump, provided with 6.10" diameter impeller and 1/4 hp motor, will deliver the required flow and head.

FIGURE B-1

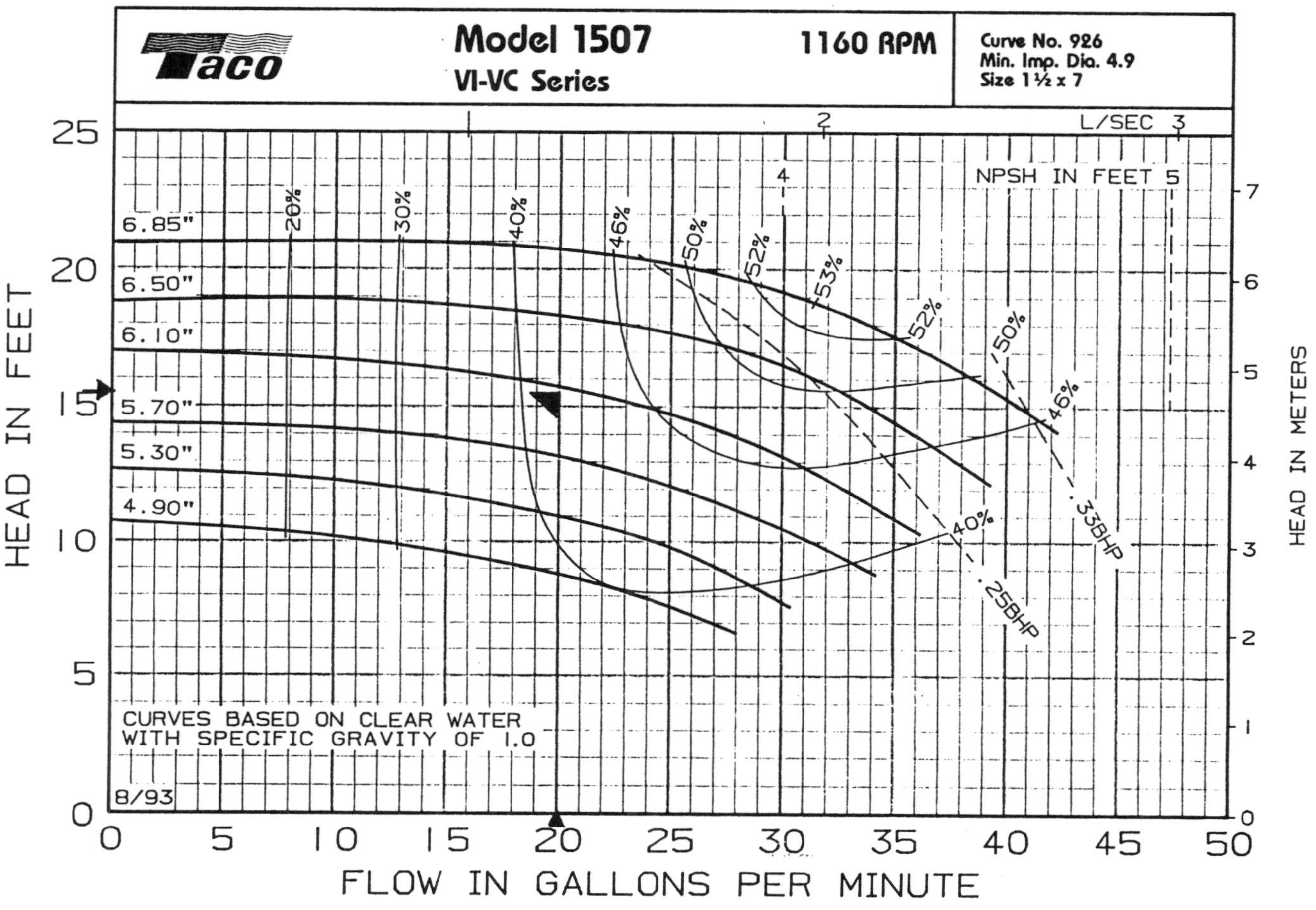

(Courtesy, Taco, Inc.)

VII. SIZING OF THE HOT WATER EXPANSION TANK

The following formula will be used to determine the minimum volume of the tank. Please note that the tank is located on the suction side of the pump.

From the piping layout it is estimated that the system contains 135 gallons of water:

$v_s = 135$ gallons

Other system parameters are:

The high point of the system is 11 ft above the boiler.
The average water temperature is t = 190°F.
The system relief valve is set at 15 psig. We will add 5 psig safety margin to prevent inadvertent opening of the valve.

$p_o = 15 + 5 = 20$ psig

System fill pressure is 5 psig.

$p_f = 5 \times 2.3 + 11 + 34 = 56.5$ ft abs

$p_o = 20 \times 2.3 + 34 = 80$ ft abs

$p_a = 34$ ft abs

$V_t = [(0.00041 \times 190 - 0.0466)135]/[(34/56.5)-(34/80)] = 23.9$ gal

A standard tank of 25 gallon capacity will be selected for this system.

CHILLED WATER SYSTEM

I. PROJECT DATA FOR COOLING SYSTEM:

Type of Building:	Commercial
Location:	Massachusetts
Space Design Temperatures:	Warehouse Area 72°F$_{db}$ / 60°F$_{wb}$ Office and Lobby Area 78°F$_{db}$ / 65°F$_{wb}$
Type of Cooling:	Chilled Water: Supply Temp. 42°F Return Temp. 52°F
Type of Cooling System:	Vapor Compression Chiller
Type of Pipes:	Copper Tubing, Type K
Number of Zones:	Two
Type of Controls:	One Circulating Pump, Two Three-Way Control Valves

II. REQUIRED CAPACITY OF THE CHILLER

Based on the heat gain calculations (not included), the capacity of the chiller and the chilled water flow (based on 10°F temperature difference) has been determined for each air handling unit.

Table B-9 summarizes cooling loads and chilled water flow requirements for each air handling unit.

TABLE B-9

AIR HANDLING UNIT	HEAT GAIN(1) (Btu/hr)	WATER FLOW (gpm(2))
E1	59,000	12
E2	59,000	12
E3	85,000	17
TOTAL	203,000	41

(1) Calculations for heat gain are not included.
(2) gpm = (Btu/hr)/(500 x 10°F)

The chiller capacity:

$$(203,000 \text{ Btu/hr}) / (12,000 \text{ Btu/ton}) = 16.9 \text{ tons}$$

For equipment selection, the chiller capacity will be rounded up to 20 tons for practical purposes.

In this capacity range, the roof-mounted, air-cooled, reciprocating chiller package will be selected.

Table B-10 represents a typical capacity rating of the air-cooled, reciprocating chillers, provided by the equipment manufacturer.

TABLE B-10

UNIT SIZE	LEAVING WATER TEMP. (°F)	AMB. TEMP. (90°F)		AMB. TEMP. (95°F)		AMB. TEMP. (100°F)	
		Tons	kW	Tons	kW	Tons	kW
022C	42	21.4	22.6	20.8	23.4	20.4	24.2
AIR-COOLED	44	22.0	22.8	21.4	23.6	20.9	24.4
CHILLER	45	22.4	22.9	21.8	23.8	21.3	24.6

The capacity of chiller size 022C, rated at 20.8 tons at an ambient temperature of 95°F and leaving chilled water temperature of 42°F, closely matches the required capacity of 203,000 Btu/hr, with approximately 20% safety margin in chiller capacity.

III. SELECTION OF THE AIR HANDLING UNIT COOLING COILS

Table B-11 represents the typical capacity of the cooling coil, installed in the air handling unit, provided by the equipment manufacturer.

TABLE B-11

UNIT SIZE	RATED FLOW (gpm)	COIL ENT. WTR (°F)	NUMBER OF ROWS	WATER TEMP. RISE (°F)	COOLING CAPACITY (Btu/hr)
1.0(1)	12	42	2	10	59,900
1.0(2)	17	42	2	10	85,500

(1) Entering air conditions from the warehouse area (Unit E1 & E2): $72°F_{db}$ / $60°F_{wb}$

(2) Entering air conditions from the office area (Unit E3): $78°F_{db}$ / $65°F_{wb}$

The capacity of the above listed coils closely matches the required capacity listed in Table B-9.

IV. DETERMINATION OF REQUIRED CHILLED WATER PUMP (CIRCULATOR) HEAD

Evaluation of Sketch No. 3 indicates that the Unit E1 has the longest pipe length and, in conjunction with the pressure loss in elbows and fittings, as shown on Sketch No. 4, will determine the selection of the circulating pump.

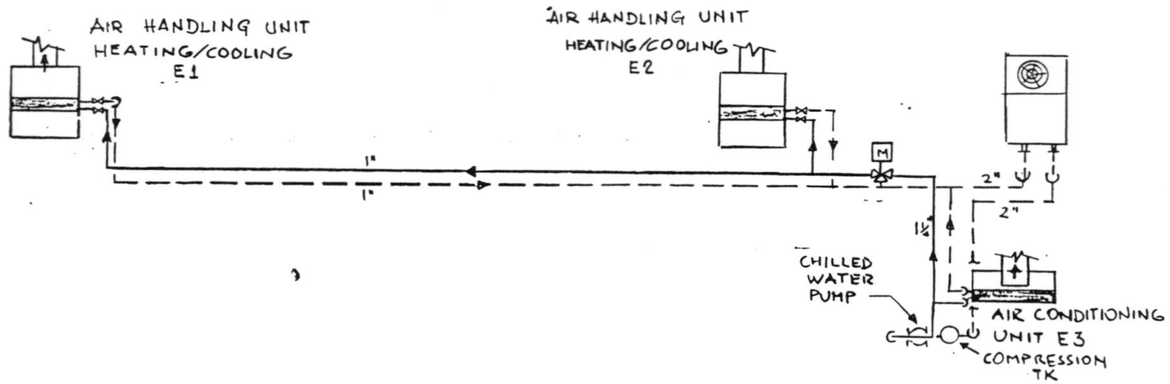

Sketch No. 3

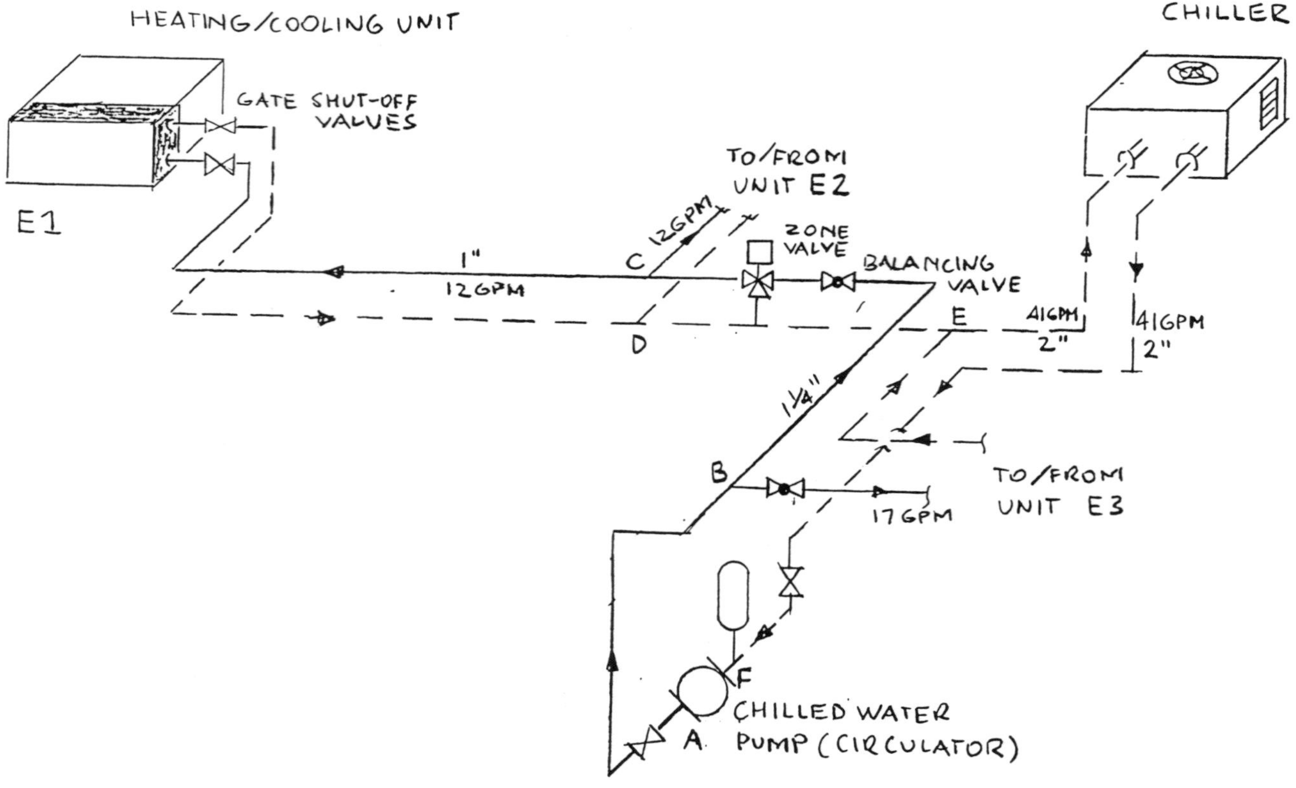

Sketch No. 4

Table B-12 summarizes the pressure loss associated with Unit E1.

TABLE B-12

Section	Flow gpm	Pipe Size in	Velocity fps	Loss ft/ln ft	Length of Pipe and Equivalent Length of Fittings, ft		Total Loss ft
A-B	41.0	2	4.1	0.034	Pipe 90^0Elbow (2) Tee-Through(1) Globe Valve(1) Total	58.50 10.34 3.50 58.60 130.94	4.45
B-C	24.0	1 1/4	4.51	0.055	Pipe 90^0Elbow(1) Tee-Through(1) Globe Valve(2) Total	15.50 3.45 2.30 78.20 99.45	5.47
C-D	12.0	1	1.8	0.093	Pipe 90^0Elbow(5) Tee-Through(1) Gate Valve(2) Cooling Coil(1) Total	38.00 13.10 1.80 1.40 11.00 65.30	6.07
D-E	24.01	1 1/4	4.51	0.055	Pipe Tee-Through(1) Total	102.00 2.30 104.30	5.74
E-F	41.0	2	4.1	0.034	Pipe 90^0Elbow (7) Tee-Through(1) Gate Valve(3) Chiller(1) Total	64.5 36.19 3.50 1.65 23.3 129.14	4.39
TOTAL							26.12

V. SELECTION OF THE CHILLED WATER PUMP

As determined in Sections II and IV, the pump must be capable of developing at least 26.12 ft of total head, while circulating 41 gpm of water through the system. For the pump selection, the total head will be rounded up to 27 ft.

Figure B-2 represents performance curves for Pump Model 1612, developed by the pump manufacturer. This pump, provided with 5.75" diameter impeller and 1/2 hp motor, will deliver the required flow and head.

VI. SIZING OF THE CHILLED WATER EXPANSION TANK

The following formula will be used to determine the minimum volume of the tank. The tank volume will be selected with 50% of the capacity determined for the water expansion between 70°F and 200°F. Please note that the tank is located at the suction side of the pump. From the piping layout, it is estimated that the system contains 125 gallons of water:

v_s = 125 gallons

FIGURE B-2

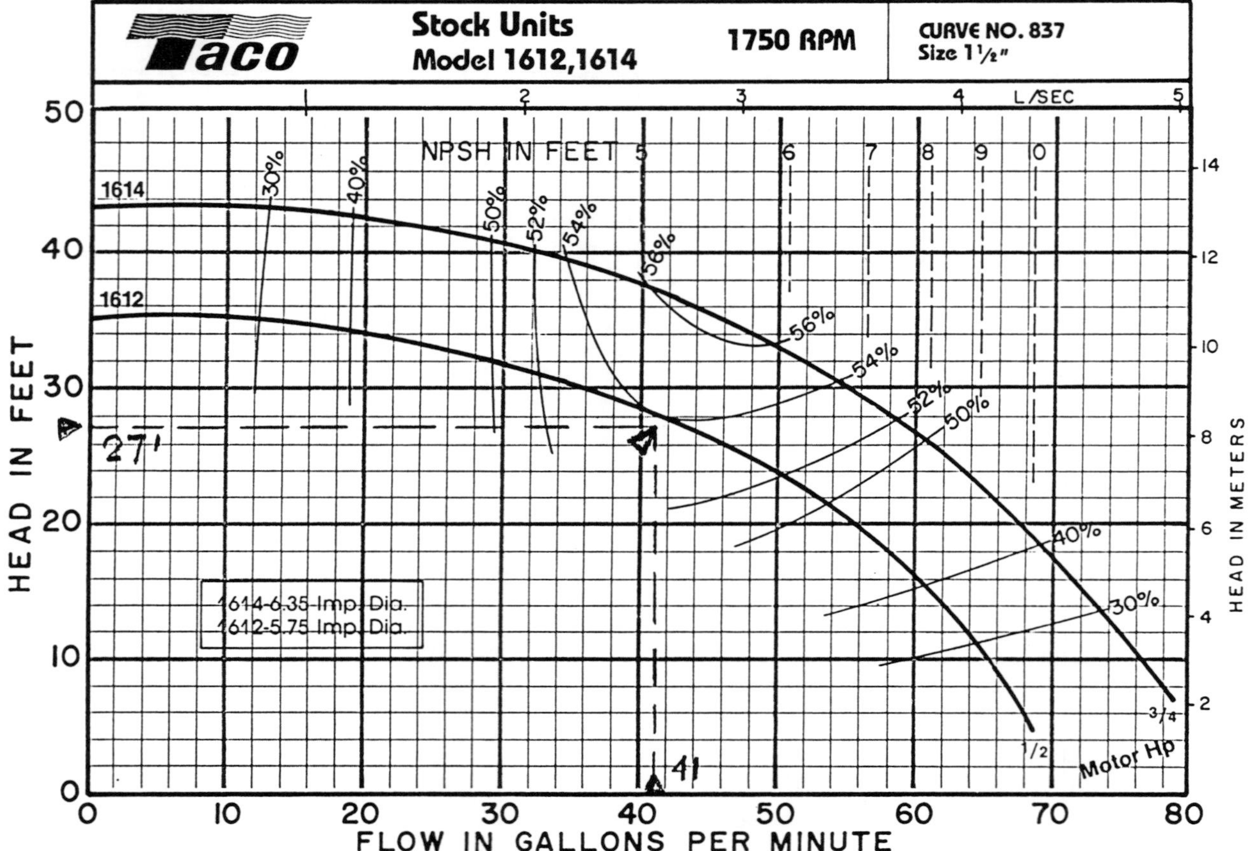

(Courtesy, Taco, Inc.)

Other system parameters are:

The high point of the system is 21 ft above the pump.
The average water temperature is 47°F.
The system relief valve is set at 15 psig. We will add 5 psig safety margin to prevent inadvertent opening of the valve.

$p_o = 15 + 5 = 20$ psig

System fill pressure is 5 psig.

$p_f = 5 \times 2.3 + 21 + 34 = 66.5$ ft abs

$p_o = 20 \times 2.3 + 34 = 80$ ft abs

$p_a = 34$ ft abs

$V_t = 0.5 \times [(0.00041 \times 200 - 0.0466) \times 125]/[(34/66.5) - (34/80)] = 25.8$ gal

A standard tank of 30 gallon capacity will be selected for this system.

SAMPLE SPECIFICATIONS FOR A HYDRONIC SYSTEM

This sample specification consists of four sections:

I. Mechanical Specification General Requirements
II. Pipe, Fittings, Valves, and Specialties: Materials and Installation
III. Hydronic Systems: Heating and Cooling
IV. Insulation

Note: These specifications shall be customized (paragraphs added, deleted, or modified) for each project. Where alternates are indicated specifier shall select.

SECTION I. MECHANICAL SPECIFICATION GENERAL REQUIREMENTS

TABLE OF CONTENTS

1 GENERAL

1.1 DESCRIPTION

This section applies to all sections of these mechanical specifications except as otherwise specified in each individual section.

1.2 SUBMITTALS

Within 30 days after receiving the notice to proceed and prior to fabrication or installation, submit shop drawings and manufacturer's data, showing type and size for equipment, materials and finish, and pertinent details for each system specified. Reference will be made to the applicable section of the specification. Obtain approval from the Engineer before procurement, fabrication, or delivery of the items to the jobsite. Partial submittals are not acceptable and will be returned without review.

A. Shop Drawings

Drawings shall not be less than 8½ inches by 11 inches in size and not more than of 36 inches by 48 inches with a minimum scale of 1/8 inch per foot. Drawings shall include floor plans and sections, wiring diagrams, and installation details of equipment. The drawings shall show location of equipment, space for removable parts, control panels, accessories, piping, and other items required to make a complete installation. For wiring diagrams, identify circuit terminals. For equipment, provide clearance for operations, maintenance, and replacement of parts as per manufacturer's recommendation. If equipment is disapproved, revise and resubmit. Verify sizes of sleeves.

B. Manufacturer's Data

Provide manufacturer's descriptive literature cataloged data including technical data sheets, equipment drawings, diagrams, performance and characteristic curves, and catalog cuts for each item. Each submittal shall include manufacturer's name, trade name, catalog model number, nameplate data, size, layout dimensions, and capacity.

C. Standard Compliances

For materials or equipment indicate conformance to, but not limited to, applicable standards such as the American National Standards Institute (ANSI), American Society for Testing and Materials (ASTM), National Electrical Manufacturer's Association (NEMA), American Society of Mechanical Engineers (ASME), American Gas Association (AGA), American Refrigeration Institute (ARI), and Underwriters' Laboratories (UL). For materials and equipment which are not regulated by an organization's standards or specifications, submit a certificate of conformance from the manufacturer for approval.

D. Certified Test Reports

Before shipment of materials and equipment, these shall be tested for compliance to the specification and the test results submitted to the Engineer.

E. Operation and Maintenance Manual
For each item of equipment furnish an operation and maintenance manual. After approval of the same, furnish three copies of the manual. In addition, furnish one complete manual prior to the time that equipment field acceptance tests are performed, and furnish the remaining manuals before the contract is completed. Inscribe the following identification on the cover: OPERATION AND MAINTENANCE MANUAL, name and location of the equipment and building, name of the Contractor, and contract number. Include names, addresses, and telephone numbers of each Subcontractor installing equipment and of the local representatives for each item of equipment. Provide a table of contents with the tab sheets placed before instructions covering the subject. Include wiring and control diagrams with data to explain detailed operation and control of each item of equipment; a control sequence describing start-up operation and shutdown; installation instructions; maintenance instructions; lubrication schedule, including type, grade, temperature range, and frequency; troubleshooting; safety precautions, diagrams, and illustrations; test procedures; performance data; and parts list and recommended spare parts to be kept in storage.

1.3 MANUFACTURER'S RECOMMENDATIONS
Manufacturer's representative and the design engineer shall be present to supervise the installation of major items of equipment or systems. Prior to start-up, certificates shall be submitted by the manufacturer that the equipment and systems have been installed in accordance with the manufacturer's requirements.

1.4 INSTRUCTIONS TO THE MANUFACTURER WITH REFERENCE TO OPERATING AND MAINTENANCE PERSONNEL
Furnish the services of competent instructors who will provide instructions to the maintenance personnel in the adjustment, operation, and maintenance of equipment and systems. Pertinent safety requirements of the equipment and systems specified shall be included. Provide instructions during the first regular work week after the equipment or systems have been accepted and turned over for regular operation.

1.5 DELIVERY AND STORAGE
Handle, store, and protect equipment and materials to prevent damage during transportation and storage before installation, in accordance with the manufacturer's recommendations and as approved. Replace damaged or defective items. Materials shall be delivered and stored in the manufacturer's original unopened protective packages, and packages shall be marked to identify the contents. Materials shall be stored with protection from the weather, humidity, temperature, dirt, dust, water, and other contaminants.

1.6 CATALOGED PRODUCTS
Provide materials and equipment that are the manufacturer's standard products.

1.7 PROHIBITED MATERIALS
No equipment, materials, or items that contain lead, mercury, asbestos, or ozone depleting substances shall be used on this project.

1.8 CONCRETE PADS
Concrete pads shall be provided for floor supported mechanical equipment and the associated floor mounted electrical and mechanical accessories. Pads shall be a minimum of four inches high and shall be sufficiently large to accept the vibration isolation device installed under the equipment.

1.9 SAFETY REQUIREMENTS
Belts, pulleys, chains, gears, couplings, projecting set screws, keys, and other rotating parts shall be fully enclosed or properly guarded in accordance with OSHA regulations. High-temperature equipment and piping located in areas where they might pose a danger for personnel or create a fire hazard must be properly guarded or covered with insulation where required for safe operation and maintenance of equipment.

1.10 WELDER'S QUALIFICATIONS
Welding operators and welders shall be qualified in accordance with the requirements of AWS D1.1 and ANSI B31.1. Contractor is to provide copies of test certificates. Welders shall have been tested within one year of working on the project.

1.11 ELECTRICAL REQUIREMENTS
Provide electrical components of mechanical equipment and systems such as motors, starters, instruments, and controls under the respective sections of the specifications and as specified and required to provide complete and operable systems. Provide interconnecting wiring for components of packaged equipment as an integral part of the equipment. Interconnecting power wiring and conduit for field erected equipment and control wiring and conduit shall be as specified in the ELECTRICAL Section (not included in this material). Motor control equipment shall be part of motor control center(s).

1.12 COORDINATION
Mechanical work shall be coordinated with that of other trades. All work shall be carefully laid out in advance. Mechanical features shall be carefully laid out in advance. Mechanical features shall be coordinated with those of architectural, structural, civil, and electrical features of construction. Where cutting and patching of the construction becomes necessary because of poor coordination, the work shall be performed in accordance with the applicable divisions of the specification and will be the responsibility of the Mechanical Contractor (at no additional cost).

1.13 MATERIALS
All materials and equipment shall be new and good for the intended use. Used material shall not be acceptable. Before and during the installation period all damaged equipment shall be replaced by the Contractor at no additional cost to the Owner.

SECTION II. PIPE, FITTINGS, VALVES, AND SPECIALTIES: MATERIALS AND INSTALLATION

TABLE OF CONTENTS

1 GENERAL

1.1 DESCRIPTION

The work covered by this specification includes the materials, tools, transportation, and supervision for furnishing, installing, and testing the piping, valves, fittings, and all associated accessories.

1.2 GENERAL REQUIREMENTS

A. Before any welding is performed, Contractor shall submit three copies of his welding procedure specification for all metals included in the work together with proof of welders' qualifications.

1.3 SUBMITTALS

A. Submit complete catalog information and shop drawings for piping materials, sleeves, valves, hangers, and all other appurtenances.

B. Submittals shall include, but are not limited to, the following:

1. Pipes, fittings, and piping layout.
2. Pipe hangers & supports: locations and types used.
3. Pipe sleeves.
4. Valves.
5. Traps, strainers, drains.
6. Escutcheons.
7. Instruments.
8. Expansion joints, guides, and anchors.
9. Air eliminators.
10. Expansion tank and fittings.

C. Submit certificates from the manufacturer stating that all material and appurtenances are designed and tested as specified herein.

D. In design and purchase of equipment, provide for the interchangeability of items of piping equipment subassemblies and parts. For standardization of products, valves shall be the product of one manufacturer where possible.

1.4 STANDARD PRODUCTS*

Material and equipment shall be the standard products of a manufacturer regularly engaged in their manufacture for a minimum period of five years. All material and equipment shall be new unless otherwise indicated. Valve ratings are for working pressure unless otherwise indicated. General requirements include those specified in other sections of these specifications and as specified herein.

*All manufacturer's designations are for information. No specific manufacturer is recommended over a similar one.

1.5 PRODUCT DELIVERY, HANDLING, AND STORAGE

A. Materials shall be delivered in the manufacturer's original unopened protective packages.

B. Materials shall be stored in their original protective packaging and protected against soiling, physical damage, or wetting before and during installation.

C. Equipment and exposed finishes shall be protected during transportation and installation against damage and stains.

D. Unloading of all materials shall be done in a manner to prevent misalignment or damage.

2 MATERIALS

2.1 GENERAL

Materials shall be suitable for the services, pressures, and temperatures specified. Pipe, valves, fittings, etc., shall conform to the published standards.

A. Pipe insulation material shall be in accordance with the specification entitled INSULATION.

B. Flanges

Flanges shall be 150 psi standard steel, screwed, slip-on or welding neck. Flanges shall be placed in all piping 3" and larger, in sufficient number to permit easy dismantlement for repairs or piping changes. Flanges (or screwed unions for smaller piping) shall also be installed at all final pipe connections to each piece of equipment.

2.2 PIPING

For Chilled Water, Condensed Water, Glycol System Water, Heating Hot Water, Fuel Oil and Gas, for pressures up to 200 psi, system materials shall be as follows:

A. Pipe shall be seamless black steel, ASTM (American Society for Testing and Materials) A53 or A106, Grade A, Schedule 40 for sizes 10 inches and smaller, 0.375 inch wall thickness for sizes 12 inches and greater.

Fuel oil piping shall have a secondary containment pipe fabricated of carbon steel. For below ground application, the secondary containment piping and fittings shall be protected with a bituminous coating. Field joints shall be covered with a heat shrinkable adhesive backed sleeve. Provision for oil leak location and monitoring shall be provided.

B. Fittings 2 inches and smaller shall be malleable iron conforming to ANSI (American National Standards Institute) B16.3, Class 150, threaded. Fittings 2-1/2 inches and larger shall be carbon steel conforming to ANSI B16.9, Class 150, butt welded. Flanges shall be welding neck type conforming to ANSI B16.5, Class 150.

C. Gate valves 2 inches and smaller shall be bronze, rising stem, solid wedge disc, threaded conforming to MSS (Manufacturers Standardizations Society) SP-80, Class 150. Gate valves 2-1/2 inches and larger shall be cast iron, bronze trim, outside screw and yoke, flanged conforming to MSS SP-70, Class 150.

D. Globe valves 2 inches and smaller shall be bronze, rising stem, bronze regrinding disc, threaded conforming to MSS SP-80, Class 150. Globe valves 2-1/2 inches and larger shall be iron body, bronze trim, flanged, tapped drains, brass plugs, outside screw and yoke, Class 150, conforming to MSS SP-85.

E. Check valves 2 inches and smaller shall be bronze regrinding disc, swing check type, bolted caps, threaded, Class 150, conforming to MSS SP-80. Check valves 2-1/2 inches and larger shall be iron, bronze trim, bolted caps, flanged, Class 150, conforming to MSS SP-70. Check valves for pump discharge shall be non-slam type, eccentric disc, bronze trim, stainless steel spring, iron body, disc guided, flanged end, and Class 150.

F. Plug valves 2 inches and smaller shall be iron body, brass plug, square head cock, Class 150 with threaded ends. Plug valves 2-1/2 inches and larger shall have semi-steel body, square head, lubricated eccentric taper plug, Class 150 with flanged ends. Valves 6 inches and smaller shall be wrench operated. Valves 8 inches and larger shall be worm gear operated.

G. Air Vent Valves

Furnish and install at all high points of the water supply and return mains and wherever else required, a manual air vent.

2.3 VENTURI FITTINGS

A. Furnish and install where shown on drawings, Venturi Fittings in the chilled and hot water lines. These fittings shall be manufactured by Illinois, Barco, Gerand, or approved equal.

2.4 DOMESTIC WATER PIPING AND FITTINGS (For Information Only) (Used for makeup water; usually installed by the Plumbing Contractor)

A. Underground plumbing pipe shall be ductile iron, bituminous coated, cement lined with "push-on" mechanical joints in accordance with AWWA (American Water Work Association) C151, Class 51.

B. Cold and hot water pipe above ground shall be Type L copper tubing, seamless, hard temper, in accordance with ASTM B88.

C. Fittings for copper piping shall be wrought copper in accordance with ANSI B16.22. Flanges shall be cast bronze in accordance with ANSI B16.24. The joining between tube and fittings shall be made with brazing filler metals complying with ANSI/AWS (American Welding Society) A58. Fluxes and brazing filler metal shall be lead-free. Threaded joints at valves shall be made with lead-free sealant.

D. Gate valves 2 inches and smaller shall be bronze, rising stem, solid wedge disc, threaded, conforming to MSS SP-80, Class 150. Gate valves 2-1/2 inches and larger shall be iron body, bronze trim, outside screw and yoke, flanged, conforming to MSS SP-70, Class 150.

E. Globe and angle valves 2 inches and smaller shall be bronze, rising stem, bronze regrinding disc, threaded conforming to MSS SP-80, Class 150. Globe and angle valves 2-1/2 inches and greater shall be iron body, bronze trim, regrinding disc, outside screw and yoke, tapped drains, brass plugs, flanged, conforming to MSS SP-85, Class 150.

F. Check valves 2 inches and smaller shall be bronze, regrinding disc, swing check type, bolted caps, threaded, Class 150, conforming to MSS SP-80. Check valves 2-1/2 inches and greater shall be iron, bronze trim, bolted caps, flanged, Class 150.

G. Balancing cocks shall be eccentric type and complete with memory stop.

H. Butterfly valves of all sizes shall be lug type with extension necks on insulated pipe. Valves 6" in size and smaller shall be provided with throttling handle and notched plate with not less than 8 intermediate positions, and valves 8" and larger shall be provided with gear drive. Where butterfly valves are used as shut-off valves on either side of equipment, care shall be taken to see that sufficient flange connections are provided so that equipment can be removed without drainage of system.

Valves shall be installed screwed to the adjacent flanges to permit dismantling of piping on the equipment side of the valve. Lug type with bronze disc and 416SS shaft EPDM seat.

2.5 STEAM PIPING

A. Pipes to be as follows:

1. 1/2" - 4" - A120/A53 C.W.

2. 5" - 10" - Standard Wall, Schedule 40

3 12" and larger - 0.375" Wall A53 E.R.W.

B. Fittings:

Standard black iron

C. Unions:

On piping smaller than 2½", unions shall be provided with malleable brass sets.

D. Flanges:

See Flanges under item 2.1 B.

E. Gaskets:

Ring type fitting inside the bolt circle, Johns-Manville Type 60 or equal.

F. Bolts:

ASTM A107 stock with threads, ANSI B1.1.

G. Gate Valves:

1. For steam distribution piping 2½" and larger to be 125# WSP IBBM, solid wedge, outside screw and yoke, flanged end, equal to Jenkins 7651-A, Hammond IR1140, or Crane 465-1/2.

2. For equipment connections 2½" and larger to be 125# WSP iron body bronze mounted, solid wedge, inside screw, non-rising stem, flanged end, equal to Jenkins 7326, Hammond IR1138, and Crane 461.

3. 2" and smaller to be 125# bronze, solid wedge, inside screw, non-rising stem, screwed end, equal to Hammond IB645, Crane 438, and Stockham B110.

H. Check Valves:

2" and smaller shall be 150# WSP bronze horizontal swing type with regrinding bronze seat and disc, screwed end, equal to Jenkins 92A, Hammond IB944 or Crane 137.

Note: It is recommended that all valves be of the same manufacturer.

I. Strainer:

To be "Y" type with stainless steel baskets having 1/16" perforations and CI body by Sarco or RP4C.

J. Traps:

Shall be F&T (float and thermostatic) as shown on drawings, equal to Barnes and Jones, Sarco, or Hoffman. All steam traps are to be sized at twice the load with a 2 psi drop.

2.6 REFRIGERATION PIPING

A. Pipe shall be seamless copper tubing, hard drawn, Type K, conforming to ASTM B88. In order to prevent the entrance of foreign materials after tubing is cleaned, it shall be sealed, capped, or plugged prior to shipment from the manufacturer.

B. Copper tubing fittings shall be wrought copper with silver soldered joints. Nitrogen shall flow through piping during brazing to prevent flaking. Flanges shall be bronze conforming to ANSI B16.24.

C. Valves shall be especially designed, manufactured, and tested for the refrigerant service. The internal parts shall be made removable for inspection or replacement purposes without applying heat or breaking pipe connections. Valves shall open when turned counterclockwise. Globe and angle valves shall have bronze alloy bodies with packed stem and seal cap. Packless type with hand wheels and bronze alloy bodies with brazing and soldered ends may be used in sizes up to and including one inch O.D. 1-1/4 inch and larger valves shall be tongue and groove flanged and shall have bolted bonnets. Refrigerant valves shall be of the backseating type, which shall make it possible to repack the valves under pressure without removal from the line.

D. Check valves shall be lift type, designed for low pressure drop. The body shall be bronze alloy with brazing ends on 2 inch size and smaller. Valve trim shall be stainless steel or bronze. Provide tight closing resilient seals for silent operation. Flow direction shall be legibly and permanently marked on the valve body.

E. Moisture indicator material shall register moisture by varying degrees of color change, based on 100°F. Indicators shall be a brass or bronze or heavily copper plated steel fitting, with the indicator material located under a bulls-eye. Indicators shall be rated for a working pressure of 350 psig.

F. Sight glass shall be installed ahead of each expansion valve and shall be of the double-port see-through type with two bull-eyes and cover caps of non-ferrous material, unless combined as a part of the moisture indicator. Sight glass indicators shall be capable of withstanding a test pressure of 350 psig without damage. Sight glass body shall be forged brass with fittings as specified for refrigerant piping.

2.7 STRAINERS

Strainers shall be installed ahead of each solenoid operated valve and shall have a means of cleaning the screen without removal of strainer from the line. Strainers shall be basket or "Y" type strainers and shall be the same as the pipelines in which they are installed. The strainer bodies shall be heavy and durable. Each strainer shall be equipped with removable cover and sediment basket.

For sizes 2½" and smaller, use threaded strainer. For sizes 3" and larger, use flanged strainer. Perforations for stainless steel screen in strainers shall be 1/16" in diameter, except for condenser water service which shall have 3/16" perforations.

2.8 FLEXIBLE CONNECTIONS

A. Furnish and install at inlet and outlet of base mounted pumps, braided stainless steel connectors with flanged ends.

B. Flexible connections shall be annual corrugations, seamless, stainless steel tubing covered with stainless steel braiding. Minimum length shall be 18 inches. For sizes 2 inches and smaller, end connections shall be threaded. End connections 2-1/2 inches and greater shall be flanged. Flexible connections shall be rated for the applicable working pressure.

2.9 MISCELLANEOUS PIPING

Piping for make-up drains, etc., are to be Type L copper with soft soldered joints. Valves for this piping are to have screwed ends, with solder adapters.

2.10 EXPANSION JOINTS, GUIDES, AND ANCHORS

A. Install where shown or required. Expansion shall be taken up by compensators and loops as shown. Compensators shall be with flanges or unions servicing.

B. Install guides and anchors to ensure proper alignment of pipe at expansion joints as shown on drawings or as may be required.

C. Locate the first guide no more than 4 pipe diameters from expansion joint. The second guide should be no more than 14 pipe diameters from the first guide. The balance of the guides should be as per manufacturer's recommendations.

D. Expansion joints shall be of the packless, corrugated type. Circumferential welded seams will not be permitted in the bellows material except for seal welds which join bellows to end connections. Bellows shall be fabricated from Type 321 stainless steel.

E Expansion joints shall be suitable for the temperatures, pressures, and thermal movements required and shall be fitted with sleeves.

2.11 BACKFLOW PREVENTERS

Backflow preventers with intermediate atmospheric vents shall be installed on the makeup water pipe for residential and small commercial applications, conforming to the applicable requirements of AWWA C506.

Reduced pressure backflow preventers shall be installed on the makeup line for industrial and large boilers.

2.12 MECHANICAL PIPE COUPLINGS

Couplings shall be self centering and shall engage and lock in place the grooved or shouldered ends of pipe and pipe fittings in a positive watertight couple. The couplings shall provide some degree of angular pipe deflection, contraction, and expansion. Special rigid couplings ("Zero-Flex" type) may be used where no flexibility is required from the joint. The coupling clamp shall be malleable iron. The gasket shall be molded rubber. Mechanical couplings and fittings shall be of the same manufacturer.

Note for installer: Before couplings are assembled, pipe ends and outsides of gaskets shall be lightly coated with lubricant approved by the coupling manufacturer to facilitate installation.

2.13 THERMOMETERS

Thermometers shall be provided where shown on the drawings. Thermometers shall have operating ranges compatible with the temperature range of the service and shall be so installed as to be conveniently read from the operating floor. Range shall be as follows:

Chilled water: 0° - 100°F

Hot water: 100° - 250°F

Steam: 100° - 300°F

Condenser water: 20° - 120°F

On thermally insulated equipment or piping, stand off mounting brackets, bases, adapters, or extended tubes shall be provided. These items shall provide clearance not less than the thickness of the insulation. These stand off mounting items shall be integral with the thermometer or standard accessories of the thermometer manufacturer. Remote type thermometers for use in pipes shall have corrosion immersion type sensing elements with corrosion resistant wells. Thermometers shall be located where shown on the drawings.

2.14 DIELECTRIC ISOLATORS

Wherever copper, brass, or bronze piping systems are connected to steel or iron piping systems, this connection shall be made with dielectric isolators. These dielectric isolators shall be so designed that no non-ferrous piping materials come in contact with ferrous materials. These materials shall be isolated by the use of Teflon or nylon isolating materials made up in the form of screwed type unions or insulating gaskets and bolt sleeves and washers for standard flanged connection. Where it will not be necessary to disconnect these piping systems, the connections may be made by the use of nylon or teflon bushings. All of these insulating units shall be selected for pressures to be encountered.

2.15 PRESSURE AND ALTITUDE GAUGES

Dial gauges for water service shall be installed where shown and shall be 3-1/2" diameter. Gauges shall have white face and black letters, cast-iron case with glass crystal, 1/4" MPT connector and calibrating device. Provide snubbers and globe shut-off valves for each gauge.

All gauges shall be provided with snubbers to eliminate needle fluctuations. All gauges in general are to be selected with a range so that the operating point is approximately at the middle of the range. Provide a compound gauge in the suction piping to the condenser water pump and oil pumps.

2.16 ACCESS PANELS

Panels shall be provided for all concealed valves, controls, or any items requiring inspection or maintenance. Minimum panel size shall be 12 by 12 inches. Access panels shall be so located that the concealed items may be serviced and maintained or completely removed and replaced.

2.17 ESCUTCHEONS

Escutcheons shall be chrome plated steel or approved cast brass, either of one piece or split type.

2.18 SLEEVES

Pipe passing through concrete or masonry walls or concrete floors or roofs shall be provided with pipe sleeves fitted into place at the time of construction. A waterproofing clamping flange shall be installed where membranes are involved. Sleeves shall not be installed in structural members except where indicated or approved. Piping passing through slab on grade shall not require sleeves, except for water piping running under the slab. Sleeves shall be #22 gauge or heavier steel, one pipe size larger than outside diameter of pipe to be sleeved. Each sleeve shall extend through its respective wall, floor, or roof and shall be cut flush with each surface. Sleeves through floors and roofs shall extend above the top surface at least 6 inches for proper flashing or finishing. Unless otherwise indicated, sleeves shall be sized to provide a minimum clearance of 1/2 inch between bare pipe and sleeves. Sleeves through outside walls shall be Schedule 40 black steel pipe with 150-pound black steel slip-on welding flanges, welded at center of sleeve, painted with one coat of bitumastic paint inside and outside. Space between sleeve and pipe shall be paceked with oakum to within two inches of each wall face. Remaining space shall be packed and made watertight with waterproof mastic. Sleeves through fire walls shall be filled smoke-tight with approved non-combustible sealant to maintain the integrity of the fire separation.

2.19 GASKETS

Gaskets shall be of non-asbestos, compressed material, suitable for the service intended, in accordance with ANSI B16.21, 1/16 inch thickness, full face or self-centering flat ring type.

2.20 HANGERS AND SUPPORTS (GENERAL)

Select, provide, fabricate, and install pipe hangers and supports complying with MSS SP-58, MSS SP-69, and MSS SP-89 (Manufacturers Standardization Society of the Valve and Fittings Industry, Inc.). Provide product which is UL-listed and FM approved.

All piping shall be supported from the building structure by means of approved hangers and supports. Piping shall be supported to maintain required grading and pitching of lines, to prevent vibration, and to secure piping in place, and shall be so arranged as to provide for expansion and contraction. Chain, perforated strap, bar, or wire hangers are not permitted.

No piping shall be hung from the piping of other trades.

Branches shall have separate supports, and no branch 5'-0" or longer shall be without support.

Do not hang pipe hangers from bottom chord of roof joists. Hangers must be installed at or near panel point of roof joists. 6" and larger pipes running parallel to roof joints must be supported by two roof joists. <u>The Contractor shall furnish and install angle iron bridging between joists of adequate size to securely support pipe hanger</u>. Bridging for pipe hangers must also be supported by the top chord of the roof joist and must be installed at or near panel point. Use approved type brackets to support piping raced along walls. Support piping running just above floor on pipe saddle supports. Pipe hangers shall not penetrate waterproof floor membrane or roof deck. Copper tubing shall be supported by copper plated hangers. Where overhead construction does not permit fastening of hanger rods in required locations, provide additional steel framing as required and approved.

Where codes having jurisdiction require closer spacing, the hanger spacing shall be as required by code in lieu of the distances specified herein.

Provide hangers at a maximum distance of 2 feet from all changes in direction (horizontal and vertical) on both sides of concentrated loads independent of the piping.

All hangers on insulated lines shall be sized to fit the outside diameter of the pipe insulation. Provide pipe covering protection saddles at all hangers on the insulated lines. Provide 16-gauge sheet metal shield, 12" long and covering 180 degrees of arc on the covering at all hangers on insulated lines.

Hangers in general for all horizontal piping shall be clevis type hangers. These hangers shall be sized to provide for insulation protectors as hereinafter specified.

Hangers for uncovered (uninsulated) copper and brass piping shall be factory-applied copper plated steel clevis hangers. Rods and nuts used with these hangers shall also be factory-applied copper plated.

Where three or more pipes are running parallel to each other, factory-fabricated gang type hangers with pipe saddle clips or rollers may be used in lieu of the hereinbefore specified clevis hangers. These hangers shall be sized to provide for insulation protectors as hereinafter specified. Pipe saddle clips shall be not less than 16 gauge metal and shall be copper when installed with uninsulated copper piping.

Trapeze type hangers shall be made of angles bolted back to back or channels for supporting parallel lines of piping. Trapeze type hangers shall be supported with suspension rods having double nuts and securely attached to construction with inserts, beam clamps, steel fish plates, cantilever brackets, lag screws, or other approved piping attached along walls.

Field painting or spraying of hangers, rods, and nuts in lieu of copper plating will not be accepted.

All suspended horizontal piping shall be supported from the building by mild steel rod connecting the pipe hanger to inserts, beam clamps, angle brackets, and lag screws as required by the construction standards in accordance with the following:

Pipe size in inches:	3/4 to 2	2-1/2 to 3-1/2	4 to 5	6	8 to 12
Rod diameter in inches:	3/8	1/2	5/8	3/4	7/8

Remove rust from all ferrous hanger equipment (hangers, rods, and bolts), and dip hangers and supports in zinc chromate primer before installation.

Piping at all equipment and control valves shall be supported to prevent strains or distortion in the connected equipment and control valves. Piping at equipment shall be supported to allow for removal of equipment, valves, and accessories with a minimum of dismantling and without requiring additional support after these items are removed.

All piping installed under this section of the specification shall be independently supported from the building structure and not from the piping, ductwork, or conduit of other trades.

All supplementary steel including factory-fabricated channels and associated accessories throughout the project for this section of the specifications, both suspended and floor mounted, shall be furnished and installed by the Plumbing Contractor and shall be subject to the approval of the Architect.

Safety retaining clips shall be installed with all beam clamps.

Support and protect underground piping so that it remains in place without settling and without damage during and from backfilling. Replace any piping which settled or was damaged.

2.21 SPECIAL HANGERS AND SUPPORTS

A. Steam Pipe:

Grinnell Roll Type Fig. 181 or Carpenter and Patterson. Rolls shall be oversized to accommodate insulation and saddles. Saddles shall be Grinnell Series 160 or equal.

B. Condensate Return Pipe:

1. Return pipe 3" and larger; same as “A” above.

2. 2½" and smaller, Grinnell Fig. 104.

C. Chilled Water Supply and Return:

Grinnell Clevis Fig. 260 or Carpenter and Patterson. Provide 14-gauge galvanized shields on bottom half of pipe, 18" long. Insulation between shield and pipe to be molded vegetable cork or Kaylo. Insulated shield units manufactured by Therman-Hanger-Shield are acceptable.

D. Condenser and Hot Water Supply and Return:

Same as “C” above, except delete shield. Clevis hanger may be installed against the pipe.

E. Use machine threaded rods attached to top chord of steel joists by either beam clamps or inserts. Do not hang piping from bottom chord of any joist or bridging. Perforated extension strap hangers are not acceptable.

F. Vertical piping is to be supported with pipe clamps in alignment and should allow expansion as well as carry the weight of the pipe. Horizontal piping is to be supported at intervals as specified in the latest ASHRAE Guide. For multiple runs of piping at the same level, use trapeze hangers.

G. Support all piping in chiller and boiler rooms from spring type isolators.

H. All piping is to be supported in such a manner that no weight is placed on equipment or on flexible connectors.

3 CONSTRUCTION METHODS

3.1 PIPING

A. Where possible, provide multiple run pipe racks. Piping shall be neatly installed in common racks and run parallel in both vertical and horizontal planes. Piping shall be pitched for venting or draining.

B. Pipe shall be thoroughly cleaned after cutting or flame beveling. Pipe ends shall be squared, reamed, and then threaded or flame beveled when welded. Threads shall be free of burrs. During construction, open ends of piping shall be protected with temporary closure plugs.

C. All pipe ends shall be reamed to full size, and all threads shall be clean cut. Joints in screwed piping shall be made with joint compound or Teflon tape.

D. Ends shall be beveled before welding. Welds shall be made without backing rings. Welds shall be clean and free of metal "icicles," loose metal, or other obstructions that can result from welding and retard flow of gases and fluids. When a leak appears in a weld, do not reweld without first chipping away the weld for about 1/2 inch each way, and then weld over the chipped surface. Slip on flanges shall be welded both front and back.

E. Branch connections shall be made with welding tees or, when the main size is a minimum of 2-1/2 times greater than the branch size, forged steel couplings, Weldolet, Threadolets, or Socolets shall be carefully welded into the piping system. Long radius welding ells shall be used at turns. Miter joints will not be acceptable.

F. Flanged joints shall be faced true and square. Flanges mating with cast iron flanges shall be flat faced.

G. On racks and in hangers, position piping requiring insulation on proper depth shoes or spacers to accommodate installation.

H. Protect against external corrosion those pipes which pass through, under, or otherwise in contact with soil, cinders, concrete, or other corrosive material. Protect by protective coating or by other approved means.

I. Connect non-ferrous piping to ferrous piping with dielectric unions/couplings.

J. Provide drain valves at low points and air vents at high points of piping.

K. Provide and install under pipe passing overhead and within 24 inches horizontally of any switch gear, motor, or controller, a stainless steel unpolished pan at least four inches wider than the outside edges of the pipe (or pipes) with a two inch turned up edge all around. Pans shall be properly stiffened, braced, and supported to prevent sagging. Run a drip pipe down to floor from the pan.

L. Install piping and piping components to ensure proper and efficient operation of the equipment and controls and in accordance with manufacturer's printed instructions. Proper supports for the mounting of vibration isolators, stands, guides, anchors, clamps, and brackets shall be provided. Piping connections to equipment shall be arranged so that removal of equipment or components of equipment including tube withdrawal from chillers, heaters, pump casing, shaft seals, and similar work can be accomplished with the least amount of disassembly or removal of the piping system. Piping connected to equipment with vibration isolators shall be provided with flexible connections which shall conform to vibration and sound isolation requirements for the system. Final connections to equipment shall be made with unions or flanges. For glycol piping screwed joints, threading joint compound used shall be for glycol water service.

M. Maximum spacing for pipe hangers shall be as follows:

Steel Pipes:

Pipe Diameter	Maximum Spacing Water	Maximum Spacing Gases
1/2 inch	5'-0"	6'-0"
3/4 inch	5'-0"	6'-0"
1 inch	7'-0"	9'-0"
1-1/4 inch	9'-0"	11'-0"
1-1/2 inch to 2 inch	10'-0"	13'-0"
2-1/2 inch	11'-0"	14'-0"
3 inch	12'-0"	15'-0"
3-1/2 inch	13'-0"	16'-0"
4 inch	14'-0"	17'-0"
5 inch	15'-0"	19'-0"
6 inch	17'-0"	21'-0"
8 inch	19'-0"	24'-0"
10 inch	22'-0"	28'-0"

Copper Pipes:

Pipe Diameter	Maximum Spacing
1/2 inch	4'-0"
1 inch	5'-0"
1-1/4 inch	6'-0"
1-1/2 inch	7'-0"
2 inch	8'-0"
2-1/2 inch	9'-0"
3 inch	9'-0"
4 inch	10'-0"
10 inch	17'-0"

3.2 TESTING

Piping, fittings, and accessories shall be hydrostatically tested and proved tight under a pressure of 1-1/2 times the working pressure. The test pressure shall be held for four hours with no allowable pressure drop.

3.3 PIPING DETAILS

A. Piping shall be installed substantially as shown. The drawings are not intended to show minor details or interferences which may exist, and location may be changed during construction as required to avoid conflict with structural work or arrangement of other equipment or architectural details.

B. No piping is to be exposed to view, except where shown exposed. Install exposed lines in neat and orderly arrangement parallel and perpendicular to structural features of building. Install in horizontal and vertical banks wherever possible.

C. (Note to Specifier: Use the following two paragraphs for steam only.)
Install free blows and drips at low points in order to permit clearing of all condensation and water from all steam piping when starting up. Install drip connections with traps at points where condensation can collect when piping is in service.

D. Piping shall conform to the best practices of the trade and shall be subject to approval by the Architect. Piping shall be free to expand without imposing excessive stresses on piping material, valves, or on apparatus to which piping is connected. Piping shall be cut short by about 1/2 the expected amount of expansion and cold-sprung into position. The amount of piping shall be cut short as follows:

Supply Steam Piping: 1/4" in 25 feet.

Return Condensate Piping: 1/4" in 30 feet.

E. Make proper provisions for expansion and contraction in all parts of piping systems, wherever possible, by means of pipe bends, pipe loops, swing connection, or changes in direction of piping.

F. Where pipe deflection cannot be employed to absorb expansion and contraction, furnish expansion joints and provide guides where necessary to confine lateral movement of piping on both sides of all expansion joints and loops.

G. An approved pipe thread compound shall be applied to the male threads only before the joints are made up. Supports shall be located near valves, fittings, and apparatus, as necessary to prevent distortion due to piping strain. Unions shall be installed near all control valves, traps, inlet and outlet connections for heaters, pumps, coils and chips, and loose scale removed from the interior, before the piping is worked into positions. Low points of all piping and where indicated on the drawings, shall be provided with a 3/4" diameter gate drain valve with a capped hose connection.

H. Install air vents at high points of hot water systems.

I. Use eccentric fittings where they are necessary to provide complete drainage of piping.

J. Provide sleeves with collars where pipes pass through walls and floors.

The refrigeration piping shall be pressure tested and leak tested in the presence of the Architect's representative. A mixture of refrigerant and dry nitrogen at 300 psi shall be charged into the system. Pressure shall hold for 24 hours. Expansion joints and compressor crankcase are not to be pressure tested.

Evacuate the system with a high vacuum pump for 24 hours, break the vacuum with the refrigerant to be used, evacuate again to 29" of mercury, and reduce the moisture content to under 100 microns. Install new cartridges in the drier, and charge the system with the required amount of refrigerant and oil.

Replace any refrigerant or oil lost from the system during the guarantee period.

3.4 IDENTIFICATION

A. Each piping valve shall have a 1-1/2 inch diameter brass tag with black filled engraved numbers and letters. Tag shall be affixed to valve by means of a brass "S" hook. Tags on different services shall be identifiable by a letter and number designation.

B. Piping identification shall be placed in clearly visible locations. Labels and tapes of general purpose type and color class may be used in lieu of painting or stenciling. Colors conforming to ANSI A13.1 shall be used. Stencil piping with approved names or code letters not less than 1/2 inch high for piping and not less than 2 inches high elsewhere. Paint arrow-shaped markings on the lines to indicate direction of flow. Spacing of identification marking shall not exceed 50 feet. Pipe markings shall be provided for risers and each change of direction for piping.

SECTION III. HYDRONIC SYSTEMS: HEATING AND COOLING

TABLE OF CONTENTS

1 GENERAL

1.1 DESCRIPTION

A. Provide all plant facilities, labor, materials, tools, equipment, appliances, transportation, supervision, and related work necessary or incidental to complete the work specified in this section and as shown on the drawings.

B. All work performed under this section of the specifications shall be subject to the requirements of the General Conditions of the Contract (as applicable).

C. The drawings indicate the extent and general arrangement of the heating and cooling systems. If any departures from the drawings are deemed necessary by the Contractor, details of such departures and the reasons therefore shall be submitted to the Owner and/or Engineer for approval. No such departures shall be made without prior written approval of the Owner and/or Engineer. Equipment and piping arrangements shall provide adequate acceptable clearances for entry, servicing, and maintenance.

The heating and cooling (hydronic system) Contractor shall be responsible for consulting with the General Contractor to indicate locations of openings for his work.

Note: Contractors must visit jobsite to acquaint themselves with existing building conditions.

1.2 WORK NOT INCLUDED

A. The following items of work to be done by others are not included in the work of this section. However, it shall be the responsibility of the Contractor to supply the Subcontractors with the necessary information, drawings, and supervision so they can properly complete their phase of the installation.

1. Electric wiring, except as noted.
2. Cutting and patching.
3. Roof equipment support.
4. Roof flashing.
5. Painting, except protective coat painting normally performed by equipment manufacturers and elsewhere as detailed herein.
6. Plumbing (except drains and final cold water makeup connection).
7. Excavation and backfilling.

1.3 SHOP DRAWINGS

A. Within 30 days after receipt of notice to proceed and prior to fabrication or installation, five complete shop drawings showing the type and size of equipment foundations, together with piping layouts for the systems shown, and a list of all Subcontractors shall be submitted to the Engineer for approval. The Contractor shall also refer to the applicable section of the specification (such as Mechanical Specification General Requirements) when submitting shop drawings.

B. As soon as is practical, the Contractor shall submit to the Engineer firm delivery dates of all equipment. This list is to be kept current, and the Engineer shall be informed immediately of any change that may affect the Contractor's completion of the project.

C. Each shop drawing shall bear the check stamp of the General Contractor or Construction Manager before it is submitted for the Engineer's approval.

D. Submit shop drawings as applicable but not limited to the following:

1. 1/4" scale plan detail and elevations of Boiler Room and Mechanical Rooms showing actual equipment to be used including ductwork and piping.
2. Boiler.
3. Centrifugal chiller.

4. Cooling tower.
5. All pumps.
6. Motors and magnetic starter schedule.

   a. Submission of shop drawings for electric motors shall include a tabulation listing the equipment the starter is intended to control, the manufacturer's number or type, starter hp, power factor, volts, phase, full load motor amperes, heater numbers and amperage, quantity of auxiliary contacts, push button arrangement, and pilot lights.
7. Radiation elements (heating).
8. All valves.
9. Steam traps.
10. Strainers.
11. Flexible connections.
12. Expansion compensators.
13. Compression (expansion) tanks.
14. Air vents.
15. Gauges.
16. Thermometers.
17. Insulation.
18. Automatic temperature controls.

    a. Shop drawings for automatic temperature controls shall include a brochure describing in detail the instruments and controls to be furnished, complete wiring and control diagrams, and a detailed description of the sequence of operation of all controls.
19. Water treatment (as applicable).
20. Oil tanks.
21. Oil pumps.
22. Electric oil heater.
23. Balancing contractor's report.
24. Operating and maintenance instructions.

1.4 AS-BUILT RECORD DRAWINGS

A. At the completion of the job and before final payment is made, the Contractor shall deliver to the Engineer a complete set of sepia drawings, including revised drawings that may have been issued, on which are to be shown the exact location of all equipment, pipes, valves, etc., as installed. These drawings shall be prepared as work progresses.

1.5 WARRANTY AND GUARANTEE

A. The entire heating and cooling systems shall be guaranteed by the Contractor against defects due to faulty materials or workmanship for a period of 12 months after written acceptance by the Owner or the Engineer.

B. The Engineer is aware and the bidders are hereby made aware that certain manufacturers' equipment guarantees are valid only for a period of one year from the date of shipment or installation and will therefore not be valid until the date of guarantee set forth herein. It shall therefore be noted that this Contractor shall be fully responsible for all material, equipment, and labor for the full guarantee period as set forth herein before, i.e., one year from date of acceptance. It shall also be noted that in the case of certain equipment guaranteed for more than one year, this Contractor shall be responsible for all material, equipment, and labor for the full one-year period as stated above, and the manufacturer shall be additionally responsible for guarantees on the equipment, only for the additional time period beyond one year, up to the limit of the guarantee.

C. The Contractor shall be responsible for servicing the system for a period of 12 months after acceptance. Warranty shall include servicing of the equipment during this time and answering any necessary trouble calls. This shall also mean the responsibility of any loss of refrigerant charge during the warranty period, regardless of cause, unless due to direct damage caused by operating personnel.

D. Test:

1. Before acceptance, Owner may call for specific tests necessary to determine if the system is functioning as intended. These tests will determine if any objectionable noise or vibration is apparent, if system is balanced, or if other objectionable items are apparent.

E. The Contractor shall be responsible for carrying out tests, securing permits, paying fees, and arranging for inspections required for work under this section.

1.6 QUALITY ASSURANCE

A. The work shall be executed in strict conformity with the latest edition of the State Building Code and all local regulations that may apply. In case of conflict between the Contract Documents and a governing code or ordinance, the more stringent standard shall apply.

B. Unless otherwise specified or indicated, materials and workmanship shall conform to the latest edition of the following standards and specifications:

1. American National Standards Institute (ANSI).
2. American Society of Heating, Refrigerating and Air-Conditioning Engineers (ASHRAE).
3. Sheet Metal and Air Conditioning Contractors National Association (SMACNA).
4. Underwriters' Laboratories, Inc. (UL).
5. American Society for Testing and Materials (ASTM).
6. National Fire Protection Association (NFPA).

2 PRODUCTS

2.1 BRAND NAME PRODUCTS

A. Bids shall be based on using equipment as scheduled on drawings and named in the specifications. Where several manufacturers of major equipment are mentioned, the Contractor has the option of using any one of these manufacturers. If the Contractor desires to use any of the other brands, he shall so state in his bid, giving details of the proposed substitution and the adjustment to be made in his bid if the substitution is approved.

B. Within two weeks after award of contracts, Contractor shall submit for approval an itemized list tabulating all equipment he proposes to use as well as the Subcontractors name list for the project.

C. Note that when manufacturers other than the ones listed on the drawing schedule are used, the cost shall include any and all charges that may be required for supports, piping, or wiring, which are additional because of the change in manufacturer.

2.2 MOTORS AND CONTROLLERS

A. All motors smaller than 1/2 hp are to be wound for 120-volts, single phase, 60-cycles.

B. All motors 1/2 hp and larger are to be wound for 460-volts, 3-phase, 60-cycles, unless otherwise noted. All 3-phase motors shall be high efficiency type, Class A or B type with power factor at full load of not less than 85%. If less than 85%, furnish power factor correction devices as per State Code to increase to 90%.

C. Motors shall be 40°C rise, normal starting torque, low starting current, squirrel cage induction type as manufactured by Century, Wagner, Reliance, Westinghouse, General Electric, or Electro Dynamic. Motors for air units are to be of one manufacturer.

D. Starters for motors 1/2 hp and larger are to be furnished by the Hydronic Contractor unless specified to be a part of the motor control center which is furnished by others. Contractor shall refer to the electrical drawings for the tabulation of equipment to be controlled from the motor control center, and he shall supply starters for all other equipment. One copy of shop drawings covering motor data as detailed above shall be submitted to the electrical contractor for his use.

E. Starters shall be magnetic, with NEMA Type 1 enclosures, with built-in transformers for 120-volt holding coils and overload protection on all three legs. See schedule on drawing for starters, HOA (Hand-Off-Auto) switches, push buttons, and pilot lights. All starters scheduled shall have pilot lights, low voltage protection, or low voltage release. Starters shall be as manufactured by Square D, Cutler-Hammer, or Allen-Bradley.

2.3 BOILERS: HOT WATER

A. Furnish and install where shown on drawings, a (two or more) hot water boiler(s), having a capacity as shown on drawings. The boiler shall be designed, constructed, tested, and installed in accordance with Section IV (Heating Boilers) of the ASME VIII Boiler and Pressure Vessel Code. Boiler shall be suitable for installation in the space shown with ample room for opening doors and cleaning and/or removal and replacement of tubes. Boiler shall be painted in accordance with manufacturer's standard requirement.

1. Each boiler shall have the output capacity in British thermal units per hour (Btuh) as shown on the drawings. Boiler and all accessories shall be designed and installed to permit ready accessibility for operation, maintenance, and service. Operating pressure for boilers supplied shall be as indicated on Contract drawings. Boiler shall be provided with all necessary connections including, but not limited to:

   a. a forced draft blower motor on the front boiler door above burner;

b. connections for water supply and water return;
c. combination thermometer and altitude gauge;
d. safety relief valve(s), (ASME rated relief valves of size and type to comply with code requirements);
e. temperature limit control, low water cutoff, drain valve, and flame safeguard controls. M&M No. 150 low water cutoff should also be included to prevent burner's operation if the water falls below a safe level. Cutoff shall have quick opening (lever operated) test valve. Temperature pressure gauges (3-1/2" dial) should be provided as well.

Boiler shall be completely pre-assembled and fire tested at the factory and have an authorized boiler inspection prior to shipment. Boiler shall have a standard nameplate containing the manufacturer's name and address, catalog number of boiler, maximum continuous capacity output in Btuh, radiant heating surface in square feet, total heating surface in square feet, and furnace volume in cubic feet. Nameplate shall be securely affixed in a conspicuous place. Boiler shall be capable of operating continuously at maximum specified capacity without damage or deterioration to the boiler, firing equipment, or auxiliaries. Boiler shall be operable automatically while burning the fuel specified. Boiler shall be of cast iron or steel construction.

2. Cast-iron boilers shall be rectangular, sectional, jacketed, and designed for use with the fuel specified.
3. Steel boilers shall be the portable, updraft, steel firebox, extended water leg, fire tube type specially designed for burning the fuel specified.

   The following features shall be included:

   a. Hinged front and rear doors permitting easy access to the flues. The rear refractory and baffling shall be completely contained in the rear door.

   b. Electric combustion control system, in compliance with I.R.I., Factory Mutual, and State codes.

   c. Combustion air damper and oil valve operating by a single damper control motor which shall regulate the fire according to the load demand.

4. Burner shall be a dual fuel forced draft type. The fuels shall be natural gas or No. 2 or No. 5 oil. (Oil No. 5 may be used but not for residential or small commercial burners.) Fuel shall burn without flame impingement on the furnace walls. The complete boiler-burner unit shall conform to I.B.R. and Factory Mutual requirements.

   The following features shall be included as required:

   a. Gas-electric ignition system utilizing propane gas in accordance with I.B.R. and Factory Mutual.

   b. Modular control panel containing the electronic programming relay, blower motor, starters and lights for low water level, flame failure, fuel valve position, and load demand shall be mounted on the front door of the boiler.

   c. The boiler shall be equipped with a low fire hold Aquastat to prevent high firing rate until boiler water is at preset temperature.

d. Boiler manufacturer shall provide terminal points and relays for operation of boiler room motorized combustion dampers. Interconnecting wire to damper motors are under automatic temperature control section.

5. The hot water heating boiler shall be as specified herein and shall be furnished complete with firing equipment, combustion chamber, insulation with steel jacket, safety and operating controls, integral electrical wiring, and all required appurtenances to make the boiler a complete, self-contained, fully automatic unit, ready for service upon completion of utility connections. Instrument panel may be free standing. Panel shall be completely wired and factory tested. Services of a manufacturer's representative who is experienced in the installation, adjustment, and operation of the equipment specified shall be provided. Representative shall supervise the installation, adjustment, and testing of the equipment.

6. Provide emergency disconnect switch located on the wall outside the boiler room entrance or just inside the door if the boiler room door is on the building exterior to allow rapid and complete shutdown of the boiler(s) in the event of an emergency. Switch shall be painted red and shall be provided with a label indicating the function of the switch.

7. Boiler relief safety valves of proper size and of the required number shall be so installed that the discharge shall be through piping extended to the nearest floor drain. Each boiler shall have at least one valve set to discharge at or below the maximum allowable working pressure of the boiler. Additional valves may be set within a range not to exceed 6 psi above the maximum allowable working pressure for boilers having working pressures up to and including 60 psi or 5 percent above the maximum allowable working pressure for boilers having working pressures exceeding 60 psi.

8. Drain valves in tandem shall be provided at each drain point of blowdown as recommended by the boiler manufacturer. Piping shall conform to ANSI B31.1 and shall be extra strong weight steel pipe conforming to ASTM A53.

B. The boiler must be guaranteed to operate at a minimum fuel to water efficiency of 84% at 100% of rating.

C. The local supplier of this equipment shall furnish the services of a factory trained engineer on his employ to check the installation of the unit, make the necessary adjustments, and train Owner's personnel in the care and maintenance of this equipment. All equipment covered under this section shall be guaranteed to be free of defects in material and workmanship for one year from date of first starting this equipment. The guarantee shall include the free service and labor necessary to effect such repairs and at no cost to the Owner.

D. Manufacturer shall be Cleaver-Brooks Kewanee, Weil McLain, or approved equal.

2.4 HOT WATER HEATING SPECIALTIES

All of the following equipment shall be furnished by one manufacturer, for example Bell and Gossett, Taco, or approved equal, and is to be ASME coded.

A. VALVES, FITTINGS, AND DEVICES

1. All radiators are to be controlled by self-contained non-electric valves. Valves shall be complete with remote sensor and control. Calibrate control as per manufacturer's recommendations.

2. Reducing valves shall be diaphragm operated with anti-siphon check valve and inlet strainer. The strainer must be easily removed without system shutdown. The valve seat strainer and stem must be removable and of noncorrosive material.

3. Relief valves shall be diaphragm operated with ASME label. The fluid should not discharge into the spring chamber. The valve should have a low blowdown differential, and the valve seat and all moving parts exposed to the fluid should be of non-ferrous material.

4. Triple duty check valve shall be furnished and installed by the Contractor as shown on plans, as well as a check valve with a spring-loaded contoured disc and a calibrated adjustment feature permitting regulation of pump discharge flow and positive shutoff. Valves shall be designed to permit repacking under full line pressure. Unit shall be installed on discharge side of pump in a vertical position with the stem up. Allow for minimum clearance of valve stem. This unit shall be cast-iron body construction suitable for maximum working pressure of 175 psig and maximum operating temperature of 300°F.

B. AIR SEPARATORS AND EXPANSION TANKS (SPECIFIER USE AS APPLICABLE)

1. Air separators and expansion tanks shall be constructed in accordance with the ASME Boiler and Pressure Vessel Codes Sections VIII and IX and stamped for 125 psig working pressure. The tanks shall be constructed of carbon steel.

   a. The tanks shall have one coat each of prime and finished paint applied at the factory.

   b. Compression tanks are to be constructed for 125 psig working pressure and guaranteed leakproof by the manufacturer. Tanks are to be stamped with the "U" symbol and Form U-1 furnished, denoting compliance with paragraph U-69 for construction of unfired pressure vessels, Section VIII, ASME. Tank shall be completed with gauge glass and cocks.

   c. The airtrol tank fitting for proper air control at the compression tank shall be constructed for 125 psig working pressure and shall include a manual vent tube for establishing the proper air volume in the compression tank on initial fill. Each unit, when installed and operated in accordance with the manufacturer's instruction, shall be furnished with a published performance guarantee.

2. The air separators shall be complete with integral strainers and collecting screen and 1 inch drain valve.

3. The units shall be of the sizes indicated on the drawings and shall be installed in strict accordance with the manufacturer's recommendations on supplementary steel supports.

4. The expansion tanks shall be the diaphragm pre-pressured type. The pressurization system shall include a diaphragm which will accommodate the expanded water of the system generated within the normal operating temperature and pressure range, limiting this pressure increase at all components in the system to the maximum allowable operating pressure necessary to eliminate system air. The air in the tanks shall be the permanently sealed-in air cushion contained in the diaphragm tanks.

5. For each tank, provide the following:

   a. Suitable approved supplementary structural steel supports and tank saddles.
   b. All necessary tappings, connections, and piping as indicated on the drawings.
   c. Approved gasketed handhole.
   d. Shut-off valves at the tanks.

2.5 COMBINATION GAS/OIL-FIRED BURNERS AND CONTROLS

A. Control systems and safety devices for automatically fired boilers shall conform to ASME CSD-1. Electrical combustion and safety controls shall be rated at 120 volts, single phase, 60 Hertz. The control panel shall have a NEMA 1a enclosure and be conveniently located for the operator. An alarm bell shall be provided with a suitable transformer and shall be located on the control panel. The alarm bell shall ring when the boiler is shut down by any safety control or interlock. Indicating lights shall be provided on the control panel. A red light shall indicate flame failure, and a green light shall indicate that the main fuel valve is open. The following shut down conditions shall require a manual reset before the boiler can automatically recycle:

   1. Flame failure.
   2. Failure to establish pilot flame.
   3. Failure to establish main flame.
   4. Low water cutoff.
   5. High temperature cutoff.

B. Burners shall be UL approved forced draft burners with all necessary air for combustion supplied by a blower which is an integral part of the burner. Boiler units shall be provided with modulating combustion controls with direct spark ignition. Burner shall be provided complete with fuel supply system in conformance with the following safety codes or standards: UL 726 and 795, and NFPA standards (as applicable), except that cast iron sectional boilers shall meet the requirements of UL 726, 795, and 296, and NFPA standards (as applicable).

2.6 PREFABRICATED CHIMNEYS

A. Furnish where shown on drawings, factory built sectional type chimneys for heating boilers, 2000°F refractory enclosed in metal jackets, as manufactured by Van Packer products, Hermit, or equal as approved by the Engineer. The chimneys shall be listed as "Model HT" by Underwriters' Laboratories, Inc., and rated as Medium Heat Appliance Chimneys.

   1. Each chimney section shall be up to __ in length with an insulating refractory wall, encased in an 11 gauge black (or specifier make selection) steel jacket. Insulating refractory shall be flush and square with ends of jackets to ensure 100 percent bearing surface between sections after erection.

2. Sections shall be joined with a high temperature, acid proof cementing compound, and section joints shall be covered with a draw band of the same material as jackets welded in the field. Joint cement and draw bands shall be supplied with the chimneys.

3. The chimney sections for heating boilers shall have an internal diameter as required and an external diameter of __ in. The chimney components shall consist of base section with cleanout and anchor lugs, special 45 degree T section for (stainless steel) flue connection, straight sections, insulated thimble, flashing and counter flashing, and stainless steel guy bands as required.

2.7 BOILER VENTS TO CHIMNEYS

A. Boiler vents shall be fabricated of 11 gauge black steel, as manufactured by U.S. Steel or an approved equal of all welded construction. Provide all necessary steel supports and hangers and make all connections between each boiler and its respective chimney as shown. Stainless steel clips shall be welded to boiler vents and shall be drilled for fastening of insulation wires.

B. Each chimney section shall be up to 4 ft in length with an insulating refractory wall, encased in a 1/4" black steel jacket suitable for field painting.

2.8 BOILER BREECHING

A. A hot water boiler shall be connected to the stack or flue by breeching constructed of steel sheets not less than 0.250 inch thick nor less than thickness of stack whichever is larger. Breeching construction shall include external perimeter angle iron welded joint reinforcement on 24 inch centers, continuously welded companion bolted flanged joints with gaskets.

B. The clear distance between any portion of the breeching surface and any combustible material shall not be less than that specified in the NFPA (National Fire Protection Association) Standard No. 211. Joints and seams shall be securely fastened and made airtight. Suitable hinged and gasketed cleanouts shall be provided at every change of direction and where they will permit cleaning of the entire smoke connection without dismantling. Gaskets shall be flexible ceramic cloth type. Cleanouts shall be raised face type for the installation of insulation. Flexible type expansion joints shall be provided as required and shall not require packing.

2.9 STEAM BOILER(S)

Where a steam boiler is specified it shall be furnished complete with the following:

A. Gauge glass and water column.

B. 3-1/2 or 4-in. steam pressure gauge with siphon and gauge cock.

C. Safety valve set at ___ psig (as applicable for the operating pressure; 15 psi for low pressure operation).

D. Combination low water cutoff and water feeder.

E. Boiler feed pump controller.

The boiler shall be erected on a concrete foundation as indicated on the drawings.

The combustion chamber shall be designed to suit the burner and boiler selected and shall be constructed of insulating firebrick with suitable insulation under the floor lining.

(Note to Specifier: Use next paragraph for gas-fired boiler.)

The Contractor shall furnish and install the collector and draft hood and breeching for the gas-fired boiler, as indicated on the drawings. The hoods and breeching shall be fabricated of aluminized sheet steel (Alclad), not lighter than No. 22 USS Gauge, and shall be amply reinforced.

(Note to Specifier: Use next paragraph for oil or oil and gas boiler.)

The Contractor shall furnish and install a smoke pipe for the boiler, as indicated on the drawings. The smoke pipe shall be fabricated of galvanized sheet steel, not lighter than No. 16 USS Gauge, and shall include an automatic barometric draft adjuster unless otherwise recommended by the boiler manufacturer.

Before acceptance by the Owner, the Contractor shall thoroughly clean out the boiler by boiling out with an acceptable cleaning compound solution.

Owner's set of suitable cleaning tools shall be furnished for cleaning the boiler.

If the boiler the Contractor proposes to furnish will not perform at the specified rating with the chimney height indicated, an induced-draft fan shall be installed in the flue to provide the necessary draft to achieve the rating.

The combination low water cutoff and water feeder shall be arranged so that water will be fed to the steam boiler automatically when the water level in the boiler drops below a predetermined point and will cause an alarm bell to ring when the water level reaches the low point. The feeder shall be arranged so that the burner will cease to operate whenever the water level drops below a dangerous low point.

The boiler feed shall be constructed so that the feedwater valve and seat will be isolated from the float chamber, thereby preventing the overheating of the feedwater and the precipitation of scale on either the valve or seat. Each float mechanism, valve, and seat shall be constructed of an acceptable, durable, and corrosion-resistant alloy. Valve seats shall be removable and renewable. The feeder shall be equipped with a large self-cleaning strainer.

The combination low water cutoff and water feeder shall be approved by Underwriters' Laboratories, Inc., and shall be suitable for not less than 35 psi steam working pressure (for low pressure operation) and shall be made by McDonnell & Miller, Inc., Chicago, IL; Magnetrol, Inc., Downers Grove, IL; or an acceptable equivalent product.

The pump controller shall be suitable for not less than 35 psi steam working pressure and shall be made by McDonnell & Miller, Inc., Chicago, IL; Magnetrol, Inc., Downers Grove, IL; or an acceptable equivalent product, and shall be approved by Underwriters' Laboratories, Inc.

The boiler makeup water piping shall include a pressure-regulating valve.

F. Steam Condensate Unit - Boiler Feed Pump Unit

Furnish a completely pre-fabricated package boiler feed system in Boiler Room located on 6" concrete pad in accordance with supplier's detailed drawings, instructions, and wiring diagrams as approved by the Engineers, for feeding the boiler(s).

1. The unit shall be of single tank horizontal construction.

2. The completely pre-fabricated and factory tested unit shall consist of condensate receiver __" diameter by __" shell length, water makeup assembly, two (2) boiler feed pumps (one for spare), one (1) NEMA II consolidated control cabinet, and accessories as specified.

3. The condensate receiver shall be of horizontal welded steel construction, 1/4" thick on heightless cradles. The receiver shall be welded inside and outside. After fabrication the inside welds shall be grounded to radius all edges and smooth all welds. The receiver interior shall be shot blasted to remove scale, dirt, and oil and shall be coated with epoxy-phenolic lining of at least 6 mil thickness. Receiver shall be equipped with 11" x 15" flanged manhole. Isolation valve shall be installed between each pump suction and receiver. Accessories shall include: water lever gauge, dial thermometer, overflow loop, low level pump cutoff float switch, and companion flanges.

4. One inlet strainer with self-cleaning bronze screen and large dirt pocket shall be mounted on the receiver. The screen shall be vertically removable for cleaning requiring no additional floor space for servicing.

5. A water makeup assembly shall be installed on the receiver of capacity equal to one (1) boiler feed pump. The assembly shall consist of a level control switch and solenoid electric valve. The valve shall be packless, piston pilot-operated type with cushioned closing feature and epoxy-resin molded waterproof coil. The valve shall be equipped with a strainer and a manual bypass provided around the valve.

G. Furnish a completely pre-fabricated package boiler feed system in Boiler Room, where shown, on 6" concrete pad in accordance with supplier's detailed drawings, instructions, and wiring diagrams as approved by the Engineers, for feeding the boiler(s).

1. The unit shall be of single tank horizontal construction, as manufactured by Domestic Pump and Manufacturing Company Type "CMHD," equal by Skidmore, Dunham Bush, or an approved equal.

2. The completely pre-fabricated and factory tested unit shall consist of condensate receiver 36" diameter by 84" shell length, water makeup assembly, two (2) boiler feed pumps (one for spare), one (1) NEMA II consolidated control cabinet, and accessories as specified.

3. A magnesium anode with tell tale weephole to inhibit electrolytic corrosion shall be mounted on the receiver by means of a 1-1/4" IPS removable plug. It shall be no less than 1-15/16" diameter by 18" long and reinforced by a steel wire core.

4. The boiler feed pumps (2) shall be two stage centrifugal design, each for __ gpm at __ psig, __ hp, permanently aligned and flange mounted for vertical operation. Two boiler feed pumps are usually installed. It is up to the specifier to decide. Each pump shall be bronze fitted with enclosed bronze impellers, axial flow bronze impeller, bronze straightening vanes, renewable bronze case rings, stainless steel shaft, and shall be coupled to a drip-proof motor, 3500 rpm, and shall deliver its full capacity at __ ft NPSH (Net Positive Suction Head).

5. The unit manufacturer shall furnish, mount on the pump unit, and wire a NEMA II control cabinet with piano hinged door enclosing the following:

   a. One combination magnetic starter (having three overload relays) with circuit breaker and cover interlock for each pump.
   b. One "Off-Lead-Lag-Continuous" selector switch for each pump.
   c. One control circuit transformer.
   d. One circuit breaker for the control circuit with cover interlock.
   e. One numbered terminal block.
   f. One alarm bell with silencer relay and indicating lights for high and low level alarms.
   g. Terminal block shall provide terminals for remote annunciator to indicate that the standby pump has been actuated.

6. The unit manufacturer shall furnish (1) each of the following for boiler: domestic pump hydraulic feed valve; McDonnel & Miller No. 150 pump control with (2) two-wire switches.

7. Control components shall be provided by the unit manufacturer for operation as follows: As the level in any boiler recedes, the pilot valve will open, releasing the pressure inside the feed valve bellows. Upon 3/4" drop in boiler level, the pump control switch will close, starting the "Lead" pump. As the level is restored, the switch will open and stop the pump. Should the level continue to recede, the second switch on the pump control will close, starting the "Lag" pump.

8. The unit shall be factory tested as a complete unit and a certified test report including NPSH characteristics shall be submitted prior to shipments. The unit manufacturer shall furnish complete elementary and connection wiring diagrams, piping diagrams, and installation and operation instructions.

Include services of factory trained engineer to supervise the installation and start the unit when ready, adjust the operation of the complete boiler feed system to the satisfaction of the Engineers, and instruct the Owner's personnel in its operation.

2.10 FUEL OIL SYSTEM

A. The Contractor shall furnish and install a complete fuel oil storage and piping system to serve the boiler, as shown on the drawings and specified herein and required for the proper operation of the oil burner specified. The system shall consist of, but not be limited to, a fuel oil storage tank with all necessary appurtenances, fuel oil pump, all complete with gauges, valves, strainers, electric heater, and other appurtenances as required and all installed in accordance with the rules and regulations of the National Board of Fire Underwriters, local authorities, and the I.R.I.

B. The fuel oil storage tank shall have a capacity of __ gallons and shall be __ diameter x __ long x __ thick. Steel tank shall be constructed and installed in accordance with the rules and regulations as stated herein. If underground tank is installed, double wall shall be required (check the State Code). Fiberglass tank may be installed instead of steel tank.

C. The excavation for the placement of (underground) tank, backfilling the hole, will be done under another section of the work. Tank shall be provided with access manhole, shelf plates, hold down straps, and all other openings as detailed on the drawings. Connections to these openings shall be as detailed. Before setting in the excavation, steel tanks shall be thoroughly cleaned and the exterior surface primed and then coated with two coats of hot coal tar enamel, applied in accordance with the manufacturer's published instructions. Furnish anchors to the General Contractor for setting in the concrete mat.

D. Furnish and install where shown on the drawings a shop built duplex fuel pump and strainer set. The pump and accessories shall be mounted on a base plate with 1" drip rim and tapped drain and shall be complete with driving motor, sheaf, belt, and belt guard. The pump shall be an iron alloy herring bone gear pump, carefully fitted to give quiet operation and smooth flow. Pump shall have a mechanical shaft seal, provided with a steel retainer and spring, and synthetic rubber seal. Each pump shall be belt-driven and have a maximum speed of 400 rpm and a capacity of not less than __ gpm at the discharge pressure required, and shall be driven by a motor of not less than 1/2 horsepower.

The pump set shall be installed on a 3" deep galvanized iron pad having a drain. Pump shall be as manufactured by Kraissl, Viking, Wayne, or approved equal.

E. Contractor shall furnish all miscellaneous valves, fittings, day tank, fill caps, sidewalk boxes, strainer, gauges, oil pressure regulating valve, as shown on drawings.

F. Oil piping: All piping throughout the building shall be constructed of Schedule 40 black steel and fittings, with welded joints, except for connections at the pump and boiler. All piping is to be in accordance with State and Local Codes. Run 1/4" copper line in conduit from oil tank to remote oil gauge as shown on the drawing. Wrap all underground piping with 3M PVC 20 mil minimum coal tar tape. Fuel oil suction line from tank to building is to be electric traced as required by electrical contractor. This pipe shall be insulated after it has been electrically traced.

1. At the Contractor's option, in lieu of the steel piping specified, copper piping with 95/5 tin-antimony soldered joints may be used, Type "K" underground and Type "L" indoors. Vent piping shall be either Schedule 40 galvanized steel with screwed galvanized malleable iron fittings or copper as detailed above.

2. The oil suction line should be tested for four hours and proven tight before placing the system in operation.

2.11 RETURN LINE HEATER

A. Furnish and install in Pump Room, an electric heater as manufactured by Chromolox or approved equal, complete with necessary safety controls and 3-valve by-pass.

B. Electric fuel oil heater in return line shall be a compact, 150 psi constructed, automatic fuel oil heater with U-shaped steel sheathed tubular elements mounted in a heavy flange. Heating surface watt density shall be a minimum of __ watts per square inch with __ kW total capacity. Heating chamber shall be completely insulated and enclosed in sheet steel housing. Provide non-indicating, direct immersion, adjustable thermostatic control device, with 50° - 250°F range.

C. Heater shall be for operation on ___volt, ___-phase, 60-cycle service. (Check with Electrical Engineer.)

2.12 FUEL OIL TANK GAUGE

A. Tank gauge shall be master Model Level meter or approved equal. The gauge shall have dimensions marked in gallons and liters. Gauge is to include closed bellows system incorporating knife edge bearings and shall contain no fluids.

2.13 HOT WATER HEATING RADIATORS AND CONVECTORS

A. Units shall be the types and sizes indicated on contract drawings. The supply connection to each unit shall contain the radiator gate valve, and the return connection shall be made with a union elbow containing a balance valve or as indicated on contract drawings. The supply and return connections shall be the same size. Nonferrous convectors shall be tested hydrostatically at the factory and proved tight at a pressure of not less than 100 psi. A certified report of these tests shall be available. Provide access doors in cover for valves. On wall to wall enclosures provide blank sections.

B. Radiators shall have large or small cast-iron tubes or cast-iron fins permanently bonded to cast-iron cores either threaded or with sweat fittings at each end for connecting to external piping. Radiators shall have capacities not less than those indicated, determined in capacities with The Hydronics Institute I=B=R Testing and Rating Code for Finned Tube (Commercial) Radiation. Radiators shall be equipped with expanded metal cover grilles fabricated from steel sheets not less than 0.0598 inch in nominal thickness, secured either directly to the radiators or to independent brackets or with solid-front, slotted sloping top cover grilles fabricated from steel sheets not less than 0.0478 inch in nominal thickness, independently secured to masonry with brackets.

C. Convectors shall be constructed of nonferrous alloys of suitable composition with fins permanently bonded to core and shall be installed where indicated. Capacities shall not be less than those indicated on contract drawings. The overall space required by convectors shall not be greater than the spaces provided and indicated. Convectors shall be complete with heating unit and enclosing cabinet which has bottom recirculating opening and top supply grille. Convector cabinet shall be constructed of sheet steel not less than 18 gauge. Cabinet shall have sloping top. Valves shall be as indicated on contract drawings.

2.14 FINNED TUBE RADIATORS

A. Furnish and install finned tube radiators, in accordance with manufacturer's instructions. Radiators shall be constructed with back plate, front enclosure, heating element, hangers, and accessories of model and size indicated.

Joints between enclosure sections shall be neat appearing without cover straps or other concealing pieces. No unfinished metal edges shall be visible in the installation.

B. Enclosure shall be constructed of furniture quality steel front panels and separate round edge bar stock steel outlet grille. Front panel shall be finished in Neutral Gray (or specific color requested by the architect) high gloss baked enamel. The grilles and accessories shall be anodized aluminum. Arrangement shall be as called for in schedules on contract drawings.

C. Accent trim shall be furnished where indicated and shall attach to enclosure without screws or other external fasteners. Accent trim shall be finished in color contrasting with (matching) enclosure.

D. Solid back plate shall be 20 gauge steel, continuous construction the full height of enclosure front panel. All mounting holes shall be provided for mounting to the wall and for attaching pipe hangers. Separate brackets shall be provided for securing front panel to back plate at bottom.

E. Pipe hangers shall be 16 gauge steel and shall be attached to back plate with self tapping screws.

F. Heating elements shall be constructed of embossed aluminum fins mechanically bonded to seamless copper tubing. One end of each element shall be expanded to receive the unexpanded end of the connecting element to make a sweat-type joint without using couplings. Elements shall be suitable for pressures up to 100 psig.

G. End caps shall be of 20 gauge steel and shall be furnished with bolts for attachment to enclosure ends.

H. Sleeves shall be of 18 gauge steel and shall telescope over enclosures for length variation. 7-1/2" sleeves shall have optional access doors for use as valve compartments.

I. Corner accessories for inside and outside 90 corners shall be of 18 gauge steel and shall mount flush with enclosure without concealing trim.

J. Filler pieces of 8" and 18" lengths shall be of 18 gauge steel and shall mount flush with enclosure without concealing trim. The 8" and 18" fillers shall have optional access doors for use as valve compartments.

2.15 RADIATION TYPE HEATERS

A. Furnish and install wall fin radiators where shown on drawings, having a capacity and dimensions as scheduled on drawings.

B. Tubing shall be threaded 1¼" with 0.123" wall thickness. Fins shall be 2½" x 5¼" and mechanically bonded to tube.

C. Heating element hangers shall be of the rolling, free expansion 2-piece, cradle type and shall be fastened to the wall not more than 4' on centers.

2.16 HOT WATER UNIT HEATERS

A. Unit heaters shall be rated and tested in accordance with ASHRAE and shall have Btu capacities not in excess of 125 percent of that indicated on contract drawings. The noise level of each unit heater shall be appropriate for the space in which the heater is installed.

B. Propeller unit heaters shall be designed for suspension. The casings shall be constructed for not less than 16 gauge steel and enameled painted. Suitable stationary or rotating air deflectors shall be provided to ensure proper air and heat penetration capacity at floor level based on established design temperature. Suspension from heating pipes shall not be permitted. Discharge shall be arranged for horizontal or vertical flow of air as indicated. Fan for vertical discharge heaters shall operate at speeds not in excess of 1200 rpm, except that units with 50,000 Btuh output capacity or less may operate at speeds up to 1800 rpm. Horizontal discharge unit heaters shall have discharge or face velocities not in excess of the following:

Unit capacity (cfm)	**Face velocity (fpm)**
Up to 1000	800
1001 to 3000	900
3001 and over	1000

2.17 DETAILS FOR VARIOUS TYPES OF HEATERS

A. Unit heater cabinets shall be constructed of highest quality furniture steel. Heat transfer portion of heating elements shall be copper tube with aluminum fins hydraulically expanded to the tubes. Heating coils and radiating fins shall be of suitable nonferrous alloys. The heating elements shall be free to expand or contract without developing leaks and shall be properly pitched for drainage. The elements shall be tested under a hydrostatic pressure for 200 psi, and a certified report of the test shall be submitted to the Engineers. Coils shall be suitable for use with water with temperatures up to 250°F.

Cabinet type heaters shall have centrifugal fans and be arranged for floor or ceiling mounting as indicated on contract drawings. The heating elements and fans shall be housed in steel cabinets with angle iron frames. The cabinets shall be constructed of not lighter than 16-gauge steel and enameled painted. Each unit heater shall be provided with an approved means of diffusing and distributing the air. The fans shall be mounted on a common shaft with one fan to each air outlet. The fan shaft shall be equipped with self-aligning vibration isolated ball or roller bearings and accessible means of lubrication. The fan shaft may be either directly connected to the driving motor or indirectly connected by an adjustable V-belt drive with belt guard rated at 150 percent of motor capacity. All fans in any one unit heater shall be the same size. Floor mounted units shall have sloping top. Provide access door for controls and valves. Provide throw-away type filters.

B. Motors shall be provided with manual selection switches with "On," "Off," and "Automatic" positions and shall be equipped with thermal overload protection.

Motors for horizontal units shall be sleeve bearing type mounted on resilient type base to isolate motor and fan from cabinets. Motors for vertical units may be sleeve bearing for all angle operation or ball bearing. Vertical unit motors shall be fully protected from radiation from unit coil by metal shield between coil and motor. Motor current characteristics shall be as covered under electrical specifications.

C. The space temperature shall be maintained automatically by stopping and starting the fan by the room thermostat. Strap on type aquastat shall be provided.

2.18 CENTRIFUGAL REFRIGERATION CHILLER (Other types of chillers are available. For appropriate selection, consultation with one of the major chiller manufacturer representatives is highly recommended.)

A. Furnish and install where shown on drawings, one electric motor driven centrifugal refrigeration machine of the hermetic type, constructed to conform to ASTM Code. Unit shall be completely factory assembled with all accessories and shall have capacity as indicated on the drawing. Maximum kW/ton is __ (specified as indicated).

B. Selection of unit shall allow for water side fouling factors of 0.0005 in the cooler tubes and 0.0005 in the condenser tubes.

C. Unit shall conform to ASA B9.1 Safety Code.

D. The cooler and condenser shall be of the shell and tube type construction. Water boxes shall be fabricated steel and securely welded to the heat exchanger tube sheets. Water box covers shall be removable to allow for tube cleaning. Water boxes shall be designed for 150 psi working pressure and shall be in accordance with the ASME Code for Unfired Pressure Vessels. Each water box shall be tapped for vent and drain piping connections. Provide flanged outlet connections at cooler and condenser. The tubes shall be seamless copper tubing with integral fins and shall be individually replaceable. Each tube is to be rolled into the tube sheets. The cooler shall be equipped with sufficient eliminator area to prevent liquid with a relief device to prevent excessive pressure in the heat exchangers.

E. The compressor shall be hermetically sealed and shall have impeller wheels of cast-aluminum alloy. The shaft bearings shall have forced feed lubrication and shall be designed to minimize thrust wear. Variable inlet guide vanes shall be provided to vary the suction gas flow entering the compressor.

F. The motor shall be of the two-pole, single-speed, non-reversing, squirrel-cage induction type, and shall be suitable for 460-volt, 3-phase, 60-cycle service. The design speed shall be 3550 rpm. The motor shall be refrigerant cooled and shall be suitable for operation in a refrigerant atmosphere. The motor in-rush current shall not exceed 150% of the maximum motor full load current.

G. A forced feed lubricating system with a 1/2 hp motor driven oil pump shall be provided. The system shall include an automatic oil heater and an oil cooler as integral parts of the compressor motor assembly. The oil pump motor shall be suitable for 460-volt, 3-phase, 60-cycle service. The starter for this pump shall be located within the main chiller starter enclosure or on the machine, with its own fused disconnect switch.

H. A refrigerant cooler purge system shall be provided to automatically remove noncondensible gases from the refrigerant system. The purge pump motor shall be rated at 1/4 hp and suitable for 115-volt, 1-phase, 60-cycle service. The purge system shall include a separation chamber to allow for the removal of water from the system.

1. An indicator light shall be mounted in a conspicuous location on the control console or panel to indicate operation of the purge pump.

I. Capacity shall be controlled by a thermostat which senses the leaving chilled water temperature and a temperature sensor controller which modulates the compressor capacity in response to a change in the leaving chilled water temperature. A device shall be provided, by which maximum current may be set at any value between 40% and 100% of rated full amperes. The machine shall be provided with the following protective devices:

1. Condenser high-pressure cut out.
2. Bearing high-temperature cut out.
3. Motor winding high-temperature cut out.
4. Refrigerant low-temperature cut out.
5. Chilled water low-temperature controllers.
6. Oil low-pressure cut out.
7. Chilled water and condenser flow switches.
8. Control to insure no-load starting for compressor.
9. Timer to prevent machine starting more than once every 20 minutes.

Items "1" through "6" shall be provided with indicator lights.

In addition, the following gauges shall be provided:

1. Suction pressure gauge.
2. Condenser pressure gauge.

J. Contractor shall furnish and install necessary auxiliary water piping for the oil cooling coil and purge condenser, starting from the chiller water line.

K. Machine shall be installed on a concrete base. Spring type isolators shall be furnished by the unit manufacturer.

L.

1. A complete charge of refrigerant and oil shall be provided for the machine, and this charge shall be guaranteed for one year's operation, after acceptance of the system.

2. Refrigerant 12 & 22 or any other containing ozone depleting fluorocarbons are not acceptable.

M. A compressor motor starter of the star delta closed transition type shall be furnished for the unit, complete with disconnect switch installed in a cabinet. An across-the-line starter for oil pump motor shall also be provided within the starter enclosure unless starter is mounted on the machine. In the cabinet shall also be installed a motor current ammeter, voltmeter, 3-phase switch, 3 kW (minimum) control transformer with disconnect for chiller auxiliaries, and ground fault protection. Ammeter shall have divisions of 10 amperes. Furnish with starter, a flush-mounted elapse time meter (unless mounted on machine) indicating the total number of hours the machine has been in operation. Lugs are to be sized in accordance with electrical feeder requirements.

N. Chiller shell, suction piping, etc., shall be factory insulated.

O. Contractor shall include in his bid, the service of a factory representative for a period of three working days who will advise on the following:

1. Installation of purge unit.
2. Testing the refrigeration system under pressure for leaks, evacuation, and dehydration of system using a high vacuum pump, as recommended by the refrigeration system manufacturer.
3. Charging the system.
4. Starting the system and instructing the operation of the machine as to proper care and operation. Furnish three copies of complete parts list of the water chilling system.

P. Manufacturer's representative is to meet with the temperature control subcontractor and the electrical contractor to coordinate the operation of the chiller and the interlocking of pumps, cooling tower, etc. Chillers must be tested in the manufacturer's plant and, after installation, in the field.

2.19 COOLING TOWER (Selection of the specific type is to be made in consultation with leading tower manufacturers.)

A. The cooling tower shall be induced draft, vertical discharge, packaged tower cross flow type with vertical air discharge and capacity and horsepower requirements as listed under equipment schedule on the drawings.

B. Casing shall be constructed entirely from hot-dip galvanized steel panels supported by heavy gauge hot-dip galvanized steel angle and channel framework, all finished inside and out with zinc chromatized aluminum.

(Note to Specifier: Casing can be wood, concrete, or other materials.)

C. Louvers shall be corrugated, hot-dip galvanized steel, finished with zinc chromatized aluminum. Louvers shall be factory installed.

D. Cold Water Basin and Other Specialties

The cold water basin shall be constructed of heavy gauge, hot-dip galvanized steel, finished inside and out with zinc chromatized aluminum. Basin shall be self cleaning and include depressed center section, with drain and clean-out connections. Suction connection shall be provided with anti-vortexing device and large area, lift out, hot-dip galvanized steel strainers. A brass float operated makeup valve shall be provided complete with large diameter plastic float, arranged for easy adjustment.

1. Distribution System Hot water distribution basins shall be open gravity type, constructed of hot-dipped galvanized steel finished with zinc chromatized aluminum.

Distribution weirs and plastic metering orifices shall be provided to ensure even distribution of water over wet deck surface.

2. Wet Deck Surface The wet deck surface shall consist of wareformed sheets of self-extinguishing polyvinyl chloride. It shall have a Flame Spread Rating of 25 per ASTM Standard E-84 and be impervious to rot, decay, fungus, or biological attack. The surface shall be manufactured and performance tested by the cooling tower manufacturer to ensure single source responsibility and control of the final product.

3. Drift Eliminators The drift eliminators shall be constructed of polyvinyl chloride. They shall have a minimum of three distinct changes in air direction and limit drift loss to less than 0.2% to 1% of the total water circulated.

4. Mechanical Equipment Fan(s) shall be fixed pitch, heavy duty, cast aluminum with a minimum of six blades. It shall discharge through a hot-dip galvanized fan cylinder designed for streamlined air entry and minimum tip loss for maximum fan efficiency.

   Fan and shaft shall be supported by heavy duty, relubricatable ball bearings with special moisture seals, slingers, and housings designed to prevent moisture accumulation.

   Fan shall be driven by a one-piece, multi-groove, neoprene/polyester Power Band designed specifically for cooling tower service.

   Fan motor shall be totally enclosed, air-over (TEAO), 1800 rpm, reversible, squirrel cage, ball bearing type designed specifically for cooling tower service. Motor shall be furnished with special moisture protection on windings; shafts and bearings pie fill line shall be external. Provide extended tube line with sight glass.

5. Access and Safety Access doors shall be provided on both sides of tower(s) for access to eliminators and plenum section. A heavy gauge, hot-dip galvanized wire fan guard shall be provided over each fan cylinder.

6. Corrosion Protection All steel components of the tower(s) shall be hot-dip galvanized, with cut edges or other exposed surfaces given a protective coat of zinc-rich compound. A final coating of zinc chromatized aluminum shall be applied after unit assembly for additional corrosion protection. The final protective finish shall be guaranteed by the tower manufacturer to permit no corrosion when exposed to a twenty percent salt spray at 95°F for 47,000 consecutive hours.

7. All bolts, nuts, and washers shall be galvanized steel. All steel used in construction of the cooling tower shall be galvanized with a least 2½ ounce coating per square foot.

8. Ladder shall be hot-dip galvanized after fabrication and shall extend from the top of the tower to roof. Ladder shall be bolted to clips furnished as a part of the tower to provide a permanent, secure installation.

9. Tower shall also be supplied with hand rail and hot water basin cover.

10. Starter is included by other in the Motor Control Center section of the specification.

11. Noise level of the cooling tower shall be within acceptable limits (to be specified based on tower location).

12. Structural supports shall be in accordance with manufacturer's recommendations.

2.20 COOLING TOWER SYSTEM

A. Recommended Water Qualities:

Phosphorate - Polyphosphate: 5 - 10 parts per million
pH: 6.5 to 8.0

Cycles of concentration are dependent on local water quality. Maintenance personnel shall be instructed to maintain cycles of concentration at the most economical and safe limit to prevent scale deposition.

Organic Growths: None

B. Feeding Equipment:

Controlled volume pump and tank packaged system, Milton Roy "Roy" pump with 55-gallon polyethylene lined tank, or equal, are to be installed as shown on drawing. Discharge of chemical feed pump shall be piped to condenser water piping returning from cooling tower, and pump motor shall be electrically interlocked with condenser water pump motor or flow switch. Bleed-off tap may be located at any point in the condenser water supply piping. Bleed-off line shall be fitted with gate valve and plug cock and shall run to an adequate drain.

Furnish and install Specific Conductivity Sensor controller complete with hinged door, mode lights indicating power on and bleed, test button, and solid state circuits; 3/4" electric solenoid valve; 3/8" strainer with Monel Screen for flow-through probe; relay; all as manufactured by Mogul, Cambridge Scientific, or approved equal.

2.21 AUTOMATIC WATER TREATMENT SYSTEM

A. Mechanical contractor shall engage a suitable water treatment subcontractor to design and install a supervised water treatment program for the following systems, for a period of one year from the date of startup.

1. Cooling tower.

2. Chilled water.

3. Boiler.

Chemical proposed to be used shall have no adverse effect on pump packing glands, seals, valves, piping, etc., and all treated water discharge shall meet all anti-pollution requirements of all Federal, State, and Local Codes having jurisdiction.

B. Water Treatment Subcontractor Shall:

1. Provide Owner with complete written instructions for chemical feeding, bleed-off, blow-down control, and testing procedures.
2. Supply a complete water treatment service to protect the water side of the boiler from problems of scale, corrosion, pitting, priming, or foaming.
3. Furnish and deliver all the chemicals required to be fed into the treatment system to maintain the following chemical conditions: chrome concentration - 500 ppm.
4. Furnish and deliver all the chemicals required to maintain specified conditions in the condenser and water systems.
5. Obtain samples, for all systems being tested, within five days after treatment has been started and at least once per month. Thereafter, analyze such samples and furnish written reports and recommendations to Owner and/or Engineer.
6. Instruct Mechanical Contractor as to the amount of feeding equipment.
7. Instruct Plant Engineer or maintenance personnel as to the amount of chemicals to be fed into the system, proper blowdown schedule as called for by analysis of the water, and testing procedures to maintain the required concentration. Water treatment company shall treat the water initially after the systems have been thoroughly flushed and cleaned as detailed before. He shall also provide a complete set of laboratory quality test equipment for the periodic testing by the operating personnel and shall thoroughly instruct the operating personnel on the use of test equipment and the application of chemicals

2.22 CHILLED WATER SYSTEM

A. Recommended Water Conditions:

Sodium nitrate-Sodium borate: 1000 parts per million

pH: 7.0 to 8.5

B. Feeding Equipment: by-pass feeder

2.23 PUMPS

A. Description

Furnish and install as shown on the drawing, pumps for the heating and/or chilled water system. Chilled and hot water pumps shall be of the single stage, double suction, centrifugal, vertical or horizontal split case type. Pumps shall be directly connected with flexible couplings, standard NEMA frame, and splash-proof motors. Pumps and motors shall be mounted on a cast-iron bed plate or base. Units shall have non-overloading curve characteristics and shall be equipped with mechanical seals.

B. General Requirements

1. Standard Products
Pumps and associated materials shall be the standard products of a manufacturer regularly engaged in their manufacture for a minimum period of ten years. All

material and equipment shall be new. General requirements include those in the Section MECHANICAL SPECIFICATION GENERAL REQUIREMENTS and as specified herein.

2. Nameplates
Pumps and motors shall have a standard nameplate securely affixed in a conspicuous place showing the manufacturer's name and address, pump type, model, and serial number. In addition, the nameplate for each pump shall show the capacity in gpm at rated speed in rpm and head in feet of water and psi. Nameplate for each electric motor shall show at least the minimum information required by NEMA (National Electrical Manufacturers Association) MG1-10.38. The nameplate may also contain such other information as the manufacturer may consider necessary for a complete identification.

3. Electrical Work
Electrical motor-driven equipment herein shall be provided complete with motors, motor starters, and controls. Motor starters shall be provided complete with properly sized thermal overload protection in each phase and other appurtenances necessary for the motor control specified. Each motor shall be of sufficient capacity to drive the pump at the specified capacity without exceeding the nameplate rating of the motor when operating at proper electrical system voltage and frequency.

4. Qualified Representative
The Contractor shall have a qualified representative of the equipment manufacturer to supervise the installation, adjustment, testing, and start-up of the equipment and associated controls.

C. Submittals

1. Submit for approval detailed shop drawings, wiring diagrams, elementary diagrams, equipment templates for foundations, supports and alignment indicating bolts, anchors, supports, concrete and steel work, and appurtenances required for proper installation. Shop drawings shall be fully dimensioned and shall indicate the intended installation.

2. Submittals shall also include product data and information on operating and maintenance instructions, pump capacity and head, motor characteristics, descriptions of controls, sensors, starters, materials, finishes, and accessories. Manufacturer's pump curves shall be submitted for each pump and motor.

   Submit pump curves showing capacity in gpm, NPSH (Net Positive Suction Head), efficiency, and pumping horsepower from 0 gpm to 110 percent of design capacity.

3. The Contractor shall submit description of operation and electrical and mechanical schematic drawings, including interconnections to all associated equipment, for approval before starting work. Bulletins describing each item of control equipment or components shall be included.

4. Operations and Maintenance Manual: See Mechanical Specification General Requirements.

5. The Contractor shall furnish detailed test procedures with test forms prior to performing the equipment tests. After completing the test, the results of the test shall be submitted to the Engineer.

D. Selection Criteria

1. Pumps shall be designed using hydraulic criteria based upon actual model developmental test data. Pumps shall be designed for a point within the maximum efficiency for a given impeller casing combination. Deviations within 3 percent of maximum efficiency is not less than the scheduled efficiency. Pumps having impeller diameters larger than 90 percent of the published maximum diameter of the casing or less than 15 percent larger than the published minimum diameter of the casing will not be accepted.

2. Pumps shall be factory tested by the manufacturer or a nationally recognized testing agency in compliance with Hydronics Institute standards. Where two or more identical pumps are specified, only one representative pump must be tested.

3. Supply drawings showing the pump characteristics. The pump characteristics shall include the head versus capacity curve, the required power characteristics, the efficiency, and the Net Positive Suction Head curves. Pumps shall operate at not less than 70% efficiency within plus or minus 20% of the design capacity. Pump characteristics shall slope gradually from the shut-off head to the design head, with the difference in head not exceeding 10% of the shut-off head.

E. Centrifugal Water Pumps

1. The pumps shall be horizontal centrifugal water pumps of the types and capacities indicated on the drawings. The driving units for the pumps and capacities shall be electric motors. Centrifugal water pumps shall be constructed in accordance with HI (Hydronics Institute) Standard 01. Pumps shall operate at optimum efficiencies to produce the most economical pumping system under the actual conditions.

2. Pump casing, shall be constructed of cast iron and shall be designed to permit replacement of wearing parts. Horizontal-split casings shall have the suction and discharge nozzles cast integrally with the lower half, so that the upper part of the casings may be removed for inspection of the rotating parts without disturbing pipe connections or pump alignment. Pump casings shall be of uniform quality and free from blowholes, porosity, hard spots, shrinkage defects, cracks, and other injurious defects. Defects in casings shall not be repaired except when such work is approved and is done by or under the supervision of the pump manufacturer, and then only when the defects are small and do not adversely affect the strength or use of the casing. Casings shall be single or double volute with flanged piping connections conforming to ANSI B16.1, Class 125. The direction of shaft rotation shall be conspicuously indicated. The casing shall have tapped openings for air venting, priming, draining, and suction and discharge gauges. Vent cock shall be furnished for venting except where automatic air vents are indicated. Drain openings in the volute, intake, or other passages capable of retaining trapped water shall be located at the low point of such passages.

3. Impellers shall be of enclosed design and shall be constructed of carefully finished bronze with smooth water passageways and shall be statically and dynamically balanced. Impellers shall be securely keyed to the pump shaft.

4. Wearing rings of bronze shall be provided for impellers. Wearing rings of a different composition or of a suitable ferrous material shall be provided for pump casings. Casing rings shall be securely fixed in position to prevent rotation. Rings shall be renewable and designed to ensure ease of maintenance.

5. Mechanical seals shall be balanced or unbalanced, as necessary to conform to specified service requirements. Mechanical seals shall be constructed in a manner and of materials particularly suitable for the temperature service range and quality of water being pumped. Seal construction shall not require external source of cooling for higher pumped-fluid service temperatures. Seal pressure rating shall be suitable for maximum system hydraulic conditions. Materials of construction shall include ANSI 300 series stainless steel, solid tungsten carbide rotating seal face, and buna-N seals. Throttling bushing shall have clearances to minimize leakage in case of complete seal failure. Mechanical seals shall not be subjected to hydrostatic test pressures in excess of the manufacturer's recommendations.

6. Couplings shall be of the heavy duty flexible type, keyed and locked to the shaft. The outside surface of the couplings shall be machined parallel to the axis of the shaft. The faces of the couplings shall be machined perpendicular to the axis of the shaft. Disconnecting the couplings shall be accomplished without removing the driver half or the pump half of the couplings from the shaft. Flexible couplings shall not be used to compensate for misalignment of pump.

7. All rotating parts of the equipment shall operate throughout the required range without excessive end thrust, vibration, or noise. Defects of this type that cannot be eliminated by installation adjustments will be sufficient cause for rejection of the equipment. Pump impeller assemblies shall be statically and dynamically balanced to within 1/2 percent of W times R squared, where W equals weight and R equals impeller radius. Shaft construction shall be substantial to prevent seal or bearing failure due to vibration. Total shaft peak to peak dynamic deflection measured by vibrometer at pump seal face shall not exceed 2.0 mils under shutoff head operating conditions. Flow from 1/4 inch iron pipe shall be provided during testing.

8. Bearings shall be ball or roller type, and the main bearings shall take all radial and end thrust. Pumps that depend only on hydraulic balance to overcome end thrust will not be acceptable.

9. Bearings shall be either oil bath type or grease type. Each oil reservoir shall be liberal in size and provided with an opening for filling, an overflow opening at the proper location to prevent overfilling, an oil level sight glass, and a drain at the lowest point. Pumps with oil lubrication systems shall be designed so that all shaft bearings will be isolated from the pumped liquid. An automatic sight feed oiler shall be provided with fittings for a grease gun and, if the bearings are not easily accessible, with grease tubing extending to convenient locations. The grease fittings shall be of a type that prevent over lubrication and the buildup of pressure injurious to the bearings.

10. Horizontal shaft centrifugal pumps with the exception of close coupled pumps shall be provided with a common base for mounting each pump and driving unit of the pump

on the same base. Each base shall be constructed of cast iron with a raised lip tapped for drainage or welded steel shapes with suitable drainage pan.

11. The pumps shall be equipped with air cocks, drain plugs, and duplex gauges indicating discharge pressure and suction pressure. Gauges, equipped with a shutoff cock and snubber, shall conform to ANSI B40.1 and shall be calibrated in pounds per square inch and feet of water in not more than 2 psi increments. Gauge ranges shall be appropriate for the particular installation. Normal operating suction and discharge pressures of the pump shall be indicated on the mid point range of the gauges. Pressures relief valve shall be furnished and installed where indicated.

12. The pump suction and discharge shall be provided with flanged connections of suitable size and arranged for piping shown. Pipe flanges shall conform to ANSI B16.1 and ANSI B16.5. Piping shall be installed to preclude the formation of air pockets.

13. Pump shall have painted or enameled finish as is standard with the manufacturer.

F. Pumps for Specific Systems: Chilled Water and Condensing Water Pumps

(Note to Specifier: Select as applicable.)

1. Chilled and condensing water pumps shall be of the single stage, double suction, centrifugal, vertical or horizontal split case type.

2. Pumps shall be bronze fitted, directly connected with flexible couplings, standard NEMA frame, and splash-proof motors. Pumps and motors shall be mounted on a cast-iron bed plate or base. Units shall have non-overloading curve characteristics and shall be equipped with mechanical seals.

3. Provide "high-efficiency" motor.

4. All pumps shall be provided with grease-lubricated ball or roller bearings with Alemite or equal grease fittings. Impellers shall not exceed 85% of the maximum possible size for the selected pump. Pumps shall be equipped with coupling guards.

5. Pumps shall be provided with a pet cock at the top of the casing and with tapped suction and discharge flanges for 1/4" gauge connections. Furnish pump performance curves. Pump manufacturer shall certify that pumps have been properly aligned, that installation does not place undue strain on the pump, and that the pumps are ready for operation.

6. Starters will be furnished by others as part of the motor control center.

7. The condensate return pump shall be multistage, if required to pump condensate at 210°F, and shall be of all iron construction and selected with mechanical seals. The receiver shall be constructed of carbon steel and shall be provided with level gauge glass, handhole, inlet removable cast iron strainers with a stainless steel basket, all necessary outlets, vent, drain, stainless steel float, and automatic float switch assembly for 120V service suitable to operate the control circuit of a magnetic contactor. Magnetic contactor should be adjusted to provide automatic alternate operation of the

pumps, simultaneous operation of both pumps under peak load conditions, automatic operation of both pumps under load conditions, and automatic operation of the standby pump if the active pump fails to operate.

G. Hot Water Pump - Base Mounted

1. The pump shall be vertical split case design, making possible complete servicing without breaking piping or motor connections.

2. The entire bearing bracket assembly shall be removable and interchangeable with a variety of pump sizes. Grease-lubricated anti-friction ball bearings shall be provided.

3. Pump shall use a mechanical rotating type seal and shall face against a diamond hard ceramic material called "Remite" or its approved equal.

4. Pump and motor shall be connected by means of a flexible coupler and guard. Motor enclosure, operating and electrical characteristics shall be as specified on the plans. Provide "high efficiency" motors as per motor section of specifications.

5. Pump and motor shaft alignment should be made before and after the grout is poured and again after the piping is connected.

6. Impellers shall not exceed 85% of the maximum possible size for the selected pump. Pumps shall be equipped with coupling guards.

7. Starters will be furnished by Others as part of the motor control center.

8. Pumps shall be provided with a pet cock at the top of the casing. Furnish pump performance curves. Pump manufacturer shall certify that pumps have been properly aligned, that installation does not place undue strain on the pump, and that the pumps are ready for operation.

9. Inline pumps (circulators) shall be directly driven by electric motors installed at any position and shall be supported from the pump casing. By construction, motors shall be protected from heat rising from the hot pump casing. Pumps shall be equipped with lubricating system and oil level indicator in bearing housing. Horizontal pumps shall be directly driven by electric motors through flexible couplings. Bearing supports and pump frame shall be of one-piece construction with three point mounting pads. Drip water shall be piped to a drain connection.

H. Spare Pump Parts

Furnish the following spare parts:

1. One __ hp motor.

2. One set of mechanical seals for each chilled, condensate pump and hot water pump.

3. One set of bearings for each chilled, condensate pump and hot water pump.

4. One complete set of gaskets for chilled, condensate pump and hot water pump.

I. Suction Diffusers

Provide at each primary heating pump, a suction diffuser of size and type noted on drawings. Units shall consist of angle type body with inlet vanes and combination diffuser-strainer-orifice cylinder with 3/16" diameter openings for pump protection. (Unit shall be equipped with disposable fine mesh start-up strainer which should be removed after 30 days of operation.)

Strainer-free area shall be no less than five times the section areas of the pump connection. Unit shall be provided with adjustable support foot to carry weight of suction piping. Provide inlet pressure gauge and pump suction compound gauge to indicate when cleaning is necessary.

Diffuser shall be as manufactured by Bell and Gossett, Taco, or approved equal.

3 CONSTRUCTION METHODS

3.1 GENERAL

A. The equipment shall be placed in accordance with the general arrangement as shown on the contract drawings. The general arrangement may be modified only as required to suit specific equipment. Such modifications shall not affect the structural design of the building or its components. Layout dimensions as shown may be modified to improve operating efficiency. Changes shall be specifically approved by the Engineer. The equipment shall be installed in accordance with the manufacturer's instructions.

B. Provide concrete pads for the proper installation of the equipment.

C. Install vibration isolators and flexible connections where shown and as required.

D. Install piping, controls, and structural supports for equipment as shown and as required.

E. Arrange belt guards to permit the use of tachometer, oiling, and testing with guard in place. Adjust equipment to operate without noticeable vibration.

F. Equipment shall be protected from dirt, foreign objects, and damage during the installation period. All damaged equipment shall be replaced at no additional cost.

G. Installation shop drawings shall be coordinated with other work and the structural conditions at the site to avoid interferences.

H. Breeching shall pitch up one inch in 10 feet in the direction of flow.

I. Breeching shall be properly supported from the structure.

J. Breeching shall be installed to allow for movement without strain on the structure and connections to equipment.

3.2 TESTING

All balancing and testing shall be completed as specified and required. The testing shall adequately show that the equipment performs as it is intended to and shall be subjected to approval.

Before final approval and after testing the Contractor shall show by in service demonstration that the equipment and all associated accessories are in good operating condition and properly performing their intended function.

A. Before any covering is installed, the entire heating system, cooling system piping (if applicable), and terminal heating/cooling units, shall be hydrostatically tested and proved tight under a pressure of 1.5 times the design working pressure. After the hydrostatic tests have been made and before the operating tests, the boilers and feed water piping shall be thoroughly cleaned by filling the system with a solution consisting of either 1 pound of caustic soda or 3 pounds of trisodium phosphate per 100 gallons of water. The water shall be heated to approximately 150°F and the solution circulated in the system for a period of 48 hours.

B. The system shall then be drained and thoroughly flushed out with fresh water. Upon completion and before acceptance of the installation, the Contractor shall subject the heating system/cooling system to the operating tests to demonstrate satisfactory functional and operational efficiency. The operating test shall cover a period of at least 24 hours for each system and shall include the following specific information, as applicable, in a report together with conclusions as to the adequacy of the system:

1. Time, date, and duration of test.
2. Outside and inside dry bulb temperatures.
3. Temperature of hot water supply leaving boiler or heater.
4. Temperature of hot water return from system at boiler or heater inlet.
5. Boiler steam pressure.
6. Temperature of condensate supply to the steam boiler.
7. Boiler or heater make, type, serial number, design pressure, and rated capacity.
8. Fuel burner make, model, and rated capacity; ammeter and voltmeter readings for burner motor.
9. Circulating pump make, model, and rated capacity; ammeter and voltmeter readings for pump motor during operation.
10. Flue gas temperature at boiler or heater outlet.
11. Percent carbon dioxide in flue gas.
12. Grade or type and calorific value of fuel.
13. Draft at boiler or heater flue gas exit.

14. Draft or pressure in furnace.

15. Quality of water circulated.

16. Quality of fuel consumed.

17. Stack emission pollutants concentration.

C. All indicating instruments shall be read at half-hour intervals unless otherwise directed. The report of the test shall be supplied in quadruplicate to the Engineer. The Contractor shall furnish all instruments, test equipment, fuel oil, and test personnel required for the tests. Gas, water, and electricity shall be furnished. Operating tests shall demonstrate that fuel burners and combustion and safety controls meet the requirements of applicable NFPA standards.

D. The entire fuel oil system shall be tested for leaks after installation, before any underground tanks, piping, and fittings are covered, and prior to the operational test of the heating system as hereinafter specified. Fuel piping shall be tested with air at not less than 1.25 times the maximum working pressure but not less than 5 psi in excess of the static pressure produced with the oil level at the highest point of the fuel system. The test shall be so made as not to impose a pressure of more than 10 psi on the tank. In lieu of the pressure test, the suction piping between underground storage tank and fuel pump may be tested under a vacuum of not less than 20 inches of mercury.

E. The gas fuel system shall be tested in accordance with the test procedures outlined in NFPA No. 54.

3.3 COMBUSTION AIR

A. Provide outside air louver, bird screen, and motorized damper for each boiler or domestic hot water heater. Motorized dampers shall be control wired to the boiler or heater panel.

B. Louvers shall be sized for 2000 Btuh input to one square inch of free louver area.

SECTION IV. INSULATION

TABLE OF CONTENTS

1 GENERAL

1.1 DESCRIPTION

The work of this Section consists of furnishing, installing, and supervising the installation of insulation for hydronic piping and equipment.

1.2 APPLICABLE PROVISIONS

In case of conflict between provisions of codes, laws, ordinances, and these specifications, including the contract drawings, the more stringent requirements will apply.

1.3 GENERAL REQUIREMENTS

A. Fire Ratings

Unless otherwise specified, installed insulation materials, adhesives, coatings, and other accessories shall have surface burning characteristics, as determined by ASTM E 84, not to exceed 25 for flame spread and 50 for smoke-developed rating where used indoors. All insulation shall be tested for the same densities and installed thickness as the material that will be used in actual construction applications. Compliance with the smoke-developed limitation is not required and a greater flame spread rating of up to 100 is permitted for insulation installed within wall assemblies or enclosures. In such installations, the interior finish materials or enclosure shall have a minimum fire retardant rating of 15 minutes by ASTM E 119 test, when tested as it will be installed in the actual construction application. Insulation that has been treated with a flame retardant additive to obtain the flame spread and smoke-developed ratings shown above are not permitted.

B. Identification of Materials

Packages or standard containers of insulation, jacket material, cements, adhesives, and coatings delivered for use and all samples required for approval shall have manufacturer's stamp or label attached giving the name of the manufacturer and brand and a description of the material.

1.4 SUBMITTAL

A. Material and equipment shall be the standard products of a manufacturer regularly engaged in their manufacture for a minimum period of ten years. All material and equipment shall be new unless otherwise indicated. General requirements include those specified in the section MECHANICAL SPECIFICATION GENERAL REQUIREMENTS and as specified herein.

2 MATERIALS

2.1 INSULATION

A. Materials shall be compatible with each other and shall not contribute to corrosion, softening, or otherwise attack surfaces to which applied in either the wet or any other state. Materials to be used on stainless steel surfaces shall meet ASTM C 795 requirements. Materials shall be new and asbestos free and shall conform to the following standards.

B. Pipe insulation shall conform to the following:
1. Mineral fiber per ASTM C 547.
2. Calcium silicate per ASTM C 533.

C. Equipment insulation shall be:
1. Rigid mineral fiber per ASTM C 612.
2. Flexible mineral fiber per ASTM C 553.
3. Calcium silicate per ASTM C 533.

D. Wire for insulation shall be ASTM A 580 stainless steel, 16 or 18 gauge.

E. All pipe insulation shall be provided with a factory applied all purpose jacket with or without integral vapor barrier as required by the service. Provide jackets on insulation in exposed locations with a white surface suitable for painting. All purpose jacket shall have a perm rating of not more than 0.02 in accordance with ASTM E 96; a puncture resistance of not less than 50 Beach units in accordance with ASTM D 781; and a tensile strength of not less than 35 lb/in/width in accordance with ASTM D 828.

F. Vapor barrier material shall conform to Fed. Specifications HH-B-100, Type I. Material shall be resistant to flame and moisture penetration and not support mold growth. Vapor barrier material shall be provided on pipe insulation as required, except that all pipe in crawl spaces shall have a factory applied vapor barrier jacket for all services. Vapor barrier shall be provided for all fittings and valves located in systems requiring vapor barrier. Provide vapor barrier on domestic cold water piping, chilled water piping, refrigerant suction piping, and dual temperature piping. Provide vapor barrier on chilled water and refrigeration equipment.

G. Pipe insulation passing through sleeves where caulking is required shall be provided with metal jackets.

H. Aluminum jackets shall conform to ASTM B 209, 0.016 inch thick, smooth.

I. Stainless steel jackets shall be Type 304, 0.010 inch thick, smooth.

J. Adhesives, sealants, and compounds shall be compatible with materials to which they are applied and suitable for the service.

K. Insulation cement shall conform to ASTM C 195 mineral fiber, thermal conductivity 0.85 maximum at 200°F mean temperature when tested per ASTM C 177.

The following are guidelines for insulation types for piping conveying various temperature fluids.

INDOOR COLD AND CHILLED WATER PIPING 35°F TO 100°F

MATERIAL: Fiberglass, sectional, preform, pipe insulation with vapor barrier jacket.

JACKET: Factory-applied, fire-resistant glass cloth jacket, presized with a vapor barrier laminate of saran or fire-resistant jacket with a vinyl coated and embossed vapor barrier laminate.

VAPOR TRANSMISSIONS: 0.02 perms minimum.

DENSITY: 6 pounds per cubic foot.

THERMAL CONDUCTIVITY: 0.30 Btu/in/sq ft/°F/hr at 200°F.

SPECIFICATION COMPLIANCE: Federal Specification HH-1558a, Form D, Type III, Class 12.

APPLICATION:

1. Apply pipe insulation, and butt adjoining sections firmly together. Longitudinal overlap and butt-joint strips of the jacket shall be sealed with fireproof vapor adhesive.

2. Fittings and valves shall be insulated with segments of molded insulation securely wired in place. Apply a skim coat of insulating cement to the insulation to produce a smooth surface. Apply a coat of vapor seal adhesive, and wrap the fitting with glass fiber fitting tape. Adhesive and tape shall overlap adjoining sections of pipe insulation by two inches. Apply a second coat of adhesive to produce a smooth surface.

3. All terminating ends of the pipe covering shall be sealed off with insulated material applied on a bevel. The insulation material and finish shall be the same as specified for fittings and valves.

4. At pipe connections to equipment, the pipe covering shall be terminated and sealed off over drip or drain pans to prevent condensate dripping.

5. All pipe connections to the water systems, such as makeup lines, drain lines, air eliminator, relief valves, instruments, etc., shall be insulated separately in the same manner as fitting. The insulation shall be applied along the connecting pipe to a distance not less than 6 inches and then sealed off.

6. When in contact with the chilled water pipe, clamp hanger, pipe anchors, etc., shall be insulated and vapor sealed as specified above.

7. Under no circumstances will stapling of the insulation and jackets be permitted.

ALTERNATE

MATERIAL: Fiberglass, one piece preform, pipe insulation with jacket.

JACKET: Factory-applied, 0.16 inch thick aluminum jacket with integral moisture barrier, self-locking weatherproof longitudinal joint, and aluminum butt-joint sealer straps with sealing mastic and stainless steel locking band. Preformed miter sealer straps shall be used on mitered sections.

DENSITY: 4 pounds per cubic foot.

THERMAL CONDUCTIVITY: 0.30 Btu/in/sq ft/°F/hr at 200°F.

SPECIFICATION COMPLIANCE: Federal Specification HH-1-558a, Form D, Type III, Class 12, ASTM C 547-67.

APPLICATION:

1. Pipe: Insulation with factory-applied aluminum jacket shall be applied in a single layer. It shall be carefully fitted to the pipes with side and end joints butted tightly. The insulation and aluminum jacket shall be held in place by a continuous self-locking longitudinal joint providing a positive weatherproof seal along the entire length of the jacket. Then a preformed strap, containing a permanently plastic weatherproof seal, shall be centered and secured over each circumferential joint and secured by a separate stainless steel banding.

2. Valves and Fittings: The bodies of fittings and the bodies and bonnets of valves shall be covered with mitered sections of factory-applied aluminum jacketed prefabricated pipe insulation of the same materials and thickness as the insulation on the adjacent piping. Insulation of 90 degree pipe elbows shall have a minimum of three mitered sections. Humped insulation fittings shall not be permitted. Valve insulation caps and end caps for valves shall be constructed of 0.032 inch thick hard aluminum and secured with snap-straps with stainless steel bands to provide a weatherproof seal. Joints shall be sealed with sealing compound and preformed miter joint aluminum bands.

3. Flanges: All flanges shall be covered with sections of factory-applied aluminum jacketed prefabricated pipe insulation of a thickness equal to that of the insulation on the adjacent piping. The insulation shall overlap the adjacent piping insulation to a minimum of 3 inches, with filler rings of the same material being provided if necessary to ensure a good bearing on the pipe insulation and on the periphery of the flanges. At the flange joint, the pipe insulation shall be tapered off at approximately a 45 degree angle on both sides of the joint for a length sufficient to permit removing flange bolts without damage to the covering. Covers shall be of a design which will permit the flanges and bolts to heat up quickly and be maintained at a temperature nearly equal to that of the pipe, thus avoiding excessive strain. End caps shall be constructed of 0.032 inch thick hard aluminum and secured with snap-straps with stainless steel bands to provide a weatherproof seal.

4. Insulation shall be omitted indoors only on the following fittings and piping:
   a. Screwed unions and exposed connections to units.
   b. Return piping from coil connection to steam trap assemblies, including trap strainers, valves, unions, and also about 5 feet of pipe downstream of trap.
   c. Vents to atmosphere, overflow, drain pipes, and blow-off pipes.
   d. Any piping with fins under radiation covers or exposed.

5. Vertical Risers: On vertical pipes exceeding 15 feet in height, intermediate support for the insulation shall be furnished by the Contractor. For carbon steel pipe, this support shall consist of angle clips or other suitable device welded to the pipe at about 15 feet on centers and concealed by the pipe covering.

   For chrome-moly alloy or stainless steel piping, no welding clips shall be used. Clamps or other approved non-welded devices shall be used.

HOT WATER PIPING 100°-199°F

MATERIAL: Hydrous calcium silicate, sectional, preform, pipe insulation with jacket.

JACKET: Factory-applied, 0.016 inch thick aluminum jacket with integral moisture barrier, self-locking weatherproof longitudinal joint, and aluminum butt-joint sealer straps with sealing mastic and stainless steel locking band. Preformed miter sealer straps shall be used on mitered sections.

DENSITY: 11 pounds per cubic foot.

THERMAL CONDUCTIVITY: 0.40 Btu/in/sq ft/°F/hr at 300°F.

SPECIFICATION COMPLIANCE: ASTM C345 - 54T, to 1200°F, Federal Specification HH-1-523, Class 2 to 1200°F.

APPLICATION:

1. Pipe: Insulation with factory-applied aluminum jacket shall be applied in a single layer. It shall be carefully fitted to the pipes with side and end joints butted tightly. The insulation and aluminum jacket shall be held in place by a continuous self-locking longitudinal joint providing a positive weatherproof seal along the entire length of the jacket. Then a preformed strap containing a permanently plastic weatherproof seal shall be centered and secured over each circumferential joint and secured by use of a separate stainless steel banding.

2. Valves and Fittings: The bodies of fittings and the bodies and bonnets of valves shall be covered with mitered sections of factory-applied aluminum jacketed prefabricated pipe insulation of the same materials and thickness as the insulation on the adjacent piping. Insulation of 90 degree pipe elbows shall have a minimum of three mitered sections. Humped insulation fittings shall not be permitted. Valve insulation caps and end caps for valves shall be constructed of 0.032 inch thick hard aluminum and secured with snap-straps with stainless steel bands to provide a weatherproof seal. Joints shall be sealed with sealing compound and preformed miter joint aluminum bands.

3. Flanges: All flanges shall be covered with sections of factory-applied aluminum jacketed prefabricated pipe insulation of a thickness equal to that of the insulation on the adjacent piping. The insulation shall overlap the adjacent piping insulation to a minimum of 3 inches, with filler rings of the same material being provided if necessary to ensure a good bearing on the pipe insulation and on the periphery of the flanges. At the flange joint the pipe insulation shall be tapered off at approximately a 45 degree angle on both sides of the joint for a length sufficient to permit removing flange bolts without damage to the covering.

Covers shall be of a design which will permit the flanges and bolts to heat up quickly and be maintained at a temperature nearly equal to that of the pipe, thus avoiding excessive strain. End caps shall be constructed of 0.032 inch thick hard aluminum and secured with snap-straps with stainless steel bands to provide a weatherproof seal.

4. Insulation shall be omitted indoors only on the following fittings and piping:
   a. Screwed unions and exposed connections to units.
   b. Return piping from coil connection to steam trap assemblies, including trap strainers, valves, union, and also about 5 feet of pipe downstream of trap.
   c. Vents to atmosphere, overflow, drain pipes, and blow-off pipes.
   d. Any piping with fins under radiation covers or exposed.

5. Vertical Risers: On vertical pipes exceeding 15 feet in height, intermediate support for the insulation shall be furnished by the Contractor. For carbon steel pipe, this support shall consist of angle clips or other suitable device welded to the pipe at about 15 feet on centers and concealed by the pipe covering. For chrome-moly alloy or stainless steel piping, no welding clips shall be used. Clamps or other approved non-welded devices shall be used.

ALTERNATE

LOCATION: Indoor.

MATERIAL: Hydrous calcium silicate, sectional, preform, pipe insulation with jacket.

JACKET: Glass cloth jacket with longitudinal and butt-end lap joints.
Burst strength: 300 psi.

DENSITY: 11 pounds per cubic foot.

THERMAL CONDUCTIVITY: 0.40 Btu/in/sq ft/°F/hr at 300°F.

SPECIFICATION COMPLIANCE: ASTM C345 - 54T, to 1200°F, Federal Specification HH-1-523, Class 2 to 1200°F.

APPLICATION:

1. Pipe: Insulation shall be applied in a single layer. All sectional coverings shall be carefully fitted to the pipes with side and end joints butted tightly. The side and laps of factory-applied glass cloth shall be pasted down smooth and coated with lagging adhesive. All segmental coverings shall be carefully fitted to pipes with side and end joints butted tightly and securely wired in place with not less than three loops of wire per section of insulation. Wire shall be 16 gauge copperweld steel. The ends of all wire loops shall be firmly twisted together with pliers, bent over and carefully pressed into the surface of the insulation. Joints between the segments shall be carefully rubbed closed with smooth tool using plastic made from the same material as the layer itself. A glass cloth jacket shall then be applied, lagged and finished as specified hereafter in this specification.

2. Valves and Fittings: The bodies of fittings and the bodies and bonnets of valves shall be covered with sections of prefabricated pipe insulation or molded pipe fitting insulation of the same materials and thickness as the insulation on the adjacent piping. At all flange joints, the

insulation shall be tapered off at approximately a 45 degree angle on both sides of the joint for a length sufficient to permit removing flange bolts without damage to the covering.

3. Insulations shall be omitted on the following fittings and piping:
   a. Screwed unions and exposed connections to units.
   b. Return piping from coil connection to steam trap assemblies, a including trap strainers, valves, union, and also about 5 feet of pipe downstream of trap.
   c. Vents to atmosphere, overflow, drain pipes, and blow-off pipes.
   d. Any piping with fins under radiation covers or exposed.

4. Vertical Risers: On vertical pipes exceeding 15 feet in height, intermediate support for the insulation shall be furnished by the Contractor. For carbon steel pipe, this support shall consist of angle clips or other suitable device welded to the pipe at about 15 feet on centers and concealed by the pipe covering. For chrome-moly alloy or stainless steel piping, no welding clips shall be used. Clamps or other approved non-welded devices shall be used.

5. Lagging and Finishing of Glass Cloth Covering: All glass cloth covering for piping and equipment shall be lagged and finished as follows, unless otherwise specified elsewhere in these specifications. Color of lagging and finishing adhesive shall be as selected by the Engineer.
   a. Factory-Applied Covering. The laps shall be sealed with a brush coat of lagging adhesive, and the entire surface of the covering coated with adhesive (brushed or sprayed) at a rate of 80 sq ft per gallon.
   b. Field-Applied Covering. A tack coat of adhesive shall be applied by brush or spray at a rate of 60 sq ft per gallon, into which the glass cloth covering shall be imbedded. Laps shall be sealed with a brush coat of adhesive and the entire surface of the glass cloth coated with adhesive (brushed or sprayed) at a rate of 80 sq ft per gallon.
   c. General. All glass cloth shall be stretched tight and smooth, with laps in least visible location. All laps shall be tightly sealed and adhered with adhesive. No metal bands shall be allowed. Lagging and finishing adhesive shall be used as furnished by the manufacturer, without adulteration by any other substance. All valves, flanges, and similar in-line items shall be finished similar to the adjacent piping or equipment.

ALTERNATE

MATERIAL: Fiberglass, sectional, preform, pipe insulation with jacket.

JACKET: Factory-applied, fire-resistant glass cloth jacket, pre-sized with a vapor barrier laminate of saran or fire-resistant jacket with a vinyl coated and embossed vapor barrier laminate.

DENSITY: 6 pounds per cubic foot.

THERMAL CONDUCTIVITY: 0.30 Btu/in/sq ft/°F/hr at 200°F.

SPECIFICATION COMPLIANCE: Federal Specification HH-1558a, Form D, Type III, Class 12, ASTM C547-67.

APPLICATION:

1. Apply pipe insulation, and butt adjoining sections firmly together. Longitudinal overlap and butt-joint strips of the jacket shall be sealed with fireproof vapor seal adhesive.

2. Fittings and valves shall be insulated with segments of molded insulation securely wired in place. Apply a skim coat of insulating cement to the insulation to produce a smooth surface. Apply a coat of mastic, and wrap the fitting with glass fiber fitting tape. Mastic and tape shall overlap adjoining sections of pipe insulation by two inches. Apply a second coat of mastic to produce a smooth surface.

3. Insulation shall be omitted on the following fittings and piping:
    a. Screwed unions and exposed connections to units.
    b. Vents to atmosphere, overflow, drain pipes, and blow-off pipes.
    c. Any piping with fins under radiation covers or exposed.

4. The insulation shall be neatly terminated, where insulation is omitted, with insulating cement troweled on a bevel. The cement and finish shall be the same as specified for fittings and valves.

5. Covering shall be neatly finished at pipe hangers, pipe anchors, and pipe covering protection saddles as specified for fittings and valves.

HOT WATER AND HIGH PRESSURE CONDENSATE RETURN PIPING 200° TO 290°F

LOCATION: Outdoor.

MATERIAL: Hydrous calcium silicate, sectional, preform, pipe insulation with jacket.

JACKET: Factory-applied, 0.016 inch thick aluminum jacket, with integral moisture barrier, self-locking weatherproof longitudinal joint, and aluminum butt-joint sealer straps with sealing mastic and stainless steel locking band. Preformed miter sealer straps shall be used on mitered sections.

DENSITY: 11 pounds per cubic foot.

THERMAL CONDUCTIVITY: 0.40 Btu/in/sq ft/°F/hr at 300°F.

SPECIFICATION COMPLIANCE: ASTM C345 - 54T, to 1200°F, Federal Specification HH-1-523, Class 2 to 1200°F.

THICKNESS: Pipes 1/2 inch to 3/4 inch use 1-1/2 inch thick insulation.
Pipes 1 inch to 1-1/2 inches use 2 inch thick insulation.
Pipes 2 inches to 2-1/2 inches use 2-1/2 inch thick insulation.
3 inch pipes use 3 inch thick insulation.
4 inch pipes use 3-1/2 inch thick insulation.
Pipes 6 inches to 8 inches use 4-inch thick insulation.
Pipes 10 inches and larger use 4-1/2 inch thick insulation

APPLICATION:

1. Pipe: Insulation with factory-applied aluminum jacket shall be applied in a single layer. It shall be carefully fitted to the pipes with side and end joints butted tightly. The insulation and aluminum jacket shall be held in place by a continuous self-locking longitudinal joint providing a positive weatherproof seal along the entire length of the jacket. Then a

preformed strap containing a permanently plastic weatherproof seal shall be centered and secured over each circumferential joint and secured by a separate stainless steel banding.

2. Valves and Fittings: The bodies of fittings and the bodies and bonnets of valves shall be covered with mitered sections of factory-applied aluminum jacketed prefabricated pipe insulation of the same materials and thickness as the insulation on the adjacent piping. Insulation of 90 degree pipe elbows shall have a minimum of three mitered sections. Humped insulation fittings shall not be permitted. Valve insulation caps and end caps for valves shall be constructed of 0.032 inch thick hard aluminum and secured with snap-straps with stainless steel bands to provide a weatherproof system.

3. The insulation shall be neatly terminated, where insulation is omitted, with insulating cement troweled on a bevel. The cement and finish shall be the same as specified for fittings and valves.

4. Covering shall be neatly finished at pipe hangers, pipe anchors, and pipe covering protection saddles as specified for fittings and valves.

3 CONSTRUCTION METHODS

3.1 GENERAL

A. Insulation for boiler flues, breaching, shall be 4 inches of calcium silicate insulation conforming to ASTM C 533. Joints in the insulation shall be filled with mineral wool or an equally suitable cement. Insulation shall be applied in double layer construction with staggered joints. Insulation at access doors shall be removable type.

B. The following pipes shall be insulated:
   1. Heating hot water lines.
   2. Chilled water lines.
   3. Dual temperature lines.
   4. Refrigerant suction lines.
   5. Glycol/water lines.

C. The following equipment shall be insulated:
   1. Pumps.
   2. Domestic hot water storage tanks.
   3. Heat exchangers.
   4. Chilled and dual temperature expansion tanks.
   5. Air separators (tank type).
   6. Coil headers and return bends.
   7. Coil casings.

D. Insulation thickness (inches for piping shall be as follows):

Service	Pipe Sizes (Inches) 1/2 to 2	2-1/2 to 5	6 to 10	Over 10
Chilled Water	1-1/2	2	2-1/2	3
Heating Hot Water	1-1/2	2	2-1/2	2-1/2
Dual Temperature	1-1/2	2	2-1/2	3
Refrigerant Suction	1	2	2	2
Glycol/Water	1	1-1/2	2	2

E. Thickness of equipment insulation shall be 2 inches. Insulation shall be rigid mineral fiber except as noted.

F. Insulation materials shall not be applied until all system tests have been satisfactorily completed and surfaces to be insulated have been cleaned and dried. Insulation shall be clean and dry when installed and during the application of any finish. Install materials neatly with smooth even surfaces with jackets drawn tight and smoothly cemented down on longitudinal and end laps. Scrap pieces shall not be used where a full length section will fit. Pipe insulation shall be continuous through sleeves, wall, and ceiling openings. Piping shall be individually insulated. A complete moisture and vapor barrier shall be provided wherever insulation terminates against metal hangers, anchors, and other projections through insulation on cold surfaces for which a vapor seal is specified. Chrome plated pipes shall not be insulated. Omit insulation from vibration isolating connections, but adjacent insulation shall be neatly terminated, and beveled. Name plates and access plates in housings and equipment shall not be insulated, but insulation must be carefully beveled and sealed around same. All piping and equipment that is heat traced shall be insulated.

3.2 PIPE INSULATION

A. Pipe insulation shall be mineral fiber or special materials function of the fluid temperature. Sections of insulation shall be placed around the pipe and tightly butted into place. The jacket laps shall be drawn tight, smooth, and secured with fire-resistant adhesive, factory-applied self sealing lap, and with non-corrosive outward clinching staples spaced not over 4 inches on centers and 1/2-inch minimum from edge of lap. Circumferential joints shall be covered with butt strips, not less than 3 inches wide, of material identical to the jacket material. Adhesive used to secure the butt strip shall be the same as used to secure the jacket laps. Staples shall be applied to both edges of the butt strips. When a vapor barrier is required, staples and seams shall be sealed with a brush coat of fire-resistant vapor barrier coating applied at all longitudinal and circumferential laps. Ends of sections of insulation that butt against flanges, unions, valves, and fittings, and joints at intervals of not more than 12 feet in continuous runs of pipe shall be coated with a vapor barrier coating. Breaks and punctures in the jacket material shall be patched by wrapping a strip of jacket material around the pipe and cementing, stapling, and coating as specified for butt strips. The patch shall extend not less than 1-1/2 inches past the break in both directions. At penetrations such as thermometers, the voids in the insulation shall be filled with vapor barrier coating and the penetration sealed with a brush coat of the same coating.

B. Calcium silicate pipe insulation shall be secured with stainless steel metal bands on 12 inch maximum centers. For high temperature piping, unless single layer insulation is recommended by the manufacturer, insulation shall be applied in two layers with the joints tightly butted and staggered a minimum of 3 inches. The inner layer of insulation shall be secured with 14 gauge soft annealed stainless steel wires on 12 inch maximum centers. The outer layer shall be secured with stainless steel metal bands on 12 inch maximum centers. Apply a skim coat of hydraulic setting cement directly to the insulation. Apply a flooding coat adhesive over the hydraulic setting cement, and while still wet, press a layer of glass cloth or tape into adhesive and seal laps and edges with adhesive. Coat cloth with adhesive cut at a ratio of one part water to five parts adhesive in color other than white for the purpose of visual inspection to ensure sizing of entire surface. When dry, apply a finish coat of adhesive at can consistency.

C. Pipe insulation shall be continuous through pipe hangers. Where pipe is supported at the insulation, shields or saddles shall be provided. Where shields are used on pipes 2 inches and larger, insulation inserts shall be provided at points of hangers and supports. Insulation inserts shall be of calcium silicate, cellular glass, molded glass fiber, or other approved material of the same thickness as adjacent insulation. Inserts shall have sufficient compressive strength to adequately support the pipe without compressing the inserts to a thickness less than the adjacent insulation. Insulation inserts shall cover the bottom half of the pipe circumference 180 degrees and shall not be less in length than the protection shield. Vapor barrier facing of the insert shall be of the same material as the facing on the adjacent insulation. Seal inserts into the insulation with vapor barrier coating. Where protection saddles are used, fill all voids with the same insulation material as used on the adjacent pipe. Where anchors are secured to chilled piping to be insulated, insulate anchors same as piping for a distance not less than four times insulation thickness to prevent condensation. Insulation around anchors shall be vapor sealed.

D. Where penetrating interior walls, extend a metal jacket 2 inches out on either side of the wall and secure on each end with a band. Where penetrating floors, extend a metal jacket from a point below the back-up material to a point 10 inches above the floor with one band at the floor and one not more than one inch from end of metal jacket.

E. When segments of insulation are used, provide elbows with not less than three segments. For other fittings and valves, cut segments to required curvature or use nesting size sectional insulation. Place and join the segments of the insulation with adhesive. After the segments are in place, apply vapor barrier coating. Where unions, flanges, and valves are to be insulated with removable sections, terminate the covering neatly at the ends with insulation cement troweled on a bevel. Apply a vapor barrier coating to the beveled ends. Cover unions, flanges, and valve bonnets with removable sections of insulation vapor barrier sealed inside and out with adjacent insulation ends neatly finished and vapor barrier sealed.

F. Metal jackets shall have side and end laps at least 2 inches wide with the cut edge of the side lap turned under one inch to provide a smooth edge. Place laps to shed water. Seal laps on chilled and cold piping with weatherproof coating. Secure jackets in place with stainless steel bands on 9-inch centers or stainless steel screws on 5-inch centers. Where pipes penetrate exterior walls, continue the increased insulation thickness required for piping exposed to weather and the metal jackets through the sleeve to a point 2 inches beyond the interior surface of the wall. For fittings, flanges, and valves in outdoor locations, secure metal covers in place with metal bands and seal with a weatherproof coating. Protect fittings, flanges, and valves with a weatherproof coating prior to installation of metal covers. Jackets covering calcium silicate insulation shall be constructed of stainless steel.

3.3 EQUIPMENT INSULATION

A. Apply equipment insulation to fit as closely as possible to equipment. Insulation shall be grooved or scored where necessary to fit the contours of equipment. Stagger end joints where possible. Bevel the edges of the insulation for cylindrical surfaces to provide tight joints. After the insulation is in place on areas to be insulated, except where metal encased, fill joints, seams, chipped edges, or depressions with bedding compound to form a smooth surface. Bevel insulation around name plates, ASME stamp, and access plates. Insulation on equipment that must be opened periodically for inspection, cleaning, and repair shall be constructed so insulation can be removed and replaced without damage.

B. For heating equipment, except pumps, secure the insulation with 16 gauge galvanized steel wire or with 3/4 inch wide 26 gauge galvanized steel bands per ASTM A641, spaced on 12-inch centers. Seal joints with insulating cement, and cover insulation with a coat of finishing cement. Provide water boxes with 20 gauge galvanized sheet metal covers internally insulated.

C. For chilled water and refrigeration equipment, secure insulation with 16 gauge, galvanized steel or copper clad wire spaced on 12-inch centers. Seal joints with vapor barrier cement. Cover irregular surfaces with a smoothing coat of insulating cement. Provide water boxes with 20 gauge galvanized sheet metal covers internally insulated.

D. Pumps shall be insulated by forming a box around the pump housing. The box shall be constructed by forming the bottom and sides using joints which do not leave raw ends of insulation exposed. Bottom and sides shall be banded to form a rigid housing which does not rest on the pump. Joints between top cover and sides shall fit tightly. The joint should form a female shiplap joint on the side pieces and a male joint on the top cover, thus making the top cover removable. The entire surface of the removable section shall be finished as specified. Caulking shall be applied at penetration.

Index

B

C

D

E

F

H

I

L

M

P

R

S

T

U

V

W

Other Titles Offered by BNP